COMPLEX ANALYSIS AND ITS APPLICATIONS
Vol.I

INTERNATIONAL CENTRE FOR THEORETICAL PHYSICS, TRIESTE

COMPLEX ANALYSIS AND ITS APPLICATIONS

LECTURES PRESENTED AT
AN INTERNATIONAL SEMINAR COURSE
AT TRIESTE FROM 21 MAY TO 8 AUGUST 1975
ORGANIZED BY THE
INTERNATIONAL CENTRE FOR THEORETICAL PHYSICS, TRIESTE

In three volumes

VOL. I

INTERNATIONAL ATOMIC ENERGY AGENCY
VIENNA, 1976

THE INTERNATIONAL CENTRE FOR THEORETICAL PHYSICS (ICTP) in Trieste was established by the International Atomic Energy Agency (IAEA) in 1964 under an agreement with the Italian Government, and with the assistance of the City and University of Trieste.

The IAEA and the United Nations Educational, Scientific and Cultural Organization (UNESCO) subsequently agreed to operate the Centre jointly from 1 January 1970.

Member States of both organizations participate in the work of the Centre, the main purpose of which is to foster, through training and research, the advancement of theoretical physics, with special regard to the needs of developing countries.

COMPLEX ANALYSIS AND ITS APPLICATIONS
IAEA, VIENNA, 1976
STI/PUB/428
ISBN 92–0–130376–9

Printed by the IAEA in Austria
September 1976

FOREWORD

The International Centre for Theoretical Physics has maintained an interdisciplinary character in its research and training programmes in different branches of theoretical physics and related applied mathematics. In pursuance of this objective, the Centre has since 1964 organized extended research courses in various disciplines; most of the Proceedings of these courses have been published by the International Atomic Energy Agency.

In 1972 the ICTP held the first of a series of extended summer courses in mathematics and its applications. To date, the following courses have taken place: Global Analysis and its Applications (1972), Mathematical and Numerical Methods in Fluid Dynamics (1973), Control Theory and Topics in Functional Analysis (1974), and Complex Analysis and its Applications (1975). The present volumes consist of a collection of the long, basic courses (Volume I) and the individual lectures (Volumes II and III) given in Trieste at the 1975 Summer Course. The contributions are partly expository and partly research-oriented – in the spirit of the very wide range of interest of the participants. The programme of lectures was organized by Professors A. Andreotti (Pisa, Italy, and Oregon, United States of America), J. Eells (Warwick, United Kingdom) and F. Gherardelli (Florence, Italy).

Abdus Salam

EDITORIAL NOTE

CONTENTS OF VOLUME I

AN INTRODUCTION TO POTENTIAL THEORY AND MEROMORPHIC FUNCTIONS

D. DRASIN
Department of Mathematics,
Purdue University,
West Lafayette, Indiana,
United States of America

Abstract

AN INTRODUCTION TO POTENTIAL THEORY AND MEROMORPHIC FUNCTIONS.

The potential theory centres on the standard decomposition and representation theorems for subharmonic functions. In addition, the first and second fundamental theorems of Nevanlinna's theory of meromorphic functions are derived and, after two applications to differential equations, some open problems are discussed. Systematic use is made of L. Schwartz's theory of distributions.

CONTENTS: I. Riemann surface of a rational function. II. Introduction to distributions: Convex functions; Introduction to distributions; Positive distributions; More on convex functions. III. Subharmonic functions – potential theory: Harmonic functions; Potentials; Subharmonic functions; Subharmonicity and convexity. IV. The first fundamental theorem – Nevanlinna's characteristic: Poisson–Jensen formula; Connection with entire functions; Cartan's identity, convexity. V. Nevanlinna's second fundamental theorem: Properties of Ψ and $[\Delta]\log\Psi$; Pullback to the z-plane; Defect relation (preliminary form); Defect relation (final form); Logarithmic derivative; Valiron deficiencies. VI. Applications and open problems: Applications to differential equations; Some open questions.

I. THE RIEMANN SURFACE OF A RATIONAL FUNCTION

Let $f(z)$ be a function which is meromorphic in the finite complex $z(= x+iy)$-plane, which we call $\mathbb{C}$. In particular f is <u>single-valued</u>, a property that will be preserved by all functions we shall study. By being meromorphic, we mean that f satisfies the equivalent conditions:

(A) $f(z) = g(z)/h(z)$, where g and h are entire, $h \not\equiv 0$, with no common roots;

(B) There exist sets

$$P = \{z_k\}$$

with no point of accumulation in the finite plane (and thus at most countable) and non-negative integers $\{n_k\}$ such that

(i) f is holomorphic in $\mathbb{C} - P$

and

(ii) in a neighborhood of each z_k, f may be written as $f(z) = (z - z_k)^{-n_k} f_k(z)$, with f_k holomorphic and non-vanishing in this neighborhood.

The <u>Riemann surface</u> of f offers one perspective to study the function f. Riemann surfaces may be defined and constructed very abstractly, but we can do better if it is known that the surface in question "arises" from a meromorphic function. Thus for our purposes the Riemann surface of the meromorphic function $f(z)$ is the set of pairs:

$$S = \{(z, f(z)); z \in \mathbb{C}\} \tag{1.1}$$

where S is endowed with a topology so that the mapping

$$f: \mathbb{C} \to S$$

(here, of course, $z \in \mathbb{C}$ is sent to $(z, f(z)) \in S$) is 1-1, onto, continuous and has a continuous inverse (i.e., $\mathbb{C}$ and S are <u>homeomorphic</u>).

Equivalently, S may be thought of as

$$S = \{(w,z); z \in \mathbb{C}, w = f(z)\} \tag{1.2}$$

where the coordinates in (1.1) have been switched. Most of the pictorial language of Riemann surfaces comes from the representation (1.2).

<u>Example</u>. Let $w = z^2$. According to (1.1), each z appears as the first coordinate of precisely one point of S, so that "above" each $z \in \mathbb{C}$ "lies" one point of S. However, if $0 \neq w$,

then $w = f(z)$ for two points z, so "above" each w (using the viewpoint of (1.2)) are two points of S, and so S is often described as a "two-sheeted covering" of $\mathbb{C}$ where the two "sheets" are joined together as a corkscrew.

Suppose first that $f(z)$ is rational and non-constant with

$$f(z) = \frac{g(z)}{h(z)} \tag{1.3}$$

where g and h are polynomials having no common factors. If

$$g(z) = g_1 z^m + g_2 z^{m-1} + \dots \qquad (g_1 \neq 0) \tag{1.4}$$

$$h(z) = h_1 z^n + h_2 z^{n-1} + \dots \qquad (h_1 \neq 0) \tag{1.5}$$

we define

$$N = \max(m,n) \tag{1.6}$$

Then N is a positive integer, and Theorem 1 below shows that N plays a role in describing the behavior of f which is similar to that of m and n in describing the polynomials g and h. For this reason N is called the degree of the rational function.

Given $a \in \mathbb{C}$, the equation $f(z) = a$ is satisfied when

$$g(z) - a\,h(z) = 0$$

and in general the left side is a polynomial of degree N and so has N solutions, with due account of multiplicity. Exceptions are $a = 0$ (when $N = n > m$) and $a = g_1/h_1$ in the case $m = n = N$. In addition, since g and h have no common factors, the poles $(a = \infty)$ of f arise precisely from the zeroes of h, of which there are n. For a given rational function exactly one of these three alternatives yields a number different from N.

The situation becomes less complicated once the behavior of f at ∞ is considered; this analysis is (as always) made by studying $f^*(z) = f(1/z)$ [also a rational function] near $z = 0$. It is easy to check that if $m < n = N$, then f has a zero of multiplicity $N-m$ at ∞ and if $n < m = N$, then f has a pole of multiplicity $N-m$ at ∞. We now give details for the case $N = m = n$, so that the equation $g(z) - (g_1/h_1)h(z) = 0$ is of degree $k(< N)$. Near ∞,

$$g(z) - (g_1/h_1)h(z) = \{\lambda + o(1)\}z^k \qquad (\lambda \neq 0)$$

i.e.

$$f(z) = \frac{g(z)}{h(z)} = h_1^{-1}[g_1 + \{\lambda + o(1)\}z^{k-N}] \qquad (\lambda \neq 0)$$

so that near $z = 0$

$$f^*(z) \equiv f(1/z) = h_1^{-1}[g_1 + \{\lambda + o(1)\}z^{N-k}] \qquad (\lambda \neq 0)$$

By our convention, this means that f assumes the value $g_1h_1^{-1}$ at ∞ with multiplicity $N-k$, and we have proved

<u>Theorem 1</u>. <u>Let</u> $f(z)$ <u>be rational of degree</u> N <u>as defined in</u> (1.3)-(1.6). <u>Then if due account is made of multiplicity</u> f <u>assumes every extended complex number precisely</u> N <u>times in the extended plane</u>.

Theorem 1 shows that the degree of a rational function shares an important property commonly associated to the degree of a polynomial: it counts the number of solutions of $f(z) = a$. From our perspective, however, it has one advantage. Since a polynomial of degree m assumes every finite value m times in the finite plane, one is tempted to ignore the behavior at ∞. However, to attain a comparable result for rational function one must consider ∞ as well.

There is another way that this example can prepare us for the Nevanlinna theory, but to do this it is necessary to investigate the structure of the surface S in more detail, where S is defined in (1.1) or (1.2). Since f is rational, it is convenient to replace (1.1) by

$$S = \{(z, f(z));\ z \in \mathbb{C}^*\}$$

(where $\mathbb{C}^* = \mathbb{C} \cup \{\infty\}$) or, equivalently,

$$S = \{(w,z);\ z, w \in \mathbb{C}^*,\ w = f(z)\} \tag{1.7}$$

One reason for this is that since $\mathbb{C}^*$ is <u>compact</u>, so is S. Associated with S is the natural projection

$$\pi\colon S \to \mathbb{C}^*, \quad \pi(w,z) = w \tag{1.8}$$

and since $\pi \circ f = f$ [here f is used in two senses; on the left as the topological map from $\mathbb{C}^*$ to S, and on the right in the "traditional manner" from $\mathbb{C}^*$ to $\mathbb{C}^*$] π is continuous. According to Theorem 1, over each $w_o \in \mathbb{C}^*$ are N points z if each point is counted with regard to multiplicity.

Suppose, for example, that $f(z_o) = w_o$ $(z_o, w_o \neq \infty)$. Then there exists $\delta > 0$ such that

$$f(z) = w_o + (z - z_o)^k f_o(z) \qquad (|z - z_o| < \delta)$$

where $k \geq 1$ and f_o is holomorphic and non-vanishing in $|z - z_o| < \delta$, or equivalently (cf. [2], pp. 131-3)

$$\begin{aligned} f(z) &= w_o + [(z - z_o)f_1(z)]^k \\ &\equiv w_o + \zeta(z)^k \end{aligned} \tag{1.9}$$

with f_1 holomorphic and non-vanishing there. Since $\zeta'(z_o) = f_1(z_o) \neq 0$, it follows that ζ is univalent for

$|z - z_o| < \delta_1$ (for some $\delta_1 < \delta$) and the image under ζ of $|z - z_o| < \delta_1$ covers a neighborhood of w_o; thus the last expression in (1.9) represents f as the simplest type of composition near z_o. In particular, to each δ_o, $\delta_1 > \delta_o > 0$ corre ponds $\varepsilon > 0$ such that each w with $0 < |w - w_o| < \varepsilon$ has exactly k distinct pre-images in the disc $0 < |z - z_o| < \delta_o$. Of course, w_o has only one pre-image from this region, namely z_o, also with multiplicity k.

In order that $k > 1$, it is necessary that $f'(z_o) = 0$, and a direct computation using (1.3)-(1.6) shows this can happe at no more than $2N - 1$ distinct points w_o. Similarly, f c assume multiple poles at no more than $\frac{1}{2}N$ distinct points in the extended plane, and we denote by R this (exceptional) set of at most $\frac{5}{2}N - 1$ points (R = <u>ramification set</u>).

Now choose $w_o \notin R$ where $w_o \neq \infty$. It then follows from (1.9) (with $k = 1$) that if ε is sufficiently small, the disc $|w - w_o| < \varepsilon$ does not meet R, and that the solutions of $f(z) = w$ with $|w - w_o| < \varepsilon$ are attained in N disjoint discs $|z - z_j| < \delta$ $(j = 1,\ldots,N)$ where $f(z_j) = w_o$. We assu for convenience that no $z_j = \infty$. The topology on S has been assigned to make f a homeomorphism from $\mathbb{C}^*$ to S. Hence, S contains N disjoint open sets which project by π (cf. (1. univalently onto a domain which contains $|w - w_o| < \varepsilon$. In the parlance of Riemann surfaces, this is expressed by saying that S contains N schlicht discs over $|w - w_o| < \varepsilon$, and if $\infty \notin R$ this also holds for $|w| > \varepsilon^{-1}$ if ε is sufficiently small.

Next, let $w_o \in R$ (we take $w_o \neq \infty$ for convenience) so that (1.9) holds with $k \geq 1$ for some z_o. Let $\{z_j\}$ be the full set of pre-images of w_o, and choose δ and then ε so that all solutions of $f(z) = w$ (with $|w - w_o| < \varepsilon$) occur in the disjoint discs $|z - z_j| < \delta$ and $\{0 < |w - w_o| < \varepsilon\} \cap R = \phi$. If $0 < |w_1 - w_o| < \varepsilon$, Theorem 1 implies that w_1 has N pre-images in the z-plane, but k of these occur in $|z - z_o| < \delta$. Since f is a homeomorphism between $\mathbb{C}^*$ and S, it follows that $(w_o, z_o) \in S$ has a <u>connected</u> open neighborhood which projects by π onto $|w - w_o| < \varepsilon$. This neighborhood forms a k-sheeted covering of $|w - w_o| < \varepsilon$, with (w_o, z_o) a <u>branch point</u> of order k-1 [when k = 1, this term is not used].

To summarize in more suggestive language, S is said to have N sheets, which are connected at the points of ramification (branch points).

We can now prove the Riemann-Hurwitz theorem.

<u>Theorem 2</u>. *Let S be the Riemann surface of the rational function f(z), with deg f = N. If $\Sigma(k-1)$ is the sum of all orders of branch points of S, then*

$$\Sigma(k-1) = 2N - 2 \tag{1.10}$$

<u>Proof</u>. A fact of combinatorial topology, due to Euler, is needed. If a surface S is divided into curvilinear triangles any two of which are either disjoint or meet at a point or along a full side, and if V is the number of vertices, E the number of edges and F the number of faces of this triangulation, then the sum

$$\chi(S) = F - E + V$$

is independent of the given triangulation (Euler characteristic). Further, if S and S' are homeomorphic, $\chi(S) = \chi(S')$.

Here S is the Riemann surface of the rational function f(z), and since f maps $\mathbb{C}^*$ homeomorphically to S, $\chi(S) = \chi(\mathbb{C}^*)$

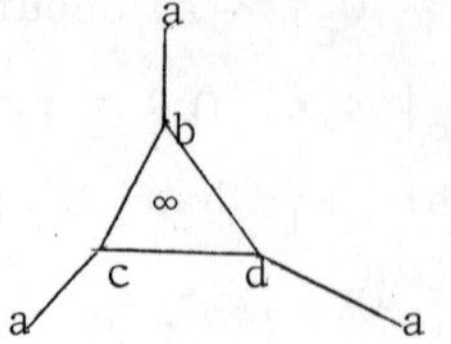

$= 2$ (to compute $\chi(\mathbb{C}^*)$ use the triangulation in the diagram: $F = 4$, $E = 6$, $V = 4$).

Let the Riemann sphere $\mathbb{C}^*$ be triangulated by a triangulat T so that the points of R are among the vertices. Then the edges, vertices and faces may be "lifted" to S, inducing a triangulation of S. Let us give some details. Suppose $E_1, \dots, E_r$ are the edges of the triangulation T, and (for $1 \le j \le r$) let E_j^o be the open edge: E_j less end points so that $E_j^o \cap R = \phi$. For each j, let $z_j \in E_j^o$ and let t_j map $[-1,1]$ topologically onto E_j with $t_j(0) = z_j$ (of course, $t_j(x) \notin R$ unless perhaps $x = \pm 1$). Let δ_j be so small that $N_j = \{|z - z_j| < \delta_j\}$ avoids R and $\bigcup_{k \neq j} E_k$. Then S contains N disjoint arcs which project onto $E_j \cap N_j$. We now apply a standard connectedness argument to show that S contains N disjoint arcs which project onto all of E_j^o.

First, let $x_j^* = \sup\{0 \le \delta \le 1;$ S contains N disjoint arcs which project onto $t_j([0,\delta])\}$. We have already shown $x_j^* \ge \delta_j$. Suppose $x_j^* < 1$, and set $z_j^* = t_j(x_j^*)$. Since we have supposed $x_j^* < 1$, it follows that $z_j^* \notin R$, and so the argument of the last paragraph may be applied at z_j^* to continu these N disjoint arcs to project onto a full neighborhood of which contradicts the definition of x_j^* if $x_j^* < 1$. Thus $x_j^* = 1$, and similarly $1 = \sup\{0 \le \delta \le 1;$ S contains N disjoint arcs which project onto $t_j([-\delta,0])\}$.

This shows that each of the open edges of T may be lifted to form N disjoint arcs in S. Next, let $w_o \in V \cap R$, let (w_o, z_o) be a branch point of order k, and let $\mathbf{U}$ be the component of $\pi^{-1}(|w - w_o| < \varepsilon)$ which contains (w_o, z_o). Then if E_j is an edge which meets $|w - w_o| < \varepsilon$, the k pre-images of E_j meet at (w_o, z_o). The reader can show that "over" each triangle $T_j \subset T$ are N subsets $T_{j,k}$ $(1 \leq k \leq N)$ of S, each bounded by arcs of $\{\pi^{-1}(E_j)\}$ which are disjoint save perhaps at the points $\pi^{-1}(w)$ $(w \in R)$, and such that π maps each $T_{j,k}$ homeomorphically onto T_j. The $T_{j,k}$ are the "lifted" triangles.

We continue the proof of Theorem 2. If V, E, F and V_1, E_1, F_1 are the number of vertices, edges and faces on $\mathbb{C}^*$ and S respectively, then we have seen that

$$F_1 = NF, \quad E_1 = NE$$

$$V_1 = NV - \Sigma(k-1)$$

where N is the degree of f and the summation is made of all the orders of the distinct branch points of S. Since $F - E + V = 2$, (1.10) now follows:

$$2 = F_1 - E_1 + V_1 = N(F - E - V) - \Sigma(k-1)$$

$$= 2N - \Sigma(k-1)$$

From our point of view, the interest of (1.10) is two-fold. It first asserts that a non-linear rational function has branch points which contribute much ramification if N is large. On the other hand, the average of the ramification per sheet behaves nicely with N:

$$N^{-1}\Sigma(k-1) = 2 - 2N^{-1}$$

which augurs the importance of the number 2 in our study.

Exercise

The function $f(z) = e^z$ has no branch points ($f' \neq 0$, $(1/f)' \neq 0$). However, the sequence $f_N(z) = (1 + z/N)^N$ of rational functions is known to tend to $f(z)$ uniformly on compact subsets of the plane. What behavior does S_N, the surface of f_N, have? As $N \to \infty$?

II. INTRODUCTION TO DISTRIBUTIONS

The calculus offers convenient ways of characterizing properties of classes of functions, but these classical methods often apply directly to artificially restricted subclasses. When considering an entire or meromorphic function $f(z)$, it will often be desirable to apply calculus methods to

$$u(z) = \log|f(z)|$$

which is <u>harmonic</u> away from the zeros or poles of f. Even near the zeros or poles, u is integrable (either in the Lebesgue sense, or as an improper Cauchy integral) but problems arise in differentiation. One of the goals of this chapter is to show that, while in the classical sense (Δ = Laplacian)

$$\Delta \log|f| = 0 \qquad \text{a.e.} \tag{2.1}$$

(again because $\log|f|$ is harmonic away from the countable number of zeros and poles of f), in the sense of distributions we have (in a sense to be made precise later)

$$\Delta(\log|f|) = 2\pi[\Sigma\delta(a_\mu) - \Sigma\delta(b_\nu)] \tag{2.2}$$

where $\{a_\mu\}$ are the zeros and $\{b_\nu\}$ the poles of f and each a_μ, b_ν appears as warranted by multiplicity. The elegant formula (2.2) is known as Poincaré's equation. One of its attractions is that it dramatically exhibits the singularities of f. The ideas given here will also simplify and clarify some of our later work with subharmonic functions.

It will be necessary to introduce definitions and conventions which may seem artificial, so in the interest of concreteness we first outline how these notions can be used to analyse convex functions. The chapter then closes with a more formal introduction to distributions.

A. CONVEX FUNCTIONS. A function $g(x)$ with second derivatives at all x $(a \leq x \leq b)$ is <u>convex</u> if

$$g''(x) \geq 0 \qquad (a < x < b) \tag{2.3}$$

It is easy to anticipate and prove most of the important properties of convex functions by manipulating (2.3), but the definition (2.3) is far too restrictive to include all <u>geometrically</u> convex functions: a function $g(x)$ on $[a,b]$ is geometrically convex if whenever

$$g(x_0) = h(x_0), \quad g(x_2) = h(x_2) \qquad (x_0 < x_2)$$

with h <u>linear</u> (and so, according to (2.3), an extremal convex function) then

$$g(x) \leq h(x) \qquad (x_0 \leq x \leq x_2)$$

Alternatively, this may be written as

$$g(tx_0 + (1-t)x_2) \leq tg(x_2) + (1-t)g(x_0) \qquad (0 \leq t \leq 1) \tag{2.4}$$

For example, let $x_0 = a < x_1 < x_2 = b$, and for real numbers a_0, a_1 let

$$g(x) = \sum_{j \le i} a_j(x_j - x_{j-1}) + a_i(x - x_i) \qquad (x_i \le x \le x_{i+1}) \tag{2.5}$$

Note that g is continuous and has left- and right-hand derivatives at all x which agree if $x \notin \{x_i\}$, and g'' exists for $x \notin \{x_i\}$. A sketch of the graph of g shows that g is convex if and only if $a_1 \ge a_0$, but it is not apparent how this is reflected in differentiability requirements.

The solution to this is to replace (2.3) by a less restrictive definition. Let $C(a,b)$ be the continuous functions on $[a,b]$ and $C^k(a,b)$ the k-times continuously differentiable functions on $[a,b]$. Then $C_o(a,b)(C_o^k(a,b))$ is the subclass of $C(a,b)(C^k(a,b))$ of functions which vanish outside some compact subset of (a,b); these functions are called of <u>compact support</u>. The proper form of (2.3) is that

$$\int_a^b g(x)\phi''(x)\,dx \ge 0 \quad \text{for all} \quad \phi \in C_o^\infty(a,b);\ \phi \ge 0 \tag{2.3'}$$

The additional requirements of g can be made very weak (so long as there is a way to assign meaning to (2.3')) but for now we assume that g is <u>continuous</u> (see Theorem 1 below).

Exercise 1. Show that (2.3') implies (2.3) if $g \in C^2(a,b)$ (integrate by parts twice, using that $\phi = 0$ if $x \notin [a',b'] \subset (a,b)$. Note that the integral in (2.3') becomes $\int_a^b g''(x)\phi(x)\,dx$ which is obviously ≥ 0 if $\phi > 0$ and (2.3) holds).

In using definition (2.3') it is important that the class of admissible ϕ's be sufficiently rich.

Lemma 1. Given $0 < \varepsilon$, there exists $\phi_\varepsilon(x)$ $(-\infty < x < \infty)$ which vanishes outside $|x| \le \varepsilon$, is nonnegative and in $C^\infty(-\infty,\infty)$ with

$$\int_{-\infty}^{\infty} \phi_\varepsilon(x)dx = 1 \tag{2.6}$$

Proof. Let

$$\phi_\varepsilon(x) = \begin{cases} 0 & |x| \ge \varepsilon \\ p(\varepsilon)\cdot\exp\left\{\dfrac{-\varepsilon^2}{\varepsilon^2-x^2}\right\} & |x| < \varepsilon \end{cases}$$

where the constant $p(\varepsilon)$ is taken so (2.6) holds. Routine use of l'Hospital's rule shows that $\phi_\varepsilon \in C^\infty$, and ϕ_ε obviously vanishes if $|x| > \varepsilon$.

Definition. Let $\varepsilon_n \downarrow 0$. Then the associated sequence of functions $\phi_n(x)$ $(= \phi_{\varepsilon_n}(x))$ are called an approximate identity. (This arises from the readily verified fact that

$$\int_{-\infty}^{\infty} g(t)\phi_n(t)dt \to g(0) \qquad (n \to \infty) \tag{2.7}$$

for any continuous g, so that the "limit" of the ϕ_n behaves as an identity).

Exercise 2. Prove (2.7).

Exercise 3. Given $0 < \varepsilon_1 < \varepsilon_2$, find a C^∞ function $\phi(x)$ which vanishes off $|x| \le \varepsilon_2$, is 1 when $|x| \le \varepsilon_1$, and $0 \le \phi(x) \le 1$ for all x.

The $\{\phi_n\}$ are very nice functions, and (2.7) shows they tend to a limit in the following sense: if $\delta(0)$ is a point-mass of measure 1 concentrated at $x = 0$, then (2.7) may be written as

$$\{\phi_n\} \to \delta(0) \qquad (n \to \infty) \tag{2.7'}$$

where the convergence in (2.7') means precisely (2.7).

The family $\{\phi_n\}$ allows us to show (2.3') sufficiently general to characterize (continuous) geometrically convex functions. For suppose g is continuous and there exist $a \leq x_0 < x_2 < b$ with, for some fixed $t \in (0,1)$,

$$g(tx_0 + (1-t)x_2) > tg(x_0) + (1-t)g(x_2) + 2p \tag{2.8}$$

where $p > 0$. This suggests that if there were a $\phi \in C_0^\infty(a,b)$ with

$$\phi''(u) = t\delta(u-x_0) - \delta(u-x_1) + (1-t)\delta(u-x_2) \tag{2.9}$$

(where we have let

$$x_1 = tx_0 + (1-t)x_2) \tag{2.10}$$

then (2.3') will fail for g, since (2.7), (2.8) and (2.9) imply

$$\int_a^b \phi''(u)g(u)\,du \leq -2p \tag{2.11}$$

It is easy to see that there is a function $\phi \in C_o(a,b)$ which satisfies (2.9), but $\phi \notin C_o^1(a,b)$, to say nothing of $C_o^\infty(a,b)$. Indeed, an integration of (2.9) gives:

$$\phi'(u) = \begin{cases} 0 & u < x_0 \\ t & x_0 \leq u < x_1 \\ (t-1) & x_1 < u < x_2 \\ 0 & x_2 < u \end{cases}$$

so

(2.12)
$$\phi(u) = \begin{cases} 0 & u \le x_0 \\ t(u-x_0) & x_0 \le u \le x_1 \\ t(x_1-x_0)+(t-1)(u-x_1) & x_1 \le u \le x_2 \\ t(x_1-x_0)+(t-1)(x_2-x_1)=0 & x_2 \le u \end{cases}$$

(where (2.10) is used to simplify the last line).

To achieve a suitable replacement p_n for this idealized solution, δ in (2.9) will be replaced by ϕ_n with n large, and then the factor $(1-t)$ slightly adjusted so that $p_n \in C_0^\infty(a,b)$.

__Lemma 2__. __Let__ n __be so large that__

$$\varepsilon_n < \frac{1}{4}\min\{(x_0-a),(x_1-x_0),(x_2-x_1),(b-x_2)\}$$

__Then there exists a unique__ t_n __such that the function__ $p_n(u)$ __with__

(2.13)
$$\begin{cases} p_n(a) = 0, \quad p_n'(a) = 0 \\ p_n''(u) = t_n\phi_n(u-x_0) - \phi_n(u-x_1) + (1-t_n)\phi_n(u-x_2) \end{cases}$$

$\in C_o^\infty(a,b)$. __Further__

$$t_n \to t \qquad (n \to \infty)$$

__Proof__. Let $h_{\lambda,n} \in C^\infty(-\infty,\infty)$ satisfy

$$h_{\lambda,n}(a) = h_{\lambda,n}'(a) = 0$$

$$h_{\lambda,n}''(u) = \lambda\phi_n(u-x_0) - \phi_n(u-x_1) + \lambda\phi_n(u-x_2)$$

Then it is clear from Lemma 1 and the choice of ε_n that

(2.14)
$$\begin{cases} h_{\lambda,n}'(u) = 0 \qquad (u \le x_1-\varepsilon_n;\ u \ge x_2+\varepsilon_n) \\ h_{\lambda,n}(u)-h_{\lambda',n}(u) = (\lambda-\lambda')\int_a^u dv\int_a^v \phi_n(s-x_o)-\phi_n(s-x_2)ds \end{cases}$$

Since $\int_{-\varepsilon}^{\varepsilon} \phi_n(u)du = 1$, the restriction on ε_n and (2.14) yield a unique t_n so that $h_{t_n,n}(x_2 + \varepsilon_n) = 0$ and, hence, that $h_{t_n,n}(u) = 0$ for $u \geq x_2 + \varepsilon_n$. Thus $p_n(n) \equiv h_{t_n,n}(u) \in C_0^\infty(a,b)$.

Finally, we observe from (2.5) and (2.10) that as $n \to \infty$

$$\begin{aligned} h_{t,n}(b) &= \int_a^b dv \int_a^v [t\phi_n(s-x_0) - \phi_n(s-x_1) + (1-t)\phi_n(s-x_2)]ds \\ &= \int_a^b (b-s)[t\phi_n(s-x_0) - \phi_n(s-x_1) + (1-t)\phi_n(s-x_2)]ds \\ &\to (b-x_0)t - (b-x_1) + (b-x_2)(1-t) = 0 \end{aligned}$$

so this with the second equation in (2.14) shows that $t_n \to t$. Thus $t_n > 0$ for all large n, and we see from the choice of ε_n that p_n first increases and then decreases. Thus $p_n(u) \geq 0$ for all u. (I thank Paul Gauthier for observing a gap in the first version of this Lemma.)

It is now very easy to show that (2.3') fails for our funct g which satisfies (2.8). According to (2.9) and (2.13),

$$\begin{aligned} \int_a^b g(u)p_n''(u)du &= \int_a^b g(u)[t\phi_n(u-x_0) - \phi_n(u-x_1) + (1-t)\phi_n(u-x_2)]du \\ &\quad + (t-t_n)\int_a^b \phi_n(u-x_2)du \end{aligned}$$

and since $t_n \to t$, we deduce from (2.7) that

$$\int_a^b g(u)p_n''(u)du \leq -p + o(1) < 0$$

As a corollary, we have that if g is continuous and

(2.15) $$\int_a^b g(x)\phi''(x)dx = 0 \quad \text{for all} \quad \phi \in C_0^\infty(a,b)$$

then g is linear. However, more is true.

Theorem 1. *Let* g *be locally integrable on* $[a,b]$ *and* (2.15) *hold. Then there is a linear function* g^* *such that* $g = g^*$ a.e.

Remarks. One has to allow an exceptional set of measure 0 since the character of (2.15) is unaffected by the behavior of g on such sets. A function $g(x)$ is *locally integrable* if each point has a neighborhood on which $\int |g|\,dx < \infty$. This definition readily extends to functions defined on m-dimensional domains.

Proof. Let $\phi(x)$ have integral 1 and vanish for $|x| \geq 1$ and set

$$g_n(x) = \int_{-\infty}^{\infty} g(x-n^{-1}y)\phi(y)\,dy$$

$$= n\int_{-\infty}^{\infty} g(y)\phi(n(x-y))\,dy$$

(where we take $g = 0$ for $y \notin [a,b]$). Then $g_n \in C^\infty[a+\delta,b-\delta]$ if $n > \delta^{-1}$, and an application of Fubini's theorem (legitimate since g is locally integrable) gives that (2.15) holds for $\phi \in C_0^\infty[a+\delta,b-\delta]$. This shows that g_n is linear (modulo definition on sets of measure 0) with slope μ_n. Finally, since $\int \phi = 1$,

$$\int_a^b |g_n - g|\,dx = \int_{-1}^{1} \phi(y)\,dy \int_a^b |g(x-n^{-1}y) - g(x)|\,dx = o(1) \tag{2.16}$$

as $n \to \infty$, and (cf. [42], p. 376-377) hence $\int_a^b |g_n - g_m|\,dx = o(1)$ $(m,n \to \infty)$. This means that the μ_n must tend to some real number μ and so (2.16) implies g is linear with slope μ.

Convexity has thus been described in terms of the differential operator g'', but the same ideas may be carried to any differential operator, and, in m-dimensions where the approximate identity Φ_n is

$$\Phi_n(x) = \frac{1}{\mu_m}\,\phi_n(||x||) \qquad (x \in \mathbb{R}^m) \tag{2.17}$$

for some μ_m. The theory of distributions formalizes this view-point.

B. INTRODUCTION TO DISTRIBUTIONS. We work in $\mathbb{R}^m$, with m usual 1 or 2. Let Ω be an open subset of $\mathbb{R}^m$. If $s = (s_1,\ldots,s_m)$ a multi-index of nonnegative integers, let

$$|s| = \sum s_j \quad , \quad s! = s_1!s_2!\ldots s_m!$$

$$D^s = \frac{\partial^{|s|}}{\partial x_1^{s_1}\ldots\partial x_m^{s_m}}$$

Then $C^k(\Omega)$ consists of these functions g for which all $D^s g$ $(|s| \le k)$ are continuous in Ω; the notions $C^\infty(\Omega)$, $C_o^k(\Omega)$, $C_o^\infty(\Omega)$ may be similarly adapted to n-dimensions.

Exercise 4. Let K be compact in Ω. Find a $C_o^\infty(\Omega)$ function which is identically 1 on a neighborhood of K. (Hint: Choose open Ω_1, cpt K_1, $K \subset \Omega_1 \subset K_1 \subset \Omega$ and consider $\psi_n(x) = \int_{K_1} \Phi_n(x-y)dy$ with n large and Φ as in (2.17). Th $\psi_n \in C_o^\infty(\Omega)$ requires careful justification.

Exercise 5. Let $g \in C_o(\Omega)$. Show that $g = \lim g_n$ with $g_n \in C_o^\infty(\Omega$ and all $g_n = 0$ off some compact subset $K \subset \Omega$.

A sequence $\{g_n\} \in C_o^\infty(\Omega)$ <u>converges to</u> $g \in C_o^\infty(\Omega)$ (in the sens of test-functions) if

(i) there is a compact K of Ω with $g_n(x) = 0$ if $x \notin K$ and $n \ge n_o$;

(ii) For any multi-index s, the sequence

$$\{D^s g_n(x)\}$$

converges uniformly to $D^s g$ in K.

<u>Definition</u>. A distribution T in Ω is a continuous linear functional on $C_o^\infty(\Omega)$. This means that

$$T(\lambda_1 g_1 + \lambda_2 g_2) = \lambda_1 T g_1 + \lambda_2 T g_2$$

and if $g_n \to g$ (in the sense of test-functions) then

$$Tg_n \to Tg$$

The most common examples of distributions used here are:

(a) Let $u(x)$ be defined in Ω so that

$$\int_K |u(x)|\,dx < \infty \qquad (dx = dx_1 \ldots dx_n)$$

for each compact K of Ω (i.e. u is <u>locally integrable</u>). Then

$$T_u g = \int_\Omega g(x)u(x)\,dx$$

defines (is) a distribution. One can show (cf [48], p. 48] that unless $u_1 = u_2$ a.e., then $T_{u_1} g$ will differ from $T_{u_2} g$ for some $g \in C_o^\infty(\Omega)$. Thus we can (and shall) identify locally integrable functions with a subset of distributions.

(b) To a discrete set $E = \{a_n\} \subset \Omega$ is associated the distribution

$$T_E = \{a_n\} = \sum \delta(a_n)$$

which means

$$T_E(g) = \sum_E g(a_n) = \sum_E g(x)\delta(a_n)$$

(a finite sum if $g \in C_0^\infty$).

(c) (generalization of (b)). To a σ-finite Borel measure m in Ω is associated the distribution

$$T_m g = \int_\Omega g(x)\,dm(x)$$

Exercise 6. Let $\Omega = \mathbb{R}^1$, and $Tg = g'(0)$. Is T a distribution? Is it one of these examples?

Distributions may be multiplied and often convolved. We ignore these possibilities since they require non-routine preparation, but we will mention some uses of these notions later. It is easier to describe how distributions are differentiated:

$$(2.18) \qquad (D^s T)g = (-1)^{|s|} T(D^s g) \qquad (g \in C_0^\infty(\Omega))$$

If $T = T_u$ (cf (A)) with $u \in C_0^\infty(\Omega)$, repeated integration by parts shows that $D^s T_u = T_{D^s u}$

Exercise 7. Show that $D^s T$ is always a distribution.

Exercise 8. Let $x_0 < x_1 < x_2$. Find a solution to the distribution equation $g'' = \delta(x_1)$. Find a solution to this equation which vanishes at x_0 and x_2.

To apply these notions in the plane, we need several analogu[es] of integration by parts. These are all consequences of

Green's theorem. Let D be a domain in the x-y plane and let th[e] boundary of D consist of piecewise smooth[1] simple closed curves $C_1,\ldots,C_n$ where C_1 includes $C_2,\ldots,C_n$ in its

[1] A smooth curve C is one representable by continuous functions $x = \phi(t)$, $y = \psi(t)$ $(a \leq t \leq b)$ with $\phi, \psi \in C^1(a,b)$ and $|\phi'(t)| + |\psi'(t)| > 0$. If ϕ and ψ are continuous on $[a,b]$ and $[a,b]$ divides into finitely many subintervals on each of which ϕ and ψ are C^1, then C is piecewise smooth.

<u>interior</u>. <u>Let</u> P <u>and</u> Q <u>be continuous with continuous partials in</u> D. <u>Then</u>

$$\oint_{C_1} Pdx + Qdy + \sum_{2}^{n} \oint_{C_j} Pdx + Qdy$$
$$= \iint_D \left(\frac{\partial Q}{\partial x} - \frac{\partial P}{\partial y}\right)dxdy$$

To state other forms of Green's theorem, let

$F, G \in C^2(D)$

η = outward normal

s = arc length

∂D = boundary of D

$\nabla F = (F_x, F_y)$ (gradient)

$\Delta F = F_{xx} + F_{yy}$ (Laplacian)

Then

(2.19) $$\int_{\partial D} \frac{\partial F}{\partial \eta}\, ds = \iint_D \Delta F\, dxdy$$

(2.20) $$\int_{\partial D} \left(F\frac{\partial G}{\partial \eta} - G\frac{\partial F}{\partial \eta}\right)ds = \iint_D (F\Delta G - G\Delta F)dxdy$$

It is now possible to prove the result outlined at the beginning of this section (cf. (2.1), (2.2)).

<u>Theorem 2</u>. <u>Let</u> f <u>be meromorphic and nonconstant in</u> $\Omega \subset \mathbb{R}^2$ <u>with zeros</u> $\{a_\mu\}$, <u>poles</u> $\{b_\nu\}$. <u>Then</u> $\log|f|$ <u>is locally integrable and as a distribution</u>,

(2.21) $$\Delta(\log|f|) = 2\pi[\sum\{a_\mu\} - \sum\{b_\nu\}] = 2\pi[\sum\delta(a_\mu) - \sum\delta(b_\nu)]$$

(For an explanation of the right side of (2.21), see Example (B) above.)

Remark. According to (2.18) it is necessary to show that

$$\iint_{\Omega} (\log|f|)(\Delta g)\,dxdy = 2\pi[\sum g(a_\mu) - \sum g(b_\nu)] \text{ [illegible]} \tag{2.22}$$

where for each g only finitely many terms appear in each sum.

Proof. If $f(z_0) \neq 0, \infty$, $|\log|f||$ is bounded and infinitely differentiable in a neighborhood of z_0. If $f(z_0) = 0$ or ∞, then $f(z) = (z - z_0)^n f_0(z)$, n an integer and $f_0 \neq 0, \infty$ in some neighborhood U of z_0; thus

$$\log|f(z)| = n \log|z - z_0| + O(1) \qquad (z \in U)$$

and so $\log|f| \in L^1(U)$. Thus $\log|f|$ induces a distribution according to example (A).

Let S be an open set with $\overline{S} \subset \Omega$ ($\overline{}$ = closure), $g = 0$ on $\Omega - S$, and ∂S disjoint from the $\{a_\mu\}$, $\{b_\nu\}$. Then if ε is sufficiently small,

$$P = \bigcup \{|z-a_\mu| < \varepsilon\} \bigcup \{|z-b_\nu| < \varepsilon\}$$

meets S in a finite union of closed disjoint discs. If ε is small,

$$\iint_{\Omega - P} (\log|f|)\Delta g \, dxdy \tag{2.23}$$

differs arbitrarily little from the left side of (2.22), but in (2.23) all functions are infinitely differentiable. Let $D = \Omega - P$, $F = \log|f|$, $G = g$ in (2.20). Then, letting

$\{a_\mu\} \cup \{b_\nu\} = \{z_n\}$ and recalling that $g = 0$ on $\Omega - S$, we have[2]

$$\iint_{\Omega-P} (\log|f|)\Delta g \, dxdy = \iint_{\Omega-P} \{\Delta \log|f|\} g \, dxdy$$

$$+ \sum_j \int_{|z-z_j|=\varepsilon} (\log|f|) \frac{\partial g}{\partial \eta} ds - \sum_j \int_{|z-z_j|=\varepsilon} g \frac{\partial \log|f|}{\partial \eta} ds$$

where the standard orientation is used in the line integrals. Now $\Delta \log|f| = 0$ on $\Omega - P$ since $\log|f|$ is harmonic, and if $|z - z_j| = \varepsilon$, $|\log|f(z)|| \left|\frac{\partial g(z)}{\partial \eta}\right| = 0(\log(1/\varepsilon))$, $g(z) = g(z_j) + o(1)$ and $\partial(\log f)/\partial\eta \sim n_j/\varepsilon$ where $n_j > 0$ for a zero, $n_j < 0$ for a pole at z_j. Thus as $\varepsilon \to 0$, we obtain (2.22).

C. POSITIVE DISTRIBUTIONS

A distribution T is positive $(T \geq 0)$ if

$$T\phi \geq 0 \qquad \phi \in C_0^\infty(\Omega), \quad \phi \geq 0$$

<u>Theorem 3.</u> <u>Every positive distribution T has the form</u>

$$T\phi = \int_\Omega \phi(x)\,d\mu(x) \tag{2.24}$$

<u>where μ is a unique positive σ-finite measure on Ω.</u>

Exercise 9. Show that the converse is true.

<u>Proof</u> (sketch). The key to the proof is the following: (I) T may be extended to a continuous functional on $C_0(\Omega)$. (A continuous functional L on $C_0(\Omega)$ assigns to each $\phi \in C_0(\Omega)$ a scalar $L\phi$ such that if $\{\phi_n\} \in C_0(\Omega)$ with all $\phi_n \equiv 0$ outside a compact subset $K \subset \Omega$ and $\phi_n \to \phi$ uniformly on K, then $L\phi_n \to L\phi$).

---

[2] The boundary of Ω-P divides into circles centred at the a_μ, b_ν and the boundary of Ω. It is possible that $\partial\Omega$ is pathological enough so Green's theorem may not be automatically applied. However, since g vanishes off S, we may, if necessary, find an open Ω_1 with $S \subset \Omega_1 \subset \Omega$ and $\partial\Omega_1$ smooth and integrate on Ω_1-P.

Exercise 10. Find a distribution which does not so extend.

We first show that T is continuous in this sense on $C_0^\infty(\Omega)$. Thus, let $\phi_n \to 0$ with all $\phi_n = 0$ outside a compact subset K of Ω. Then if $\psi \in C_0^\infty(\Omega)$, $\psi \equiv 1$ on K (cf Ex.4), $\psi \geq 0$, the functions $\phi_n \psi \to 0$ uniformly and vanish outside K. Thus given $\varepsilon > 0$

$$-\varepsilon\psi \leq \phi_n \leq \varepsilon\psi$$

holds for large n; i.e.

$$|T\phi_n| \leq \varepsilon T\psi \qquad \text{(n large)}$$

or $T\phi_n \to 0$.

This reasoning shows that if $\phi_n \in C_0^\infty$, $\phi_n \equiv 0$ outside a a compact K of Ω, and $\{\phi_n\} \to \{\phi\}$ uniformly (with now $\phi \in C_0(\Omega)$), then the numbers $T\phi_n$ form a Cauchy sequence, and the definition $T\phi = \lim T\phi_n$ extends T to ϕ unambiguously. According to Ex. 5, any $\phi \in C_0^\infty(\Omega)$ may be so represented and thus (I) follows.

It is obvious that T is linear on $C_0(\Omega)$ since T is continuous and linear on $C_0^\infty(\Omega)$.

The proof is completed by an appeal to the Riesz representat theorem (cf. [22], §56): every continuous linear functional on $C_0(\Omega)$ is of the form (2.24).

Remark. One advantage of (2.24) is that now standard measure theory allows T to be defined on some discontinuous functions so long as the expression (2.24) makes sense with ψ in place of g (i.e. $\psi \in L^1(d\mu)$). For example, this will provide the rationale for equation (5.35) which plays a key role in the Nevanlinna theory.

D. MORE ON CONVEX FUNCTIONS. As an application of these ideas, we derive some basic facts about convex functions; these will also be used later. We formally define a convex function to be a locally integrable function $g(x)$ for which $g'' \geq 0$ in the sense of distributions. It is also not hard to show that a function which satisfies (2.3') meets these criteria.

Exercise 11. Let g satisfy (2.4) with $-\infty \leq g(x) < \infty$. Show g is locally integrable.

Theorem 4. *Let g be convex on $a < x < b$ $(|a|,|b| < \infty)$. Then g is continuous and has right and left derivatives at each point, which agree off an at most countable set. Further, if g' denotes either of these derivatives and $a_1 < x_1 < b$*

$$g(x) = g(x_1) + \int_{x_1}^{x} g'(t)\,dt$$

Proof. Theorem 3 implies that (in the sense of distributions) g'' is a measure μ on the domain of g. Since $\int_a^b d\mu$ may be infinite, we consider any $x_1 \in (a,b)$, and set

$$\mu^+(x) = \mu[x_1,x] \qquad (x > x_1) \tag{2.25}$$
$$= -\mu(x,x_1] \qquad (x \leq x_1)$$

$$\mu^-(x) = \mu[x_1,x) \qquad (x > x_1) \tag{2.26}$$
$$= -\mu[x,x_1] \qquad (x \leq x_1)$$

(thus μ^+, μ^- are increasing functions, μ^+ is right-continuous and μ^- left-continuous). Then we formally introduce a function g_1 with $g_1(x_1) = 0$ whose second derivative is also μ:

$$(2.27)\qquad g_1(x) = \int_{x_1}^{x} g_1'(y)\,dy = \int_{x_1}^{x} dy \int_{x_1}^{y} d\mu(t)$$

$$= \int_{x_1}^{x} (x-t)\,d\mu(t) \qquad (a < x < b)$$

The last equation provides a proper definition. Note that $d\mu$ can be replaced by $d\mu^{+}$ or $d\mu^{-}$ without changing the values of g_1. It is routine to check that

$$(2.28)\qquad \int g_1\phi''dx = \int \phi d\mu (= \int g\phi'')$$

for all test functions ϕ. For example, if ϕ is 0 outside $[a_1,b_1] \subset (a,b)$ and, say, $a_1 < x_1 < b_1$, the left side of (2.28) is computed by dividing the range of integration into $[a_1,x_1]$ and $[x_1,b_1]$ and we see that

$$\int_{a_1}^{x_1} g_1\phi''dx = -\int_{a_1}^{x_1} \phi''(x)\,dx \int_{x}^{x_1} (x-t)\,d\mu(t)$$

$$= \int_{a_1}^{x_1} d\mu(t) \int_{t}^{x_1} (x-t)\phi''(x)\,dx$$

$$= \int_{a_1}^{x_1} \phi(t)\,d\mu(t) \qquad \text{(Integrate by parts)}$$

According to Theorem 2 $g-g_1$ is linear and so

$$(2.29)\qquad g(x) = g_1(x) + g(x_1) + A(x-x_1) \qquad (a < x < b)$$

for some constant A. It now follows directly from (2.27) and (2.29) that g is continuous. Straightforward computations using (2.25), (2.26), (2.27) and (2.29) also show

$$(2.30)\qquad g'_{+}(x) = A + \mu^{+}(x) \qquad (a < x < b$$

$$(2.31)\qquad g'_{-}(x) = A + \mu^{-}(x) \qquad (a < x < b$$

and since $\mu^+ = \mu^-$ off a countable set, g' exists off such a set. The final assertion is a consequence of (2.27) and (2.29).

Corollary 1. Let g *be as in Theorem* 3. *Given* $x_1 \in (a,b)$, *there is a* k *such that*

$$g(x) \geq g(x_1) + k(x-x_1) \qquad (a < x < b) \tag{2.32}$$

Proof. Recall the function g_1, which was defined by (2.27). Clearly $g_1(x_1) = 0$, and (2.27) readily gives

$$g_1(x) \geq g_1(x_1) + 0\cdot(x-x_1) \qquad (a < x < b)$$

The Corollary is proved with $k = A$, with A the constant of (2.29).

Corollary 2 (Minkowski). *Let* g *be convex on* $(-\infty,\infty)$ *and* f *integrable on* $[\alpha,\beta]$. *Then*

$$\frac{1}{\beta-\alpha}\int_\alpha^\beta g(f(x))dx \geq g(m) \tag{2.33}$$

where m *is the mean value*

$$m = \frac{1}{\beta-\alpha}\int_\alpha^\beta f(x)dx \tag{2.34}$$

Proof. Choose k so

$$g(y) \geq g(m) + k(y-m) \qquad (-\infty < y < \infty)$$

Then for each $x \in (\alpha,\beta)$, we take $y = f(x)$ and obtain the Corollary upon integrating both sides.

III. SUBHARMONIC FUNCTIONS: POTENTIAL THEORY

Here the analysis takes place in $\mathbb{R}^2$, although much extends to $\mathbb{R}^n$.

A region Ω in $\mathbb{R}^2$ (or $\mathbb{R}^n$) is an open, connected set.

A good reference to much of this theory and more is [8].

A. HARMONIC FUNCTIONS

Definition. Let Ω be a region in $\mathbb{R}^2$. A function $u \in C(\Omega)$ is C^∞-harmonic if $u \in C^\infty(\Omega)$ and

$$\Delta u = u_{xx} + u_{yy} = 0 \tag{3.1}$$

Example. If u is the real part of an analytic function f, then u is C^∞-harmonic. If Ω is simply-connected this is the only example. Indeed, a harmonic conjugate $v(z)$ is given by choosing $v(z_0)$ arbitrarily and setting

$$\begin{aligned} dv &= v_x dx + v_y dy \\ &\equiv -u_y dx + u_x dy \end{aligned} \tag{3.2}$$

where the latter expression depends only on u. Since Ω is simply-connected, Green's theorem (cf. Ch. II) yields that $\int_{z_0}^{z} d$ depends only on z for $z \in \Omega$, so $f = u + iv$ is single-value and analytic in Ω.

Upon taking real parts of Cauchy's formula

$$f(z_0) = (2\pi i)^{-1} \int_{|\zeta - z_0| = r} f(\zeta)(\zeta - z_0)^{-1} d\zeta \qquad (r < R)$$

we obtain

$$u(z_0) = \frac{1}{2\pi} \int_0^{2\pi} u(z_0 + re^{i\theta}) d\theta \qquad (r < R) \tag{3.3}$$

(Mean-value property)

and the usual maximum principle applied to $g(z) = \exp f(z)$ giv

Theorem 1. Let u be C^∞-harmonic in $\{|z - z_0| < R\}$, and

$$M(r) = \max_{|z - z_0| = r} u(z) \qquad (r < R) \tag{3.4}$$

Then

$$u(z) \leq M(r) \qquad (|z-z_0| < r)$$

and if equality holds for a single z *in* $|z-z_0| < r$, u *is constant.*

Exercise 1. Show if u is harmonic in $\{0 < |z-z_0| < \varepsilon\}$ and bounded as $z \to z_0$, then u extends to be harmonic at z_0. (Hint: now (3.2) does not determine a single-valued function v. However, Green's theorem shows that $\alpha = \int_{|z-z_0|=h} dv$ is independent of h for $0 < h < \varepsilon$. Apply the theorem on removable singularities of analytic functions to $g(z) = \exp\{\lambda f(z)\}$ for a good choice of λ).

Theorem 2 (Poisson's formula). Let $w = Re^{i\phi}$, $z = re^{i\theta}$,

$$(3.5) \qquad P(w,z) = \mathscr{R}\{\frac{w+z}{w-z}\} = \frac{R^2-r^2}{R^2+r^2-2Rr\cos(\phi-\theta)} \qquad \text{(Poisson kernel)}$$

and $V(\phi)$ *a* 2π*-periodic continuous function. Let*

$$(3.6) \qquad v(z) = \frac{1}{2\pi}\int_0^{2\pi} V(\phi)P(Re^{i\phi},z)\,d\phi$$

Then v *is* C^∞*-harmonic in* $\{|z| < R\}$ *and*

$$(3.7) \qquad v(z) \to V(\alpha) \qquad (z \to Re^{i\alpha},\ |z| < R)$$

Remarks. v is called the *Poisson integral* of V, and is often written $v = P_V$.

The importance of Theorem 2 is that it gives a concrete way to construct harmonic functions. More precisely, formula (3.6) solves the *Dirichlet problem* for $\{|z| \leq R\}$: it extends any continuous function on the boundary of this set to a C^∞-harmonic function inside. This Dirichlet problem makes sense in more general settings although we do not need this fact here (cf. Chapter 1 of

[43]): let D be a bounded region, and suppose each $\zeta \in \partial D$ belongs to a continuum contained in ∂D. Then to each continuou[s] function f on ∂D may be associated a (C^{∞}-) harmonic function $u(z)$, $(z \in D)$ such that

$$u(z) \to f(\zeta) \qquad (z \to \zeta,\ z \in D,\ \zeta \in \partial D)$$

Exercise 2. Let $D = \{0 < |z| < 1\}$, let $f(0) = 1$, $f(e^{i\theta}) = 0$, $0 \le \theta \le 2\pi$.

Does this Dirichlet problem have a solution?

<u>Proof</u>. One useful aspect of the representation (3.6) is that z appears only in the Poisson kernel P. Since V is continuous, it follows that all derivatives of v (for $|z| < R$) may be computed by differentiating under the integral sign. Thus $v \in C^{\infty}(\{|z| < R\})$ and since P is a harmonic function of z for each fixed w $(|z| < |w| = R)$

$$\Delta v(z) = 0 \qquad (|z| < R)$$

Hence v is C^{∞}-harmonic.

The proof of (3.7) uses arguments akin to those of Ch. II, especially in the proof of (2.7). All that is required is (whe[n] $|z| < |w| = R$)

(i) $P(w,z) \ge 0$

(ii) $\dfrac{1}{2\pi}\displaystyle\int_0^{2\pi} P(Re^{i\phi},z)\,d\phi = \mathscr{R}\{\dfrac{1}{2\pi}\displaystyle\int_{|w|=R} \dfrac{w+z}{w-z}\,\dfrac{dw}{iw}\} = 1$

(iii) $P(Re^{i\phi},re^{i\theta}) = P(Re^{i(\phi-\theta)},r) = P(R,re^{i(\theta-\phi)})$

(iv) $\dfrac{R-|z|}{R+|z|} \le P(Re^{i\phi},z) \le \dfrac{R+|z|}{R-|z|}$

(v) <u>for each</u> $\delta > 0$, $\displaystyle\max_{\delta \le |\phi| \le \pi} P(Re^{i\phi},r) = o(1)$ $(r \to$

The routine verification of these facts is left to the reader.

To prove (3.7), we first extend V to $|\phi-\alpha| \leq \pi$ by periodicity so that V remains continuous. According to (ii),

$$v(z) - V(\alpha) = \frac{1}{2\pi}\int_{\alpha-\pi}^{\alpha+\pi} P(Re^{i\phi},z)\{V(\phi)-V(\alpha)\}d\phi$$

The range of integration is divided into $|\alpha-\phi| \leq \delta$ and $\delta \leq |\alpha-\phi| \leq \pi$, with resulting integrals I_1 and I_2. For small δ, I_1 is small since V is continuous at α, and P satisfies (i) and (ii); that I_2 is small then follows from boundedness of V, (iii) and (v).

Exercise 3. Solve the analogue of the Dirichlet problem on a region in $\mathbb{R}^1$, with $\Delta u = u''$.

__Corollary 1__. __Let__ u __be__ C^∞-__harmonic in__ Ω, __and__ $\{|z-z_0| \leq R\} \subset \Omega$. __Then__

$$(3.8) \qquad u(z) = \frac{1}{2\pi}\int_0^{2\pi} u(z_0+Re^{i\phi})P(Re^{i\phi},z-z_0)d\phi \qquad (|z-z_0| < R)$$

__Proof__. Apply the maximum principle to $u-P_u$.

__Remark__. It is possible to obtain (3.8) from Cauchy's integral formula, but in a manner less direct than merely taking real parts. This is done in ([2], pp. 165-7).

Exercise 4. What are the boundary values of the harmonic function $P(R,z)$ for $|z| \leq R$? Does this contradict Theorem 1?

The next result is of the same character as Weierstrass's theorem on uniform convergence of analytic functions (in fact, a proof can be based on this Weierstrass theorem).

__Theorem 3__. __Let__ $\{u_n\}$ __be__ C^∞-__harmonic functions on__ Ω, $u_n \to u$ __uniformly on compacta__. __Then__ u __is__ C^∞-harmonic.

Proof. Let $\{|z-z_0| \leq R\} \subset \Omega$. Then by (3.8)

$$u_n(z) = \frac{1}{2\pi}\int_0^{2\pi} u_n(z_0+Re^{i\phi})P(Re^{i\phi},z-z_0)\,d\phi \quad (|z-z_0| < R)$$

so by dominated convergence (or more elementary methods)

$$u = P_u \quad (|z-z_0| < R)$$

Theorem 2 now shows that u is C^∞-harmonic.

Remark. If $\{|z-z_0| \leq R\} \subset \Omega$, and $z_0 = x_0 + iy_0$, then a C^∞-harmonic function u has a power-series expansion:

$$u(z) = \sum_{m,n\geq 0} a_{mn}(x-x_0)^m(y-y_0)^n \quad (|z-z_0| < R,\ z_0 = x_0+iy_0)$$

which converges uniformly on compact subsets of $\{|z-z_0| < R\}$; i.e. u is real-analytic. This may be proved, for example, by taking the real part of the corresponding power series for $f = u + iv$, or expanding the Poisson kernel in such a power seri and integrating term by term.

It is now possible to consider harmonic functions in the spirit of Ch. II.

Definition. A locally-integrable function u in Ω is weakly harmonic if

$$\iint_\Omega u\Delta\phi\, dxdy = 0 \qquad (\phi \in C_0^\infty(\Omega)) \tag{3.9}$$

Theorem 4. *If u is weakly harmonic, then $u = u^*$ a.e. where u^* is C^∞-harmonic.*

Remark. The proof is essentially that used to achieve Theorem 1 of Chapter II.

Proof. If u is also C^2, then (2.20) implies that (3.8) may be written as

(3.10) $$\iint (\Delta u)\phi \, dxdy = 0 \qquad (\phi \in C_0^\infty(\Omega))$$

so, since Δu is continuous, we may choose $\phi(z) = \Phi_n(z-z_0)$ for each $z_0 \in \Omega$ (where $\iint \Phi_n dxdy = 1$ with Φ_n described in Lemma 1 of Chapter II and (2.1)) and easily verify that (3.1) holds at z_0.

In general let Φ be one of these functions $\{\Phi_n\}$ with $\Phi(z) = 0 \quad (|z| \geq 1)$, and set

(3.11) $$\begin{aligned} u_n(z) &= \iint_{\mathbb{R}^2} u(z-n^{-1}\zeta)\Phi(\zeta)d\xi d\eta \\ &= n^2 \iint_{\mathbb{R}^2} u(\zeta)\Phi(n(z-\zeta))d\xi d\eta \end{aligned}$$

so that if

$$\Omega_\delta = \{z \in \Omega;\ \mathrm{dist}(z,\partial\Omega) \geq 2\delta\}$$

then $u_n \in C^\infty(\Omega_\delta)$ when $n^{-1} < \delta$. It is also easy to check that (3.9) holds for each $\phi \in C_0^\infty(\Omega_\delta)$ if $n^{-1} < \delta$ when u_n is used in place of u. Thus u_n $(n^{-1} < \delta)$ is C^∞-harmonic in Ω_δ, and

$$\begin{aligned} \iint_{\Omega_\delta} |u_n(z)-u(z)|dxdy &\leq \iint_{|\zeta|\leq 1} \Phi(\zeta)d\xi d\eta \iint_{\Omega_\delta} |u(z-n^{-1}\zeta)-u(z)|dxdy \\ &= o(1) \qquad (n \to \infty) \end{aligned}$$

(this is a standard application of Lusin's theorem (cf. [28], p. 159); for a sketch of the proof in $\mathbb{R}^1$ cf. [42], p. 376-7).

Thus $u_n \to u$ (L_1) in Ω_δ, so we are assured of the following: if $\{|z-z_0| \leq (1+2\delta)r\} \subset \Omega_\delta$, then there is an r', $(1+\delta)r \leq r' \leq (1+2\delta)$, such that $u_n \to u$ a.e. on $|z-z_0| = r'$ (relative to linear Lebesgue measure). Let P_{u_n} be the Poisson integral of u_n computed on $|z-z_0| < r'$ so that, according to Corollary 1 of Theorem 2, $u_n(z) = P_{u_n}(z)$ $(|z-z_0| < r')$. Then since $u_n \to u$ (L^1) on $|z-z_0| = r'$, it is easy to see from the definition (3.6) and property (iv) of the Poisson kernel that th u_n (= P_{u_n}) converge uniformly in $|z-z_0| \le r$. Hence there is a C^∞-harmonic function u^* with $u^* = \lim u_n$ (pointwise limit in $|z-z_0| < r$), and since for each $\delta > 0$, $\iint_{\Omega_\delta} |u^*-u|\,dxdy = 0$, $u^* = u$ a.e.

<u>Remark</u>. This shows that a weakly harmonic function is (modulo sets of measure 0) C^∞-harmonic.

The ideas of this proof, with some further technical preparati show that a harmonic distribution T (i.e. $T(\Delta\phi) = 0$ for all test functions ϕ) is a harmonic function. What is needed is a method to approximate T by smooth functions in the sense of the $\{u_n\}$ in Theorem 3. For details cf. [39], [48].

<u>Theorem 5</u> (Harnack). <u>Let $u_1 \le u_2 \le \ldots$ be an increasing sequen of harmonic functions in Ω, and for some $z_0 \in \Omega$ suppose $u_n(z_0) \to \alpha \ne \infty$. Then the sequence $\{u_n\}$ converges uniformly (to a harmonic function) on compact subsets of Ω.</u>

<u>Remark</u>. An analogous conclusion holds if $u_1 \ge u_2 \ge \ldots$ and $u_n(z_0) \to \alpha > -\infty$ for some $z_0 \in \Omega$.

<u>Proof</u>. Suppose $z_0 \in \Omega$ and $\{|z-z_0| \le R\} \subset \Omega$. Then if $|z-z_0| < R$ and $n > m$ we have (according to (3.8), (3.3) and properties (i) and (iv) of the Poisson kernel)

$$0 \le u_n(z)-u_m(z) = \frac{1}{2\pi}\int_0^{2\pi} \{u_n(z_0+Re^{i\phi})-u_m(z_0+Re^{i\phi})\}P(Re^{i\phi},z-z_0)\,d\phi$$

$$\le \frac{R+|z-z_0|}{R-|z-z_0|} \cdot \frac{1}{2\pi}\int_0^{2\pi} \{u_n(z_0+Re^{i\phi})-u_m(z_0+Re^{i\phi})\}\,d\phi$$

$$= \frac{R+|z-z_0|}{R-|z-z_0|}\,\{u_n(z_0)-u_m(z_0)\}$$

Thus the $\{u_n\}$ converge uniformly on compact subsets of $\{|z| < R\}$ and the theorem follows by covering any compact subset of Ω by a well-overlapping chain of such discs.

The final result of this section is

<u>Theorem 6</u>. *Let* u *be harmonic in* $\mathbb{R}^2$ *and define* $M(r) = M(r,u)$ *as in* (3.4). *If*

$$M(r) = O(r^k) \qquad (r = r_n \to \infty) \tag{3.12}$$

then u *is a polynomial of degree* $\le k$.

<u>Proof</u>. Let $|w| = 2R$, $|z| < R$. Then a direct computation with (3.5) shows that (differentiating with respect to x,y)

$$|D^s P(w,z)| \le s! \frac{R}{(R-|z|)^{|s|+1}} \qquad (s = (s_1,s_2))$$

If this estimate is used in (3.8) as $R \to \infty$ we find that $D^s u = 0$ for $|s| > k$.

<u>Remark</u>. Another way to obtain this fact is by writing $w = \mathscr{R} f$ (f entire) and applying the Borel-Caratheodory inequality (cf. [42], p. 175) to show $|f(re^{i\theta})| = O(r^k)$ for large r. Thus u is the real part of a complex polynomial.

B. POTENTIALS. The function $\log|z|$ plays a basic role in the study of Laplace's equation in the plane (in $\mathbb{R}^n$, $n \ge 3$ the corresponding function is $|x|^{2-n}$).

<u>Lemma 1</u>. $\Delta(\log|z|) = 2\pi\, \delta(0)$ (*distribution equation*).

<u>Proof</u>. This is the content of Theorem 2, Chapter II.

<u>Remark</u>. $\log|z|$ is called the *fundamental solution* of the Laplacian since it satisfies the elegant equation described in Lemma 1. The

search for fundamental solutions is a basic theme in partial differential equations (cf. [12], [30]).

Exercise 5. What is the fundamental solution in $\mathbb{R}^1$ of Δ?

We want to study the equation:

$$\Delta f = g \tag{3.13}$$

where g belongs to some class of functions or objects. Such an equation does not have a unique solution: adding a harmonic f_1 to f will also work (this is essentially the only reason for non-uniqueness, cf. Theorem 13). A very general result is

Theorem 7. *Let μ be a measure (positive measure) which is 0 off a compact set. Then a solution in the sense of distributio of the equation*

$$\Delta f = \mu \tag{3.14}$$

may be expressed in terms of a potential by

$$f(z) = -\frac{1}{2\pi} P^{\mu}(z) = \frac{1}{2\pi} \int \log|z-\zeta| \, d\mu(\zeta) \tag{3.15}$$

Remark. Equation (3.15) defines the logarithmic potential P^{μ} of the measure μ. P^{μ} (and hence f) is locally integrable si P^{μ} is clearly bounded below on each compact subset K and

$$\iint_K P^{\mu}(z)\,dxdy = \int d\mu(\zeta) \iint_K \log \frac{1}{|z-\zeta|}\,dxdy$$
$$< 0(1)$$

on such a set (interchange justified since the integrand is bou below for $\zeta \in$ support of μ and $z \in K$).

Proof. Let $\phi \in C_0^\infty$. Then according to Lemma 1,

$$-\iint P^\mu \Delta\phi \, dxdy = \iint \Delta\phi(z)\,dxdy \int \log|z-\zeta|\,d\mu(\zeta)$$

$$= \int d\mu(\zeta) \iint \log|z-\zeta|\,\Delta\phi(z)\,dxdy$$

$$= 2\pi \int \phi(\zeta)\,d\mu(\zeta)$$

where the interchange is justified since all functions are bounded above on the compact set where $\Delta\phi \neq 0$. This proves the Theorem.

Exercise 6. Prove that a potential is lower-semicontinuous; i.e. if P^μ is given by (3.15) and $z_n \to z_0$, prove that

$$P^\mu(z_0) \leq \liminf_{n\to\infty} P^\mu(z_n) \tag{3.16}$$

Exercise 7. Find a measure μ so that strict inequality holds in (3.16).

Exercise 8. Find a measure of total mass 1 (without compact support, of course!) such that P^μ is not locally integrable.

The logarithmic potential is not always ≥ 0 (the analogous potentials $P^\mu(z) = \int |z-\zeta|^{2-m} d\mu(\zeta)$ in $\mathbb{R}^m$ is positive). For this reason, if μ is concentrated in a ball $\{|\zeta| \leq R\}$ it is often convenient to replace P^μ by the Green potential:

$$G^\mu(z) = \int_{|\zeta|\leq R} \log\left|\frac{R^2-\bar{\zeta}z}{R(z-\zeta)}\right| d\mu(\zeta) \qquad (|z| < R) \tag{3.17}$$

It is clear that $P^\mu - G^\mu$ is harmonic in $|z| < R$ (the difference is easy to write explicitly!).

Theorem 8. *Let G^μ be the Green potential of a (finite, positive) measure on $\{|z| \leq R\}$ as in (3.17). Then*

$$G^\mu(z) \geq 0 \qquad (|z| < R) \tag{3.18}$$

$$(3.19) \qquad \lim_{r\to R} \frac{1}{2\pi}\int_0^{2\pi} G^{\mu}(re^{i\theta})\,d\theta = 0 \qquad (r \to R)$$

Proof. The function

$$w = \frac{R(z-\zeta)}{R^2-\bar{\zeta}z} \qquad (|z| < R,\ |\zeta| < R)$$

maps $\{|z| < R\}$ conformally to $\{|w| < 1\}$ (cf. [42], p. 194) the integrand in (3.17) is non-negative and (3.18) follows at o

The proof of (3.19) is trivial if μ is concentrated in $\{|\zeta| < (1-\varepsilon)R\}$ for any $\varepsilon > 0$: in fact, then $G^{\mu}(z) \to 0$ as $|z| \to R$. The general case requires more care, and depends on

Lemma 2. Let $r > 0$. Then

$$(3.20) \qquad \frac{1}{2\pi}\int_0^{2\pi} \log|re^{i\theta}-\zeta|\,d\theta = \max\{\log|\zeta|,\ \log r\}$$

Proof. If $|\zeta| > r$, this is (3.3) (with $z_0 = 0$) since log| is harmonic in $\{|z| \le r\}$. For $|\zeta| < r$, we write the integrand a $\log|e^{-i\theta}| + \log|r-\bar{\zeta}e^{i\theta}|$, and apply a similar analysis. The re for $|\zeta| = r$ follows by continuity.

Completion of proof of Theorem 8. Since an inspection of (3.17 and (3.19) suggests an eventual appeal to Fubini's theorem, we first consider

$$(3.21) \qquad S(r,\zeta) \equiv \frac{1}{2\pi}\int_0^{2\pi} \log\left|\frac{R^2-\bar{\zeta}re^{i\theta}}{R(re^{i\theta}-\zeta)}\right| d\theta \qquad (|\zeta| \le R,\ r < R)$$

as a function of ζ. If $|\zeta| > r$, (3.3) applies and so

$$(3.22) \qquad S(r,\zeta) = \log\left|\frac{R}{\zeta}\right| = \log R - \log|\zeta| \qquad (|\zeta| > r)$$

When $|\zeta| < r$, the function $\log|R^2-\bar{\zeta}re^{i\theta}|$ is still harmonic $|z| \le r$, and (3.3) with (3.20) readily give

$$S(r,\zeta) = \log R - \log r \qquad (|\zeta| \le r) \tag{3.23}$$

(a continuity argument yields (3.23) when $|\zeta| = r$).

Now according to Fubini's theorem, (3.17) and (3.21)-(3.23),

$$\frac{1}{2\pi}\int_0^{2\pi} G^{\mu}(re^{i\theta})\,d\theta = \frac{1}{2\pi}\int_{|\zeta|\le R} S(r,\zeta)\,d\mu(\zeta)$$

$$= \int_{r<|\zeta|\le R} \{\log R - \log|\zeta|\}\,d\mu(\zeta) + (\log R - \log r)\mu\{|\zeta| \le r\}$$

Both terms tend to 0 as $r \to R$ since $\mu\{|\zeta| \le R\} < \infty$ and (3.19) is proved.

C. SUBHARMONIC FUNCTIONS.

(Much of our approach is based on the ideas in [31]). Subharmonic functions are analogues in higher dimensions of convex functions. A function $u(z)$ defined on a region $\Omega \subset \mathbb{R}^2$ with $-\infty \le u(z) < \infty$ is <u>subharmonic</u> in Ω if

(i) $u \not\equiv -\infty$

(ii) u is upper-semi-continuous (i.e. $\{z;\, u(z) < \alpha\}$ is open for each real α)

(iii) if the disc $K_\delta = \{|z-z_0| \le \delta\} \subset \Omega$ and h is continuous on K_δ, harmonic in the interior of K with $h \ge u$ on ∂K_δ then $h \ge u$ in K_δ.

<u>Example 1</u>. Let $f(z)$ be holomorphic in a region Ω. Then $\log|f|$ is subharmonic. Indeed (i), (ii) are obvious. Finally, given z_0, K_δ and h as in (iii), let k be a harmonic conjugate of h and $g = h + ik$. Then $|e^g| = e^h \ge e^{\log|f|} = |f|$ on ∂K_δ, so the inequality is preserved inside (maximum principle).

<u>Example 2</u>. Let $f(z)$ be an entire function. Then for $r > 0$, $0 < \theta < \pi$ let

(3.24) $$u^*(r,\theta) = \sup_E \int_E \log|f(re^{i\phi})|\,d\phi$$

where the sup is taken over all θ-sets of measure 2θ. (Alternativ E could be restricted to be intervals, or be at most n-connecte etc). A. Baernstein has introduced this function u^* (and analogues thereof for meromorphic functions and functions meromorph in an annulus $r_1 < |z| < r_2$) and proved it is subharmonic. This is of great importance in current research, but beyond the scope of these lectures; we can only refer the reader to [3], [4], [5].

Exercise 9. Let u be C_0^∞ with $\Delta u \geq 0$. Show u is subharmonic. (Hint let z_0, K_δ and h be as in condition (iii). The theor of Chapter II suggests that $h(z_0) - u(z_0)$ may be recovered fro $\iint \Delta(\log|(z-z_0)/\delta|)\{h(z)-u(z)\}dxdy$. Use (2.20) to make sense of double integral).

Exercise 10. Show that the negative of a potential p^μ is subharmon

Remark. The interest of Ex. 10 is that it with Ex. 7 shows that a subharmonic function need not be continuous. Thus (ii) is not a specious assumption.

Exercise 11. (Structure of semi-continuous functions). Let $u(z) <$ in Ω. Show u is upper-semi-continuous iff on each compact K of Ω, u is the limit of a decreasing sequence of continuou functions $f_n(z)$. (Hint: take $f_n(z) = \sup_{\zeta \in K}\{u(z)-n|z-\zeta|\}$).

We collect some elementary properties of subharmonic functions in

Theorem 9. *Let* Ω *be a region in* $\mathbb{R}^2$

(i) *if* u *is subharmonic in* Ω, *then for* $c > 0$, cu *is subharmonic*;

(ii) *if* $\{u_\alpha\}$ $(\alpha \in A)$ *is subharmonic, then* $u = \sup u_\alpha$ *is subharmonic if* $u < \infty$ *and upper-semi-continuous*;

(iii) *if* $\{u_i\}$ $(1 \geq i)$ *is a decreasing sequence of subharmonic* functions, then $u = \lim u_n$ is $\equiv -\infty$ *or subharmonic*;

(iv) *if* u *is subharmonic and* $\{|z-z_0| \leq R\} \subset \Omega$, *then*

$$u(z) \leq P_u(z-z_0) \qquad (|z-z_0| < R) \tag{3.25}$$

where the Poisson integral is given in (3.7):

(v) *if* u *is subharmonic in* Ω *and* $\{|z-z_0| \leq R\} \subset \Omega$, then

$$\int_0^{2\pi} u(z_0+re^{i\theta})d\theta \leq \int_0^{2\pi} u(z_0+Re^{i\theta})d\theta \qquad (0 \leq r \leq R)$$

(vi) *if* u_1, u_2 *are subharmonic then* $u = u_1 + u_2$ *is subharmonic.*

Note that P_u makes sense since u is upper-semi-continuous on $\{|z| = R\}$.

Proof. The first two statements are immediate. To prove (iii) we observe that, since $\{z \in \Omega;\, u < s\} = \bigcup_j \{z \in \Omega;\, u_j < s\}$, u is upper-semi-continuous. Also, if h satisfies condition (iii) of the definition of subharmonicity, then for each $\varepsilon > 0$ the sets $\{z \in K_\delta;\, u_j \geq h+\varepsilon\}$ are compact, nondecreasing, with empty intersection. Thus each of these sets is empty for large j, and so $u_j \leq h+\varepsilon$ in K_δ for all large j. This means $u \leq h$ in K_δ.

We next consider (iv). For each n, we form P_{f_n}, where is continuous and $f_n \downarrow u$ on $|z-z_0| \leq R$ (cf. Ex. 11). Let

$$h_n(z) = P_{f_n}(z-z_0) \qquad (|z-z_0| < R)$$

be the Poisson integral of f_n. Then condition (iii) in our definition shows that $h_n \geq u$ for all m, and so, according to Theorem 5, $h = \lim h_n$ is harmonic and $h \geq u$. By monotone convergence, $h = P_h$, and (3.25) follows. Finally (v) is an integration of (3.25), and (vi) also follows from (3.25) (that $u_1 + u_2 \not\equiv -\infty$ follows from the next corollary).

<u>Corollary</u>. <u>A subharmonic function is locally integrable</u>.

<u>Proof</u>. Let $u(z_0) > -\infty$, and $K_\delta = \{|z-z_0| \leq \delta\} \subset \Omega$. For each $r \in [0,\delta]$, (3.25) yields that

$$u(z_0) \leq (2\pi)^{-1}\int_0^{2\pi} u(z_0+re^{i\theta})\,d\theta \qquad (0 < r \leq \delta) \tag{3.26}$$

so

$$-\infty < \iint_{K_\delta} u(z)\,dxdy$$

Since u is bounded above in K_δ (for u is u.s.c. and $< \infty$), u is integrable in a neighborhood of z_0. Thus if $E = \{z \in \Omega;\ u$ integrable in a neighborhood of $z\}$ we see that $u = -\infty$ in a neighborhood of each point in $\Omega - E$ so $\Omega - E$ is ope However, E is also open, and since $u \not\equiv -\infty$, the connectedness of Ω gives that $E = \Omega$.

<u>Theorem 10</u> (<u>Maximum principle</u>). <u>Let</u> $u(z)$ <u>be subharmonic and</u> $h(z)$ <u>harmonic in the region</u> Ω <u>with</u>

$$\limsup_{\substack{z\in\Omega \\ z\to\partial\Omega}}\{u(z)-h(z)\} \leq 0$$

Then

$$u(z) \leq h(z) \qquad (z \in \Omega) \tag{3.27}$$

and if equality holds in (3.27) *for some* $z_0 \in \Omega$, *then* $u \equiv h$.

Proof. Let $M = \sup u\text{-}h \ (z \in \Omega)$. If $u(z_0)\text{-}h(z_0)$ for some $z_0 \in \Omega$, then (3.25) applied to $(u\text{-}h)$ shows that $u\text{-}h = M$ on a dense set of $\{|z-z_0| \leq R\}$ whenever $\{|z-z_0| \leq R\} \subset \Omega$. By semi-continuity, $u\text{-}h \equiv M$ on a ball about z_0, and by connectedness $u = M+h$ in all of Ω, and (3.27) is obvious with $M \leq 0$. If $u\text{-}h < M$ in Ω, it follows that $M = \limsup\{u(z)-h(z)\}$ $(z \in \Omega, \ z \to \partial\Omega) \leq 0$, and (3.27) follows again.

Remark. Theorem 10 can often be mis-stated and mis-interpreted. Thus, the harmonic function $x = \mathcal{R}e\, z$ in $\Omega = \{\mathcal{R}e\, z > 0\}$ has "boundary values" zero, but obviously is not non-positive in Ω.

Corollary (*Hadamard three-circle theorem*). *Let* $f(z)$ *be holomorphic in* $0 < r_1 \leq |z| \leq r_2$ *and*

$$M(r) = \max|f(z)| \qquad (|z| = r)$$

Then $\log M(r)$ *is a convex function of the real variable* $\log r$.

Proof. We prove an inequality of the nature of (2.4) Chapter II. The function

$$h(re^{i\theta}) = \log M(r_2)\, \frac{\log(r/r_1)}{\log(r_2/r_1)} + \log M(r_1)\, \frac{\log(r_2/r)}{\log(r_2/r_1)}$$

is of the form $A \log r + B$, so harmonic in $r_1 \leq |z| \leq r_2$. The result follows from Theorem 10, with

$$u(z) = \log|f(z)| \ \ (\leq M(|z|))$$

We next consider Δu in the sense of distributions.

<u>Theorem 11(A)</u>. <u>If</u> u <u>is subharmonic in a region</u> Ω <u>then</u>

$$\iint u\Delta v \, dxdy \geq 0 \qquad v \in C_0^\infty(\Omega),\ v \geq 0 \tag{3.28}$$

<u>Proof</u>. Choose δ such that $v(z) = 0$ if $\operatorname{dist}(z,\partial\Omega) \leq 2\delta$. We consider (3.20) with $r = \delta$ for each $z_0 \in \Omega$, and upon multiplying this by $v(z_0)$ and integrating over Ω obtain that

$$2\pi\iint_\Omega u(z)v(z)\,dxdy \leq \iint_\Omega \{\int_0^{2\pi} u(z+\delta e^{i\theta})\,d\theta\}v(z)\,dxdy$$

$$= \iint_\Omega u(z)\{\int_0^{2\pi} v(z-\delta e^{i\theta})\,d\theta\}\,dxdy$$

hence

$$\iint_\Omega u(z)\{\int_0^{2\pi} v(z-\delta e^{i\theta})\,d\theta - 2\pi v(z)\}\,dxdy \geq 0 \tag{3.29}$$

for all v under consideration.

The Taylor theorem in several variables gives for small δ

$$\begin{aligned} v(z-\delta e^{i\theta}) &= v(z) - v_x(z)\delta\cos\theta - v_y(z)\delta\sin\theta \\ &+ v_{xy}(z)\delta^2\cos\theta\sin\theta + \tfrac{1}{2}[v_{xx}(z)\delta^2\cos^2\theta \\ &+ v_{yy}(z)\delta^2\sin^2\theta] + 0(\delta^3) \end{aligned}$$

where the 0 is independent of z. Hence

$$\int_0^{2\pi} v(z-\delta e^{i\theta})\,d\theta - 2\pi v(z) = \pi\delta^2\Delta v(z) + 0(\delta)^3$$

and when this is divided by δ^2 $(\delta \to 0)$ and substituted in (3.29), (3.28) follows.

<u>Remark</u>. Theorem 11A provides one way to attack Ex. 9. Together with Theorem 2 of Chapter II it provides an "obvious" proof of the subharmonicity of $\log|f(z)|$ when f is holomorphic in Ω.

Theorem 11(B). *Let* u *be locally integrable in* Ω *and assume* (3.28) *holds. Then there is a subharmonic function* u^* *in* Ω *with* $u = u^*$ a.e. (Compare with Theorem 4).

Proof. If $u \in C^2(\Omega)$, (3.28) may be integrated by parts (cf. (2.20)) and is equivalent to

$$\iint \Phi \Delta u \; dxdy \geq 0$$

for all non-negative test functions ϕ; i.e. $\Delta u \geq 0$ and u is subharmonic.

In general, let Φ and Ω_δ be as in Theorem 4, and define u_n according to (3.11) for $n^{-1} < \delta$. Then $u_n \in C^\infty(\Omega_\delta)$, $u_n \to u(L_1)$, and (3.28) holds for u_n in Ω_δ, so u_n is subharmonic in Ω_δ. In particular

$$\iint u_n(z - m^{-1}\zeta)\Phi(\zeta)d\xi d\eta$$

is a decreasing function of m with m a positive integer (cf. (v) in Theorem 9). When $n \to \infty$, it then follows from the representation (3.11) that

$$u_m(z) = \iint u(z - m^{-1}\zeta)\Phi(\zeta)d\xi d\eta$$

also decreases with m. The subharmonic function u^* is then obtained by appeal to Theorem 9 (iii).

Corollary. *If* u *is subharmonic in* Ω, *then* Δu *is a measure which is finite on each compact* K *of* Ω.

Proof. See Theorem 3 of Chapter II and (3.28).

Theorem 12 (Jensen's inequality). *Let* g *be a non-decreasing real convex function, and set* $g(-\infty) = \lim_{x \to -\infty} g(x)$. *If* u *is subharmonic in* Ω *so is* $g(u)$.

Proof. Theorem 4 of Ch. II implies g is continuous, so $g(u)$ is upper-semicontinuous. Let $x = u(z_0 + re^{i\theta})$ in (2.32) and integrate with respect to θ: if x_1 is in the domain of g,

$$(2\pi)^{-1}\int_0^{2\pi} g(u(z_0 + re^{i\theta}))d\theta \geq g(x_1) +$$
$$k\{(2\pi)^{-1}\int_0^{2\pi} u(z_0 + re^{i\theta})d\theta - x_1\} \qquad (k = k(x_1))$$

when $\{|z-z_0| \leq r\} \subset \Omega$. Since u is locally integrable, we may choose x_1 so that the last term vanishes. Then since u is subharmonic and g is nondecreasing we obtain that

$$g(u(z_0)) \leq g((2\pi)^{-1}\int_0^{2\pi} u(z_0 + re^{i\theta})d\theta)$$
$$\leq (2\pi)^{-1}\int_0^{2\pi} g(u(z_0 + re^{i\theta}))d\theta$$

Exercise 12. If g is continuous on $[a,b]$ and $g(u)$ is subharmonic for all subharmonic functions in a bounded domain Ω with range in $[a,b]$, show g is nondecreasing and convex.

The main structure theorem on subharmonic functions is

Theorem 13 (Riesz decomposition theorem; local form). *Let u be subharmonic in a domain Ω, and Ω_1 a subdomain of Ω with $\overline{\Omega}_1$ compact in Ω. Then there exists a (unique) measure μ on Ω_1 such that*

$$u(z) = v(z) - P^{\mu}(z) \qquad (z \in \Omega_1) \tag{3.30}$$

with v harmonic in Ω_1.

Proof. We have from the Corollary to Theorem 11B that $\Delta u = \mu$ is a finite positive measure when restricted to Ω_1. Let μ_1

be the measure defined by $\mu_1(E) = \mu(E \cap \Omega_1)$ and let P^{μ_1} be the associated potential. Then $u + P^{\mu_1}$ is locally integrable in Ω_1 and $\Delta(u + P^{\mu_1}) = 0$. The result now follows from Theorem 4 (perhaps after redefining u on a set of measure zero). (I thank Dr. L. Riihentaus for improving my original argument here). For a global form of this theorem cf. ([8], p. 42).

The form of Theorem 13 which is most useful here is

Theorem 14. *Let u be subharmonic in a region Ω and $\{|z-z_0| \leq R\} \subset \Omega$. Then*

$$u = -G^{\mu}(z-z_0) + P_u(z-z_0) \qquad (|z-z_0| < R) \tag{3.31}$$

where G^{μ} is the Green potential (3.17) *of the positive measure $\mu(= (2\pi)^{-1}\Delta u)$ and P_{μ} is the Poisson integral* (3.6) *of $u(z-z_0)$ computed on $\{|z-z_0| = R\}$.*

Remark. Theorem 14 is sometimes more useful than Theorem 13. Although it applies only to discs, the right side of (3.31) is now explicitly determined.

Proof. Let us assume $z_0 = 0$. According to Theorem 13 and the discussion in the paragraph which contains (3.17), we have

$$u(z) = -G^{\mu}(z) + w(z) \qquad (|z| < R) \tag{3.32}$$

where w is harmonic in a neighborhood of $|z| \leq R$. We claim that if P_u is the Poisson integral computed on $|z| = R$, then

$$w - P_u \geq 0 \qquad (|z| \leq R) \tag{3.33}$$

For according to (3.18), $w \geq u$ in $|z| \leq R$. If $\{f_n\}$ is a family of continuous functions which decreases to u on $|z| \leq R$ and

$g_n = \min(f_n, w)$ we see that $w \geq P_{g_n} \to P_u \geq u$ in $|z| \leq R$ (the last inequality is on account of (3.25)). Let

$$h(z) = w(z) - P_u(z) \qquad (|z| \leq R) \tag{3.34}$$

Then h is harmonic, and

$$2\pi h(0) = \int_0^{2\pi} \{w(re^{i\theta}) - P_u(re^{i\theta})\}d\theta$$

$$\leq \int_0^{2\pi} \{w(re^{i\theta}) - u(re^{i\theta})\}d\theta = -\int G^{\mu}(re^{i\theta})d\theta$$

$$(0 < r < R)$$

As $r \to R$, (3.19) now shows that $h(0) \leq 0$, but since (3.33) asserts that $h \geq 0$ in $|z| \leq R$, the maximum principle (Theorems 1 or 10) implies $h \equiv 0$. This, (3.32) and (3.34) yie[l]d the Theorem.

D. SUBHARMONICITY AND CONVEXITY

__Theorem 15.__ *Let $u(z)$ be subharmonic in $\{|z| < R\}$ and set*

$$h(r) = \frac{1}{2\pi}\int_0^{2\pi} u(re^{i\theta})d\theta \qquad (0 < r < R)$$

Then h is a convex function of $\log r$.

__Proof.__ The corollary to Theorem 9 implies that h is continu[ous] on $0 < |z| < R$. Now given $\phi \geq 0 \in C_0^{\infty}(0,R)$ set

$$\psi_1(re^{i\theta}) = \phi(r) \qquad (0 \leq \theta \leq 2\pi) \tag{3.35}$$

Then $\phi_1 \geq 0 \in C_0^{\infty}(\{|z| \leq R\})$, ϕ_1 depends only on r and

$$2\pi \int_0^R \frac{\partial^2\phi}{\partial(\log r)^2} h(r)d\log r = \int_0^{2\pi}\int_0^R \{r^{-1}\frac{\partial}{\partial r}(r\frac{\partial\phi}{\partial r})\}u(re^{i\theta})rd[r d\theta]$$

$$= \iint_{|z|<R} (\Delta\phi_1)u \, rdrd\theta \tag{3.36}$$

Since u is subharmonic, the last expression is ≥ 0 (cf. (3.28)) and hence h satisfies (2.3') of Chapter II and is convex.

Exercise 13. Let f be holomorphic in $\{|z| < R\}$. Show that

$$\frac{1}{2\pi}\int_0^{2\pi} |f(re^{i\theta})|^p d\theta \qquad (p > 0)$$

is a convex function of $\log r$. Generalize.

Exercise 14. Show that the function $h(r)$ in Theorem 15 is increasing. What happens if u is only subharmonic in $0 < |z| < R$?

IV. THE FIRST FUNDAMENTAL THEOREM: NEVANLINNA'S CHARACTERISTIC

All functions under analysis henceforth are assumed non-constant. If f is a general (non-rational) meromorphic function, we ask whether analogues of the results of Chapter I are valid. That some theory is possible was shown almost a century ago by E. Picard, who proved that a meromorphic function which omits 3 extended complex numbers is constant (the exponential function shows 3 may not be replaced by 2). Nevanlinna's investigations, now half a century old, give a far-reaching generalization of this. In a series of remarkable papers, Nevanlinna developed his theory by developing and applying potential-theoretic methods to $u(z) = \log|f(z)|$ in a way that even now appears modern.

A. POISSON-JENSEN FORMULA

The basic tool in Nevanlinna theory is

<u>Theorem 1</u> (<u>Poisson-Jensen formula</u>). <u>Let</u> $f(z)$ <u>be meromorphic in</u> $|z| < \rho \leq \infty$ <u>with zeros</u> $\{a_\mu\}$, <u>poles</u> $\{b_\nu\}$, <u>each</u> a_μ, b_ν <u>repeated as warranted by multiplicity</u>. <u>Then if</u> $|z| = r < R < \rho$,

$$(4.1) \qquad \log|f(z)| = \frac{1}{2\pi}\int_0^{2\pi} \log|f(Re^{i\phi})|P(Re^{i\phi},z)\,d\phi$$

$$+ \sum_{|a_\mu|<R} \log\left|\frac{R(z-a_\mu)}{R^2-\bar{a}_\mu z}\right| \qquad \sum_{|b_\nu|<R} \log\left|\frac{R(z-b_\nu)}{R^2-\bar{b}_\nu z}\right|$$

where $P(w,z)$ is Poisson's kernel.

<u>Remark</u>. The theorem includes the case $\log|f(z)| = \pm\infty$, as then both sides are infinite.

<u>Proof</u>. If f is analytic, then $\log|f|$ is subharmonic, and $\Delta(\log|f|) = \{a_\mu\}$, in which case (4.1) is a restatement of Th. 1 Ch. III. In general, write $f = g|h$ and apply the above reasoning to g and h.

<u>Remark</u>. The special case

$$(4.2) \quad \log|f(0)| = \frac{1}{2\pi}\int_0^{2\pi} \log|f(Re^{i\phi})|\,d\phi$$

$$+ \sum_{|a_\mu|<R} \log\left|\frac{a_\mu}{R}\right| - \sum_{|b_\nu|<R} \log\left|\frac{b_\nu}{R}\right|$$

is <u>Jensen's formula</u>, which was obtained in 1899. Formula (4.1) w first written down by Nevanlinna. If $f(0) = 0,\infty$, (4.2) may be in a more useful form by letting $f_1(z) = (z/R)^k f(z)$ replace f where k is chosen so that $|\log|f_1(0)|| \neq \infty$. As noted in [25] p. 3, this modification is routine but also annoying, and we usu write our formulae with the understanding that none of the terms th in is infinite, knowing that the exceptional cases may be so trea

In order to study the value-distribution of f we need additional notation. For $|a| \leq \infty$ and $r < \rho$, let

$$n(r,a) \quad (= n(r,a,f))$$

be the number of solutions of the equation $f(z) = a$ $(|z| < r)$ where an a-point of multiplicity k contributes k, and let

$$\bar{n}(r,a) \quad (= \bar{n}(r,a,f))$$

be the number of such solutions with each multiple root counted once.

The formulas of the Nevanlinna theory are expressed in terms of certain means of n, $\bar{n}$. For example, if $f(0) \neq a$, let

$$(4.3) \qquad N(r,a) = \int_0^r t^{-1} n(t,a)\,dt = \log(r/t)n(t,a)\Big|_0^r - \int_0^r n(t,a)\,d\log(r/t) = \sum_{|b|<r} \log(r/|b|)$$

where $\{b\}$ is the solution-set of $f(z) = a$ in $\{|z| < r\}$ (when $f(0) = a$, this becomes

$$N(r,a) = \int_0^r t^{-1}\{n(t,a)-n(0,a)\}dt + n(0,a)\log r$$

which can be readily manipulated as in (4.3)). If $n(t,a)$ is replaced by $\bar{n}(t,a)$ $(t \geq 0)$, these formulas define $\bar{N}(r,a)$.

Exercise 1. Let $f_n(z) = 1/(z-n^{-1})$, $f_0(z) = 1/z$. For $r > 0$ compute $n(r,f_n)$, $n(r,f_0)$, $N(r,f_n)$, $N(r,f_0)$.

The first fundamental theorem is little more than an insightful rewriting of (4.2): let

$$\log^+ u = \max(0, \log u) \qquad (u \geq 0)$$

so that

$$\log u = \log^+ u - \log^+(1/u) \qquad (u > 0)$$

Then if we set for $r < \rho$

$$m(r,\infty) \ (= m(r,\infty,f)) = \frac{1}{2\pi}\int_0^{2\pi} \log^+|f(re^{i\phi})|\,d\phi$$

$$m(r,a) \ (= m(r,a,f)) = \frac{1}{2\pi}\int_0^{2\pi} \log^+|1/(f(re^{i\phi})-a)|\,d\phi \qquad (a \neq \infty)$$

it follows at once from (4.2) that

$$(4.4)\quad m(r,\infty) + N(r,\infty) = m(r,0) + N(r,0) + \log|f(0)| \, (f(0) \neq 0,\infty$$

Exercise 2. Write out the proper form of (4.4) when $|\log|f(0)|| = \infty$

<u>Definition</u>. For $0 < r < \rho$, the <u>Nevanlinna characteristic</u> $T($
is defined by

$$(4.5)\qquad T(r) \; (= T(r,f)) = m(r,\infty) + N(r,\infty)$$

Thus Jensen's formula asserts that $T(r,f) = T(r,f^{-1}) + \log|f(0)|$ when $\log|f(0)| \neq \pm\infty$.

It is easy to see that

$$\log^+(uv) \leq \log^+u + \log^+v \qquad (u,v > 0)$$

$$\log^+(u+v) \leq \log^+(2\max(u,v)) \leq \log^+u + \log^+v + \log 2 \qquad (u,v \geq 0)$$

In particular, for any finite a,

$$\log^+|f-a| \leq \log^+(|f|+|a|) \leq \log^+|f|+\log^+|a|+\log 2$$

$$\log^+|f| = \log^+|f-a+a| \leq \log^+|f-a|+\log^+|a|+\log 2$$

We thus deduce:

$$(4.6)\qquad m(r,\infty,f-a) - m(r,\infty,f) = E^*(a,r)$$

where

$$|E^*(a,r)| \leq \log^+|a| + \log 2$$

It is obvious that $N(r,f) = N(r,f-a)$ so (4.4) (with $f-a$ in place of f) and (4.6) give

<u>Theorem 2</u> (<u>first fundamental theorem of the Nevanlinna theory</u>).

Let f *be meromorphic in* $|z| < \rho \leq \infty$. *Then if* $r < \rho$ *and* a *is any complex number,*

$$m(r,a) + N(r,a) \quad (= T(r,(f-a)^{-1})$$
$$= T(r,f) + E(r,a) \qquad (f(0) \neq \infty, a)$$

where

$$|E(a,r)| \leq \log^{+}|a| + \log 2 + |\log|f(0)-a|| \tag{4.7}$$

and a similar formula holds when $f(0) = a$ *or* ∞.

Remark. $m(r,a)$ is large when there is some non-negligible set of ϕ on which $f(re^{i\phi})$ is close to a; $m(r,a)$ is called the *proximity function*, and $N(r,a)$ the *counting function*.

Since we shall see that $T(r)$ is unbounded if, for example, f is nonconstant and meromorphic in the plane, (4.7) shows that $E(r,a)$ plays the role of an unimportant error term. Thus the sum $m + N$ is essentially independent of a.

Finally, we note that since $m(r,a) \geq 0$, the first fundamental theorem gives an upper bound for $n(r,a)$ (for example in terms of $T(2r)$ and $|f(0)-a|$).

Exercise 3. Let $f(z)$ be a rational function. Discuss $m(r,a)$, $N(r,a)$, $T(r)$ as $r \to \infty$ (cf. Theorem 1 of Chapter I and the accompanying discussion). Same for e^z.

Exercise 4. For $r < \rho$ show:

$$T(r,1/f) = T(r,f) + O(1)$$

$$T(r,f_1 f_2) \leq T(r,f_1) + T(r,f_2) + O(1)$$

$$T(r,f_1+f_2) \leq T(r,f_1) + T(r,f_2) + O(1)$$

$$T(r,cf) = T(r,f) + O(1) \qquad (c \neq 0)$$

Exercise 5. Let $ad-bc \neq 0$ and $g = (af+b)/(cf+d)$. Show

$$T(r,g) = T(r,f) + 0(1)$$

(write a Möbius transformation as a composition of simpler tran formations).

Exercise 6. Let f be meromorphic and $f = g/h$, g and h entire Show

$$T(r,f) = \frac{1}{2\pi}\int_0^{2\pi} \max(\log|g(re^{i\theta})|, \log|h(re^{i\theta})|)d\theta$$

up to a constant term. For applications and generalizations cf.[

B. A CONNECTION WITH ENTIRE FUNCTIONS

Theorem 3. If f is entire, then

(4.8) $$\max_{|z|\leq r} \log|f(z)| \leq \frac{R+r}{R-r} T(R,f) \qquad (0 \leq r < R < \rho)$$

Proof. This follows from (4.1) (since f has no poles) and properties of the Poisson kernel: if $|z| \leq r$

$$\log|f(z)| \leq (2\pi)^{-1} \int_0^{2\pi} \log|f(Re^{i\phi}) P(Re^{i\phi},z)d\phi$$

$$\leq \frac{R+r}{R-r} T(R,f)$$

Remark. This shows that

(4.9) $$M(r) = M(r,f) = \max_{|z|\leq r} \log|f(z)|$$

also reflects important properties of f. According to the viewp of (3.4), we should refer to (4.9) as a functional of the subharmonic function $\log|f|$ rather than f itself, but this notation is more convenient.

As a simple example of how $T(r)$ reflects value-distribut properties of f, we prove

__Theorem 4.__ _If f is meromorphic in the plane and for some constant_ A,

$$(4.10) \qquad T(r,f) \le A(\log r)$$

for an unbounded r-_set, then f is rational._

Proof. If f takes on the complex number $a \ge q$ times, then

$$N(r,a) \ge q \log r + O(1)$$

so (4.10) and the first fundamental theorem show $q \le A$.

Let $b_1,\ldots,b_n$ be the (finite number of) poles of f and set

$$g(z) = f(z)\Pi(z-b_i)$$

Then g is entire and so Ex. 3 and Theorem 3 yield

$$\log M(r,g) \le 3T(2r,g) \le 3[T(2r,f) + q \log r + O(1)] \le k \log r$$

for an unbounded r-set and some k. By an extension of Liouville's Theorem, g is a polynomial of degree $\le k$, and this clearly implies f is rational.

To more precisely measure the growth of f, we set

$$(4.11) \qquad \lambda = \limsup_{r\to\infty} \frac{\log T(r,f)}{\log r}$$

as the _order_ of f (replacing lim sup by lim inf gives the lower order; there is an analogous definition when $\rho < \infty$).

A rational function is of order 0, $\exp(z^k)$ of order k.

Exercise 7. Show $\lambda = \infty$ for $f = \exp(\exp z)$. One way to do this is to estimate $N(r,1)$. Or use Ex. 8.

Exercise 8. Show that if $T(r,f)$ is replaced by $\log M(r,f)$ in (4.11), the corresponding λ agrees with that given by (4.11).

C. CARTAN'S IDENTITY, CONVEXITY

Theorem 5. Let f be meromorphic in $\{|z| < \rho\}$. Then

$$T(r,f) = \frac{1}{2\pi}\int_0^{2\pi} N(r,e^{i\theta})\,d\theta + \log^+|f(0)| \qquad (r < \rho) \tag{4.12}$$

Proof. According to (4.2),

$$\log|f(0)-e^{i\theta}| = \frac{1}{2\pi}\int_0^{2\pi}\log|f(re^{i\phi})-e^{i\theta}|\,d\phi + N(r,\infty) - N(r,e^{i\theta})$$

We integrate both sides with respect to θ, interchange and recall (3.20) of Chapter III:

$$\log^+|f(0)| = \frac{1}{2\pi}\int_0^{2\pi}\log^+|f(re^{i\phi})|\,d\phi + N(r,\infty) - \frac{1}{2\pi}\int_0^{2\pi} N(r,e^{i\theta})\,d\theta$$

Exercise 9. If $f(0) = \infty$, write $f(z) = cz^k f_1(z)$ with $f_1(0) = 1$. Show that (4.12) becomes

$$T(r,f) = \frac{1}{2\pi}\int_0^{2\pi} N(r,e^{i\theta})\,d\theta + \log|c| \qquad (r < \rho) \tag{4.12'}$$

Check that (4.12) is valid if $f(0) = 0$ or $f(0) = e^{i\phi}$ for some ϕ.

There is an interesting geometric interpretation of Cartan's identity. Indeed,

$$\int_0^{2\pi} N(r,e^{i\theta})\,d\theta = \int_0^r t^{-1}dt \int_0^{2\pi} n(t,e^{i\theta})\,d\theta$$

and the argument principle shows that the inner integral represents the total length $\ell(t)$ of the arcs on $|w| = 1$ covered by f in $\{|z| < t\}$, each arc counted as frequently as it is covered. In the parlance of Ch. I, $\ell(t)$ is the length of the arcs on the Riemann surface of f (restricted to $\{|z| < t\}$ which project onto the unit circle $|w| = 1$ of the plane. Thus, (4.12), (4.12') yield

$$T(r) = (2\pi)^{-1}\int_0^r \ell(t)t^{-1}dt + O(1)$$

Corollary 1. $T(r)$ *is an increasing convex function of* $\log r$.

Proof. $N(r,e^{i\theta})$ is increasing and log convex for each θ, so clearly T increases. Also, if $\phi(\geq 0)$ is a test function (4.12) yields

$$\int_0^\infty T(r)\frac{d^2\phi(r)}{d(\log r)^2}d\log r = (2\pi)^{-1}\int_0^{2\pi}d\theta\int_0^\infty N(r,e^{i\theta})\frac{d^2\phi(r)}{d(\log r)^2}d\log r \geq 0$$

Corollary 2. $\frac{1}{2\pi}\int_0^{2\pi} m(r,e^{i\theta})d\theta \leq \log 2$.

Proof. $T(r,f) = m(r,e^{i\theta})+N(r,e^{i\theta}) + \log|f(0)-e^{i\theta}| + G(\theta)$, with $|G(\theta)| \leq \log 2$. If this is integrated with respect to θ, then (4.12) with (3.20) of Chapter III gives

$$T(r,f) = \frac{1}{2\pi}\int_0^{2\pi} m(r,e^{i\theta})d\theta + \{T(r,f)-\log^+|f(0)|\} + \log^+|f(0)| + \frac{1}{2\pi}\int_0^{2\pi} G(\theta)d\theta$$

q.e.d.

The proper generalization of Cartan's identity is due to O. Frostman, who recognized that (4.12) depends on properties of the potential $P^\mu(z)$ with μ Lebesgue measure on $|w| = 1$.

Definition. A compact set E is called of *positive capacity* if there exists a (unit)positive measure μ in E having

(4.13) $$P^\mu(z) \leq K \qquad (z \in C)$$

for some $K < \infty$; otherwise E is of *capacity zero*.

Corollary 3. *Let* E *be a compact set of positive capacity, and* μ *a measure on* E *such that* (4.13) *holds. Then*

$$|T(r) - \int_E N(r,a)d\mu(a)| = O(1) \tag{4.14}$$

Proof. When Jensen's formula (4.2) is applied to the function $\log|f(re^{i\theta})-a|$ $(a \in E)$ we obtain after an appeal to Fubini's theorem that

$$P^{\mu}(f(0)) = \frac{1}{2\pi}\int_0^{2\pi} P^{\mu}(f(re^{i\theta}))d\theta + \int_E N(r,a)d\mu(a) - N(r,\infty) \tag{4.15}$$

Now $P^{\mu}(f(0)) \leq K$ according to (4.13), and since E is compac and μ vanishes off E we have

$$|P^{\mu}(z) + \log^+|z|| \leq K_1 \qquad (z \in C^*) \tag{4.16}$$

(where $K_1 = K_1(E)$). Hence, (4.13), (4.15) and (4.16) give (4.1

Exercise 10. If E has positive capacity, (4.13) holds, and f is meromorphic in the plane, show that

$$\int_E m(r,a)d\mu(a) = O(1) \qquad (r \to \infty)$$

Remark. If E has capacity zero there exists an entire functic $f(z)$ with

$$m(r,a) \to \infty \qquad (r \to \infty)$$

for each $a \in E$ [14].

As a final application, let E be the extended plane and the Riemann sphere of unit diameter resting at the origin of E Given $z_1, z_2 \in E$, let Z_1, Z_2 be associated by stereographic projection, and let $\chi[z_1,z_2] = |Z_1 - Z_2|$ (3-dimensional distanc This induces the *chordal metric* μ in E and (cf. [29], §2.5)

$$d\mu = \pi^{-1}(1+r^2)^{-2}r\,dr\,d\theta \tag{4.17}$$

According to (3.20) of Chapter III and routine computations,

$$P^{\mu}(\zeta) = \int \log\left|\frac{1}{\zeta-a}\right| d\mu(a)$$

$$= \pi^{-1}\int_0^{|\zeta|} \frac{r}{(1+r^2)^2}dr\int_0^{2\pi} \log\left|\frac{1}{\zeta-re^{i\theta}}\right| d\theta + \pi^{-1}\int_{|\zeta|}^{\infty} \frac{r}{(1+r^2)^2}dr \int_0^{2\pi} \log\left|\frac{1}{\zeta-re^{i\theta}}\right| d\theta$$

$$= -2\int_0^{|\zeta|} \log|\zeta| \frac{r}{(1+r^2)^2}dr - 2\int_{|\zeta|}^{\infty} \log r \frac{r}{(1+r^2)^2}dr$$

(4.18) $$= -\log(1 + |\zeta|^2)^{1/2}$$

(cf. [36], p. 178). Then the lines used to obtain Theorem 5 and Corollary 2 (especially (4.15)) show that

$$\frac{1}{2\pi}\int_0^{2\pi} \log(1 + |f(re^{i\theta})|^2)^{1/2}d\theta + N(r,\infty)$$

$$= \int_0^{\infty} A(t)t^{-1}dt + \log(1 + |f(0)|^2)^{1/2}$$

where, if $\mathscr{F}(t)$ is the (homeomorphic) image of $\{|z| \le t\}$ onto the Riemann surface of f,

(4.19) $$A(t) = \pi^{-1}\int_{\mathscr{F}(t)} \frac{du dv}{(1+|w|^2)^2}$$

$$= \pi^{-1}\int_{|z|<t} \frac{|f'(z)|^2}{(1+|f(z)|^2)^2} dx dy$$

is the spherical area of $\mathscr{F}(t)$. Note that

(4.20) $$|\log(1+|w|^2)^{1/2} - \log^+|w|| \le \log 2 \qquad (w \in \mathbb{C})$$

so one can replace $T(r)$ by

(4.21) $$T^{\#}(r) = \int_0^r A(t)t^{-1}dt$$

with a bounded error. The definition (4.21) is due to Ahlfors and Shimizu.

The advantage of the chordal metric is that ∞ no longer plays a special role. We remark that

$$[z_1,z_2] = \frac{|z_2-z_1|}{(1+|z_1|^2)^{1/2}(1+|z_2|^2)^{1/2}}$$

(4.22)

$$|z,\infty| = (1+|z|^2)^{-1/2}$$

(compare with (4.18)).

V. NEVANLINNA'S SECOND FUNDAMENTAL THEOREM

In this chapter it is convenient to compute with the chordal metric on the w-plane where $w = f(z)$ with f meromorphic. Thus we take

(5.1) $$T(r) = \int_0^r A(t)t^{-1}dt$$

(where

(5.2) $$A(r) = \pi^{-1} \iint_{|z|\leq r} \frac{|f'|^2}{(1+|f|^2)^2}\, dxdy\ )$$

as the characteristic of f; it was noted in (4.21) that this revision changes the classical $T(r)$ by a bounded term, and so will not affect our results here. In this spirit, we redefine the classical proximity function $m(r,a)$ of Chapter IV by

(5.3) $$m(r,a) = -\frac{1}{2\pi}\int_0^{2\pi} \log[f(re^{i\theta}),a]d\theta$$

where $[f(re^{i\theta}),a]$ is defined in (4.22), again with bounded error (cf. (4.20)).

Some of our arguments need slight modification when f is rational, but the results are trivial in that case; thus we assume (only for convenience) that

(5.4) $\quad f$ is not rational: $(\log r)^{-1}T(r) \to \infty$

(cf. Theorem 4 of Chapter IV).

Since $m(r,a) > 0$ and (5.4) holds, it follows from the first fundamental theorem that

(5.5) $$\limsup \frac{N(r,a)}{T(r)} \leq 1$$

for all $a \in \mathbb{C}^*$, but results such as (4.12) and (4.14) suggest that equality must hold in (5.5) for most a.

The proper formulation of this is the defect relation (which is essentially the second fundamental theorem).

<u>Theorem 1</u> (Defect relation) <u>Let f be meromorphic in the plane and for each value</u> a $(|a| \leq \infty)$ <u>define</u>

(5.6) $$\delta(a) = 1 - \limsup \frac{N(r,a)}{T(r)} = \liminf \frac{m(r,a)}{T(r)} \qquad (r \to \infty)$$

(5.7) $$\theta(a) = \liminf \frac{N(r,a) - \bar{N}(r,a)}{T(r)} \qquad (r \to \infty)$$

<u>Then</u>

(5.8) $$\delta(a) \geq 0,\ \theta(a) \geq 0,\ 0 \leq \delta(a) + \theta(a) \leq 1 \qquad (|a| \leq \infty)$$

<u>and</u>

(5.9) $$\sum_{|a| \leq \infty} \delta(a) + \theta(a) \leq 2$$

Exercise 1. Show that Picard's theorem $(f \neq 0,1,\infty \to f$ constant) is a consequence of the defect relation.

Exercise 2. We call a value <u>totally ramified</u> if each solution of the equation $f(z) = a$ has multiplicity ≥ 2. Use (5.7) and (5.9) to prove that a meromorphic function can have at most 4 totally ramified values. Any improvement possible if f is rational?

There is a formal connection between the defect relation and Theorem 2 of Chapter I (cf. [36] Ch. XII). If f is rational and $|a| \leq \infty$, let $n(a)$ be the number of points on the Riemann surface S of f "over" a (with regard to multiplicity) that are situated in the finite plane (so that $n(a) = n$ = degree of f for all but one value of a; cf. Theorem 1 of Ch. I) and $\bar{n}(a)$ the number of such distinct points over a. Then if $\sum(k-1)$ is the sum of all orders of branch points of S, we have

$$\sum(k-1) = \sum(n-n(a)) + \sum(n(a)-\bar{n}(a))$$

When both sides are divided by n with $n \to \infty$, we see that the Riemann-Hurwitz formula (1.10) suggests (5.9). In some special cases, this formalism can be made precise; cf. the Ex. at the end of Ch. I and [35].

Exercise 3. Compute $\delta(a)$, $\theta(a)$ for a polynomial or rational functi Can equality ever hold in (5.9)?

We now consider the proof of Theorem 1. Nevanlinna's origi proof is based on the differentiated Poisson-Jensen formula. Sc after, his brother Frithiof gave another demonstration which developed a connection with the methodology used originally to prove Picard's theorem. It depends on boundary properties of automorphic functions that map the disc $\{|w| < 1\}$ conformally onto the universal cover of the sphere with q (≥ 3) points deleted. An account of this proof is beyond the scope of these notes; for a stimulating account cf. Ch. IX, §4 of [36].

At first glance, it would seem that such a point of view would not be amenable to generalization, but Ahlfors salvaged the principle of F. Nevanlinna's argument by introducing almost ad-hoc metrics on the w-sphere which have carefully controlled singularities. Our proof will be based on this approach, but will follow the lines of [9], [20].

Now given q (≥ 3) distinct points $a_1,\ldots,a_q$ (with no loss of generality we take $a_q = \infty$) and a sufficiently small $\varepsilon > 0$, we set

(5.10) $$g_j = (\varepsilon[w,a_j])^{-2}$$

(where $[w,a_j]$ is defined by (4.22)) and consider the <u>volume form</u>

(5.11) $$\Psi(w)\,dudv = \left(\prod_{j=1}^{q} g_j\{\log g_j\}^{-2}\right) \frac{dudv}{(1+|w|^2)^2}$$

which is singular at the a_j.

Now Ψ is defined on the full Riemann sphere, and the notation $\Delta \log \Psi$ is subject to two distinct interpretations. In general, if Φ is a function which is smooth in a (perhaps deleted) neighborhood of w_0, $|w_0| \leq \infty$, we let

(5.12) $$[\Delta]\log \Phi(w)$$

be the standard Laplacian operator in this neighborhood. The unadorned expression $\Delta \log \Phi$, when computed in such a neighborhood, will always be interpreted in the sense of distributions and is influenced by singularities if $w_0 \in \{a_j\}$. For example, if $\Phi(z) = \log|z-a|$, (2.1) and (2.2) show $[\Delta]\Phi = 0$, $\Delta\Phi = 2\pi\delta(a)$.

A. PROPERTIES OF Ψ AND $[\Delta]\log\Psi$

<u>Lemma 1</u>

(5.13) $$[\Delta]\log(1+|w|^2) = 4(1+|w|^2)^{-2}$$

Proof. A convenient way to compute the left side of (5.13) is t introduce the formal operators $u_z = \frac{1}{2}(u_x - iu_y)$, $u_{\bar{z}} = \frac{1}{2}(u_x + iu_y$ then if $u \in C^2$,

$$\Delta u = 4u_{z\bar{z}} = 4u_{\bar{z}z} \tag{5.14}$$

A direct computation using (5.14) yields (5.13).

Theorem 2. Let Ψ be given by (5.10), (5.11). Then if ε in (5.10) is sufficiently small

$$[\Delta] \log \Psi > 0 \tag{5.15}$$

$$[\Delta] \log \Psi > \Psi \tag{5.16}$$

$$\iint [\Delta] \log \Psi (1+|w|^2)^{-2} dudv < \infty \tag{5.17}$$

Proof. We first work in the finite plane. According to the conventions described in (5.12) and Lemma 1,

$$\begin{aligned} [\Delta] \log \Psi &= \sum [\Delta] \log g_j - 2\sum [\Delta] \log (\log g_j) - 2[\Delta] \log (1+|w|^2 \\ &= 4(q-2)(1+|w|^2)^{-2} - 2\sum [\Delta] \log (\log g_j) \end{aligned} \tag{5.18}$$

and the significance of $q > 2$ for (5.15) is now apparent.

Formula (5.14) implies that

$$\begin{aligned} -2[\Delta] \log \log g_j &= -8\{\log \log g_j\}_{\bar{w}w} \\ &= -8\{\frac{(\log g_j)_{\bar{w}}}{\log g_j}\}_w \\ &= -2 \frac{[\Delta]\log g_j}{\log g_j} + 8 \frac{(\log g_j)_w (\log g_j)_{\bar{w}}}{(\log g_j)^2} \end{aligned}$$

Again, by Lemma 1 and (5.12), $-2[\Delta] \log g_j = -2[\Delta] \log(1+|w|^2)$ $= -8(1+|w|^2)^{-2}$. Thus if ε is sufficiently small in (5.10), term $[\Delta] \log g_j/(\log g_j)$ may be absorbed in the term $4(q-2)(1+|w|^2)^{-2}$ of (5.18) so that, given $\eta > 0$,

$$4^{-1}[\Delta] \log \Psi \geq (q-2-\eta)(1+|w|^2)^{-2} + 2\sum \frac{(\log g_j)_w(\log g_j)_{\overline{w}}}{(\log g_j)^2} \tag{5.19}$$

We now check (5.15), (5.16), (5.17) in a neighborhood of each $w \in \mathbb{C}^*$. The analysis is easy unless $w \in \{a_j\}$. First suppose $w_0 = a_j$ with, for now, $j < q$. Then there is a neighborhood U_j about a_j such that the g_k $(k \neq j)$ are smooth and large in U_j, and $\log g_j = \log b_j + \log|w-a_j|^{-2}$ where b_j $(\geq 0) \in C^\infty(U_j)$, and is uniformly large on U_j when ε is small. Then

$$\begin{aligned}(\log g_j)_w(\log g_j)_{\overline{w}} &= (\log g_j)_{w-a_j}(\log g_j)_{\overline{w}-\overline{a}_j}\\ &= |w-a_j|^{-2} + \rho\end{aligned}$$

where $(b = b_j)$

$$\rho = (b_w b_{\overline{w}} b^{-2} - b_w b^{-1}(\overline{w}-\overline{a}_j)^{-1} - b_{\overline{w}} b^{-1}(w-a_j))$$

has the property that $|w-a_j|^2\rho$ is bounded in U_j and tends to zero as $w \to a_j$. When these estimates are substituted into (5.19), it follows that ($\eta > 0$ arbitrary, ε small in (5.10))

$$\begin{aligned}4^{-1}[\Delta] \log \Psi &\geq (q-2-\eta)(1+|w|^2)^{-2} + 2(1-\eta)(w-a_j)^{-2}(\log g_j)^{-2}\\ &\geq c|w-a_j|^{-2}(\log|w-a_j|)^{-2} \qquad (w \in U_j,\ c > 0)\end{aligned}$$

This gives (5.15) in U_j, and

$$[\Delta] \log \Psi \geq c\Psi \qquad (w \in U_j) \tag{5.20}$$

for some $c > 0$. It is also clear that

$$[\Delta] \log \Psi \leq O(1)|w-a_j|^{-2}(\log|w-a_j|)^{-2} \qquad (w \in U_j$$

and since $\int_0^{\frac{1}{2}} (\log t)^{-2}t^{-1}dt < \infty$, this means $[\Delta] \log \Psi \in L_1(U_j)$

The analysis near other points in the finite plane is made the same way. Near $w_0 = \infty$ the argument proceeds along similar lines. Now

$$\log g_q = \text{const} + \log(1+w\bar{w})$$

so near ∞

$$(\log g_q)_w(\log g_q)_{\bar{w}} = \frac{|w|^2}{(1+|w|^2)^2}$$

$$\log g_j = \text{const} - \log(w-a_j)(\bar{w}-a_j) + \log(1+w\bar{w}) \qquad (j < q)$$

and a careful computation (leading terms cancel!) yields

$$(\log g_j)_w(\log g_j)_{\bar{w}} \leq A|w|^{-4} \qquad (j < q)$$

for large w and some constant A. By taking ε small, the factors $(\log g_j)^{-2}$ $(j \leq q)$ in (5.19) are diminished. Thus in a neighborhood U_q of ∞

$$[\Delta] \log \Psi \geq (q-2-\eta)(1+|w|^2)^2 + 2(1-\eta)\frac{|w|^2}{(1+|w|^2)^2}(\log g_q)^{-2}$$

$$> c|w|^{-2}(\log|w|^2)^{-2} \qquad (w \in U_q,\ c >$$

This gives (5.15) and (5.20) in U_q, and it is now easy to check that $\iint_{U_q} [\Delta] \log \Psi\ (1+|w|^2)^{-2}dudv < \infty$ since also $[\Delta] \log \Psi < O(1)|w|^{-2}(\log|w|^2)^{-2}$ in U_q.

We now prove Theorem 1. By compactness of $\mathbb{C}^*$, there is a choice of ε in (5.10) so that (5.15), (5.17) and (5.20) hold for all w. To obtain (5.16), we replace Ψ in (5.11) by $c\Psi$ if necessary for a smaller $c > 0$, since $[\Delta] \log (c\Psi) = [\Delta]\Psi$.

Remark. The metric $\lambda(w)$ introduced by F. Nevanlinna was defined on the w-sphere, and satisfied

$$[\Delta] \log \lambda(w) = 4\lambda(w)^2$$

The weaker (5.15)-(5.17) are adequate here.

B. A PULLBACK TO THE z-PLANE

Let $w = f(z)$ be the meromorphic function under study and let Ψ_f be the metric on the z-plane induced by Ψ:

$$\Psi_f(z) = \Psi(w)|f'(z)|^2 \qquad (w = f(z)) \tag{5.21}$$

so that $|\Psi_f(z)||dz|^2 = |\Psi(w)||dw|^2$.

Lemma 2. $[\Delta] \log\{\log(\varepsilon[w,a_j])^2\}^2 = \Delta \log\{\log(\varepsilon[w,a_j])^2\}^2$.

Proof. (The Laplacian on the right is in the sense of distributions.) We check only in the finite plane; the modifications needed when $j = q$ follow by the techniques used in Theorems 2 & 3. Since g_q is smooth in the plane, we need only consider g_j $(j < q)$. According to the definitions (5.10), (5.11) and (5.12), it must be checked that

$$\iint \log\log g_j(w)\Delta\phi(w)\,dudv = \iint \phi(w)[\Delta]\{\log\log g_j(w)\}dudv \tag{5.22}$$

$$(\phi \in C_0^\infty)$$

since then $[\Delta]\{\log(\log g_j)\}$ is also the distributional Laplacian. Now $\log(\log g_j)$ is locally integrable and C^∞ away from a_j, so the left side of (5.22) is the limit as $\varepsilon \to 0$ of

$$\int_{D_\varepsilon} \log(\log g_j(w))\Delta\phi(w)\,dudv$$

where $D_\varepsilon = \{|w-a_j| \geq \varepsilon\}$. To check that this is the same as the right side of (5.22), we need (Green's theorem) that

(5.23) $$\int_0^{2\pi} \phi(a_j+\delta e^{i\theta})\frac{\partial}{\partial\eta}\,[\log\log g_j(a+\delta e^{i\theta})\}]\,\delta d\theta = o(1) \qquad (\delta \rightarrow$$

Near a_j, $(j \neq q)$ $g_j = b(w)|w-a_j|^{-2}$, where $\log b(w)$ is smoo

so it suffices to suppose $b(w) \equiv 1$.

Exercise 4. Justify this last assertion (cf proof of Theorem [illegible]).

Then with $g_j = |w-a_j|^{-2}$ and $w-a_j = \rho e^{i\theta}$, the left side of (5.23) becomes

$$\int_0^{2\pi} \phi(a_j+\delta e^{i\theta})\frac{\partial}{\partial\rho}\,[\log\log\rho^{-2}]_{\rho=\delta}\,d\theta$$

$$2(\log\delta^{-2})^{-1}\int_0^{2\pi}\phi(a_j+\delta e^{i\theta})\,d\theta = o(1) \qquad (\delta \rightarrow 0)$$

and (5.23) (and also Lemma 2) now follows.

Theorem 3. *Let Ψ_f be given by (5.21), and let*

$$D_j = \{z;\ f = a_j\} \qquad (1 \leq j \leq$$

$$D_0 = \{z;\ f' = 0\},\quad D_\infty = \{z;\ f' = \infty\}$$

where each point in these sets is repeated as warranted by multiplicity. Then Ψ_f is ≥ 0, locally integrable in the plane, and C^∞ on $\mathbb{C}-\{\bigcup_0^\infty D_j\}$. Further

(5.24) $$[\Delta]\log\Psi_f = [\Delta]\log\Psi_w \qquad (w = f(z))$$

and (in the sense of distributions)

(5.25) $$(4\pi)^{-1}\Delta\log\Psi_f = \{2D_q - D_\infty + D_0\} - \bigcup_1^q D_j + (4\pi)^{-1}[\Delta]\log\Psi_f$$

Remarks. The second fundamental theorem will be the twice integrated form of (5.25).

The term

(5.26) $$R = 2D_q - D_\infty + D_0 \qquad (a_q = \infty)$$

measures the ramification of f and vanishes if f has no multiple values.

Proof. The statements which precede (5.24) are readily checked, and (5.24) follows from (5.21) and the convention (5.12).

Away from $\bigcup_0^q D_j$, $\log \Psi_f$ is C^∞, and $\Delta \log \Psi_f = [\Delta] \log \Psi_f$. Now let $z_0 \in D_j$ $(1 \leq j \leq q-1)$. Then near z_0, $f-a_j \sim \lambda(z-z_0)^k$ $(k \geq 1, \lambda \neq 0)$ so (5.11) and (5.21) yield

$$\Psi_f(z) = \Psi(f(z))|f'(z)|^2 \sim A|z-z_0|^{-2k}|z-z_0|^{2(k-1)}\{\log|z-z_0|\}^{-2}$$

$$(5.27) \qquad\qquad (z \text{ near } z_0)$$

If $f(z_0) = \infty$ so $z \in D_q$ (or D_∞) we use the invariance of chordal metric and compute using $h = \frac{1}{f}$. Thus

$$(5.28) \qquad [f,\infty]^{-2}\{\log[f,\infty]^2\}^{-2} \frac{|df|^2}{(1+|f|^2)^2}$$

$$= \{\log \frac{|h|^2}{(1+|h|^2)}\}^{-2}|h|^{-2} \frac{|dh|^2}{(1+|h|^2)^2}$$

so if near z_0 $f(z) \sim A(z-z_0)^{-k}$, $|dh| \sim |f|^{-2}|df|$, we see that

$$(5.29) \qquad \Psi_f(z) \sim |h'|^2|h|^{-2}\{\log|h|^2\}^{-2}$$

$$\sim A|z-z_0|^{-2}\{\log|z-z_0|\} \qquad (z \text{ near } z_0)$$

Lemma 2 shows that the log factors in (5.28) and (5.29) are already accounted for in $[\Delta]\log \Psi$. Thus Theorem 1 of Chapter II with (5.21), (5.27) and (5.29) yield (5.25). [Note that according to (5.29) each pole contributes $-\delta(z_0)$ to $(4\pi)^{-1}\Delta \log \Psi_f$, independent of multiplicity, which agrees with its contribution to $2D_q - D_\infty - D_q = D_q - D_\infty$.]

The function $f(z)$ under study is henceforth assumed to have $f(0) \notin \{a_j\}$, $f'(0) \neq 0,\infty$ in addition to (5.4).

To analyse (5.25), we introduce

$$\text{(5.30)} \qquad F(r) = \frac{1}{4\pi}\int_0^r t^{-1}dt\{\iint_{|\zeta|\leq t}[\Delta]\log\Psi_f\, d\xi d\eta\}$$

$$\text{(5.31)} \qquad \mu(r) = \frac{1}{2\pi}\int_0^{2\pi}\log\Psi_f(re^{i\theta})d\theta$$

$$\text{(5.32)} \qquad N_1(r) = \sum_{a\in R}\log\frac{r}{|a|} \qquad \text{(ramification term)}$$

<u>Theorem 4</u>. <u>For</u> $r > 0$

$$\text{(5.33)} \qquad F(r) + N_1(r) = \sum_1^q N(r,a_j) + \frac{1}{4\pi}\mu(r) + 0(1)$$

<u>where the bound in</u> $0(1)$ <u>may depend on</u> D <u>but not on</u> r.

<u>Proof</u>. The first integration of (5.25) is the more delicate. Theorem 3 shows that $\Delta \log \Psi_f$ is a (signed) measure σ, so (cf. the Remark following Theorem 3 of Chapter II) $\Delta\Psi_f$ may be extended to functions integrable with respect to this measure. Our goal is to compute

$$\iint_{|z|\leq r}\Delta\log\Psi_f = \iint\Delta\log\Psi_f\chi_r$$

where χ_r is the characteristic function of $\{|z| \leq r\}$. Decom[illegible] measure σ as the difference of positive measure: $\sigma = \sigma^+ - \sigma$ (cf. [22], p. 121). Then since $\log\Psi_f$ is locally integrable, arguments of the Riesz decomposition Theorem (Theorem 13, Chapt III) yield that

$$\log\Psi_f = P^{\sigma-} - P^{\sigma+} + \text{harmonic function}$$

We thus deduce from Theorem 15 of Chapter III and Theorem 4 of Chapter II that μ has right- and left-hand derivatives for

$r \geq 0$ and, for example, if we interpret $\mu'(r)$ to be the <u>left-hand</u> derivative[3] then μ is left-continuous and

$$\mu(r) = \mu(0) + \int_0^r \mu'(t)dt \tag{5.34}$$

Further, μ has second-order distributional derivatives and (up to linear terms) may be recovered from two integrations of these derivatives.

The assumption that $f(0) \notin \{a_j\}$, $f'(0) \neq 0$ also implies after a glance at (5.21) and (5.31) that

$$r\mu'(r) \text{ is continuous on } [0,\varepsilon],\ r\mu'(r) \to 0 \quad (r \to 0)$$

for sufficiently small ε. If we take C^∞ functions ϕ_n which are 1 in a neighborhood of the origin and tend monotonically to X_r, and then define Ψ_n as in (3.35), we see from (3.36) that

$$\iint_{|z|\leq r} \Delta \log \Psi_f = \int_0^r \frac{d^2\mu}{d(\log t)^2}\, d \log t = r\frac{d\mu}{dr} \qquad (r > 0) \tag{5.35}$$

When (5.35) is divided by $4\pi r$ and integrated from 0 to r [all integrands vanish at $r = 0$] we obtain (5.33) from (5.25), (5.30), and (5.34), with $0(1)$ in (5.33) equal to $\log \Psi_f(0) = \mu(0)$.

C. DEFECT RELATION (PRELIMINARY FORM).

(The qualification is that the bound will involve $F(r)$ rather than $T(r)$).

Let

$$F^{\#}(r) = \int_0^r \{\iint_{|\zeta|\leq t} \Psi_f(\zeta)d\xi d\eta\}t^{-1}dt \tag{5.36}$$

[3] The choice of left-hand derivative in (5.34) is arbitrary and does not affect (5.33). It is used here since n(r, a) in Chapter IV is left-continuous.

Since Ψ_f is locally integrable, $F^{\#} \in C^2$, and (5.16) with (5.
show that

$$F^{\#}(r) \leq F(r) \qquad (r > 0) \tag{5.37}$$

<u>Lemma 3.</u> <u>If</u> $\mu(r)$ <u>and</u> $F^{\#}(r)$ <u>are given by</u> (5.31) <u>and</u> (5.37), <u>t</u>

$$\mu(r) \leq \log \left\{\frac{d^2 F^{\#}(r)}{d(\log r)^2}\right\} + 0(1) \tag{5.38}$$

<u>Proof.</u> To prove this requires (2.33) and (2.34)(with $g = \exp$, $f = \log \Psi_f$), (5.30), (5.31) and (5.36)

$$\begin{aligned} \mu(r) &\leq \log\left\{\frac{1}{2\pi}\int_0^{2\pi} \Psi_f(re^{i\theta})\,d\theta\right\} \\ &= \log\left\{\frac{1}{2\pi} r \frac{d}{dr} \iint_{|\zeta| \leq r} \Psi_f(\zeta)\,d\xi\,d\eta\right\} \\ &= \log\left\{\frac{1}{2\pi}\left(r \frac{d}{dr}\right)^2 F^{\#}(r)\right\} \end{aligned}$$

In order to convert (5.38) to a more useful form, a growth lemma is needed.

<u>Lemma 4.</u> <u>Let</u> f, g, α <u>be increasing continuous functions of</u> r $(r > 0)$ <u>with</u> g' <u>continuous</u>, f' <u>piecewise continuous and</u>

$$\int^{\infty} \frac{dr}{\alpha(r)} < \infty$$

<u>Then for each</u> $r_0 > 0$

$$f'(r) \leq g'(r)\alpha(f(r)) \tag{5.39}$$

<u>for all</u> $r > r_0$ <u>save for a set</u> $I \subset (r_0, \infty)$ <u>such that</u>

$$\int_I dg < \int_{x_0}^{\infty} \frac{dr}{\alpha(r)} \qquad (x_0 = f(r_0)) \tag{5.40}$$

Proof. If $I \neq \phi$ and an interval $(a,b) \subset I$, the negation of (5.39) gives

$$\alpha(f(a))\{g(b)-g(a)\} \leq (f(b)-f(a))$$

By taking b near a, we achieve

$$dg \leq \frac{df}{\alpha(f)}$$

and the bound (5.40) follows on integration.

Remark. A qualified inequality

$$a(r) \leq b(r) \qquad ||g$$

will mean that $a(r) \leq b(r)$ save on a set I such that $\int_I dg < \infty$. When $g(x) = x$, the reference to g will be ignored.

Now take $f(r) = F^{\#}(r)$, $g(r) = \beta^{-1}r^{\beta}$, $\alpha(r) = r^{\alpha}$ with $\alpha, \beta > 1$. Then (5.39) shows

$$\frac{dF^{\#}(r)}{d \log r} \leq r^{\beta}(F^{\#})^{\alpha} \qquad ||g$$

and when (5.39) is applied again with the same $g(r)$ and $\alpha(r)$ and $f = \{dF^{\#}(r)/d \log r\}$, we obtain

$$\frac{d^2F^{\#}(r)}{d(\log r)^2} \leq r^{\beta}\left(\frac{dF^{\#}(r)}{d \log r}\right)^{\alpha} \qquad ||g$$

If α and β are taken close to one, it follows that

$$\frac{d^2F^{\#}(r)}{d(\log r)^2} \leq r^{2+\varepsilon}(F^{\#})^{2+\varepsilon} \qquad ||g \tag{5.41}$$

Hence if (5.38) and (5.41) are used with (5.37) in (5.33), it follows that

$$F(r) + N_1(r) \leq \sum N(r,a_j) + O(\log F(r)) + O(\log r) \qquad ||g \tag{5.42}$$

D. DEFECT RELATION (FINAL FORM)

It is only necessary to relate $F(r)$ to the Nevanlinna characteristic. According to (5.1), (5.2), Lemma 2, (5.30) and the computation (5.18),

$$F(r) = (q-2)T(r) - \frac{1}{2\pi} \sum \int_0^r t^{-1} dt \int_{|\zeta| \leq t} \Delta \log[\log g_j(f)] d\xi d\eta \tag{5.43}$$

We check only for $j < q$. Then the steps used to give (5.35) lead to

$$\int_0^r t^{-1} dt \int_{|\zeta| \leq t} \Delta \log[\log g_j(f)] d\xi d\eta$$

$$= \int_0^r dt \int_0^{2\pi} \frac{\partial}{\partial t} \{\log[\log g_j(f(te^{i\theta}))]\} d\theta$$

$$= \int_0^r dt \frac{d}{dt} \int_0^{2\pi} \log[\log g_j(f)] d\theta$$

$$= \int_0^{2\pi} \log[\log g_j(f(re^{i\theta}))] d\theta + \text{const} \tag{5.44}$$

If ε is taken sufficiently small in (5.10) we see that each term of the sum in (5.43) is ≥ 0. An appeal to (5.3) (5.10) and (2.33), (2.34) then yields that

$$\int_0^{2\pi} \log[\log g_j(f)] d\theta \leq 2\pi \log \int_0^{2\pi} \log g_j(f) d\theta$$

$$\leq O(1) \log m(r, a_j) \leq O(1) \log T(r) + O(1)$$

and so

$$0 \leq (q-2)T(r) - F(r) \leq A(\log T(r) + \log r) \tag{5.45}$$

When this is substituted in (5.42) we achieve

<u>Theorem 5</u> (<u>Second fundamental theorem</u>). <u>Let</u> f <u>be meromorphic in the plane, and</u> $a_1, \ldots, a_q$ $(= \infty)$ <u>distinct extended complex members. Then given</u> $\beta > 1$,

(5.46) $$(q-2)T(r) \le \sum N(r,a_j) - N_1(r) + 0(\log T(r) + \log r) \qquad (r \notin I)$$

outside a set I *such that*

(5.47) $$\int_I d(r^\beta) < \infty$$

Remark. Nevanlinna's own proof of the second fundamental theorem (cf. Ch. II of [25]) shows we may let $\beta = 1$ in (5.47).

The defect relation (5.9) is a simple consequence of Theorem 5. For (5.46) and the first fundamental theorem give

(5.48) $$\sum_1^q m(r,a_j) + N_1(r) \le 2T(r) + 0(\log r\, T(r)) \qquad (r \in I)$$

for $r \notin I$. Now a point of multiplicity k contributes $(k-1)$ to $N_1(r)$ (cf. (5.26) and (5.32)); similarly a pole of multiplicity $k \ge 1$ contributes $2k - (k+1) = k - 1$ to $N_1(r)$. Hence, if $r \to \infty$ in (5.48) while avoiding I we achieve (5.9) for any finite set of a's, and hence for all a's.

Remark. Recent examples of Hayman [24] show that an exceptional set must be allowed in (5.48).

E. ON THE LOGARITHMIC DERIVATIVE

Nevanlinna's proof of the second fundamental theorem relied on a delicate estimate for $m(r,f'/f)$. The methods used here give another approach, and sharpen slightly the conclusion of Proposition 4.3 of [20].

Theorem 6. *Let* f *be meromorphic and nonrational*. *Then*

(5.49) $$m(r,f'/f) = 0 \log(rT(r)) \qquad ||$$

Exercise 5. If f is rational, show that $m(r,f'/f) = o(1)$ $(r \to \infty)$.

Exercise 6. Choose a careful gappy series $\sum c_n z^{\lambda_n}$ to be entire and have

$$T(r) = o(m(r,f'/f))$$

on an unbounded r-set (cf. [23]).

The proof of Theorem 6 imitates that used for the second fundamental theorem; consequently some details will be omitted. If $a_1 = 0$, $a_2 = \infty$ and g_1, g_2 are as in (5.10) we let

$$\sigma(w) = (g_1 g_2)^{1/2} = \varepsilon^{-1}(1+|w|^2)|w|^{-1} \tag{5.50}$$

and consider the volume form

$$\Psi(w)\,dudv = \sigma^2(\log \sigma^2)^{-2}(1+|w|^2)^{-2}dudv \tag{5.51}$$

on the sphere.

Lemma 5. *If ε is sufficiently small in (5.50), then there is an M such that*

$$\iint [\Delta]\log \Psi(1+|w|^2)^{-2}dudv < \infty \tag{5.52}$$

$$\Psi \le [\Delta]\log \Psi + M \tag{5.53}$$

Proof. Away from $w = 0, \infty$ everything is C^∞, so we study Ψ in a neighborhood of $w = 0$. According to (5.50) we have

$$\begin{aligned}[\Delta]\log \Psi &= 2[\Delta]\log \sigma - 2\Delta\log(1+|w|^2) \\ &\quad - 2[\Delta]\log(\log \sigma^2) \\ &= -2[\Delta]\log(\log \sigma^2)\end{aligned}$$

and the analysis used in the proof of Theorem 2 shows

(5.54) $$[\Delta]\log \Psi \geq c|w|^{-2}(\log|w|)^{-2} \qquad (w \text{ near } 0)$$

if ε is sufficiently small, and the same estimate holds near ∞. This proves (5.53). Inequalities reversing (5.54)(with a different c) similarly hold near 0, ∞ and (5.52) follows. (We replace Ψ by $k\Psi$ if necessary).

The pullback Ψ_f of Ψ induced by the meromorphic function f is given by (5.21), and it follows directly from (5.50), (5.51) that

(5.55) $$\log^+|f'/f| \leq 2\log^+|\Psi_f| + \log(\log \sigma^2)^2$$

(recall that $|\sigma| < 1$ since ε is small). Each term on the right side of (5.55) must now be estimated.

According to (2.33), (5.3) and the first fundamental theorem

(5.56) $$\begin{aligned}(2\pi)^{-1}\int_0^{2\pi}\log(\log \sigma^2)^2 d\theta &\leq 2 \log \int_0^{2\pi}\log(\sigma^{-2})d\theta \\ &\leq 8\pi \log\{m(r,0) + m(r,\infty) + 0(1)\} \\ &\leq A \log T(r) + 0(1)\end{aligned}$$

Thus, (5.49) depends on a bound for

(5.57) $$s(r) = (2\pi)^{-1}\int_0^{2\pi}\log^+|\Psi_f(re^{i\theta})|d\theta$$

Now $\log \Psi_f(r)$ is locally L_1 and satisfies the distribution equation (compare with (5.25))

(5.58) $$(4\pi)^{-1}\Delta \log \Psi_f = R - \{D_1+D_2\} + (4\pi)^{-1}[\Delta]\log \Psi_f$$

where R is defined in (5.26). In terms of Ψ_f we define functionals $F(r)$, $\mu(r)$ and $F^{\#}(r)$ as in (5.30), (5.31) and (5.36). Formula (5.58), with (5.52), shows that $\Delta \log \Psi_f$ as a distribution extends to χ_r (characteristic function of

$\{|z| < r\}$). Hence two integrations of (5.58) (as in Theorem 4) with the first fundamental theorem and a glance at (5.57) give

$$(5.59)\qquad F(r) = N(r,0) + N(r,\infty) - N_1(r) + \frac{1}{4\pi}\mu(r) + O(1)$$
$$\leq 2T(r) + \frac{1}{4\pi}s(r) + O(1)$$

Lemma 6. Let M be the bound obtained in Lemma 5. Then

$$(5.60)\qquad F^{\#}(r) \leq F(r) + MT(r)$$

Proof. This is a replacement for (5.37). We have from (5.53) that

$$(5.61)\qquad \iint_{|\zeta|\leq t} \Psi_f(\zeta)\,d\xi d\eta \leq \iint_{\mathscr{F}(t)} \{[\Delta]\log\Psi + M\}(1+|w|^2)^{-2}dudv$$

where $\mathscr{F}(t)$ is the portion of the surface $\mathscr{F}$ of f which is the image of $\{|z| \leq t\}$. The result thus follows from integrati and the Ahlfors-Shimizu formula (4.19), (5.1), (5.2) since

$$M\iint_{\mathscr{F}(t)} (1+|w|^2)^{-2}dudv \leq M\frac{d}{d\log t}T(t)$$

Now

$$(5.62)\qquad \exp(\log^+ u) \leq u + 1 \qquad (u > 0)$$

so after (5.61) is written in polar coordinates and integrate we obtain from (5.53), (5.59) and (5.62) that

$$(5.63)\qquad \int_0^r t^{-1}dt\{\int_0^t sds\int_0^{2\pi} \exp[\log^+\Psi_f(se^{i\theta})]d\theta\}$$
$$\leq Ar^2 + AT(r) + As(r)$$

The integrals in (5.63) are eliminated by Lemma 4. With $g = r$, $\alpha = r^\beta$ $(\beta > 1)$, (5.39) becomes

(5.64) $$f'(r) \leq [f(r)]^{\beta} \qquad ||$$

Let f be the left side of (5.63); then from (5.64) we have

$$\int_0^r t\,dt \int_0^{2\pi} \exp(\log^+ \Psi_f)\,d\theta \leq Ar(r^2 + T(r) + s(r))^{\beta} \qquad ||$$

and (5.64) again yields

$$\int_0^{2\pi} \exp(\log^+ \Psi_f)\,d\theta \leq Ar^{\beta-1}(r^2 + T(r) + s(r))^{\beta^2} \qquad ||$$

Thus we may appeal to (2.33) and (2.34) again, and then achieve

$$\int_0^{2\pi} \log^+ \Psi_f\,d\theta \leq A(\log rT(r)) \qquad ||$$

and this with (5.56) in (5.55) gives (5.49).

Remark. If f has finite order, the qualification "||" in (5.49) may be dropped (cf. Ch. II of [25]).

F. ON VALIRON DEFICIENCIES. The number

(5.65) $$\Delta(a) = 1 - \liminf \frac{N(r,a)}{T(r)} \qquad (r \to \infty)$$

is called the Valiron deficiency of a and in analogy with (5.8) satisfies $0 \leq \Delta(a) \leq 1$. The set

(5.66) $$V_f = \{a;\ \Delta(a) > 0\}$$

is substantially larger than the countable set $N_f = \{a;\ \delta(a) > 0\}$. Indeed, a very interesting construction of Hayman [24] shows that V_f can be any countable union of compact sets of capacity zero. This is essentially best-possible (cf. [36], p. 276). Here we prove a weaker positive result whose argument is in the same spirit.

Theorem 7. *The set V_f has Lebesgue (planar) measure zero.*

<u>Proof</u>. We consider a fixed positive h and a sequence $\{r_n\}$ $(n \geq 1)$ with

$$r_1 > 1, \quad T(r_1) > 1$$

$$T(r_{n+1}) = (1+h)T(r_n) \tag{5.67}$$

so that $r_n \to \infty$ $(n \to \infty)$. Let $I_n = (r_n, r_{n+1})$

The proof is based on two observations. First, if

$$N(r,a) < cT(r) \tag{5.68}$$

for some $r \in I_n$, $c < 1$ with $c(1+h) > \frac{1}{2}h$, it follows from (5.6 that

$$N(r_n,a) < N(r,a) < cT(r) < c(1+h)T(r_n) \tag{5.69}$$

We apply Corollary 3 to Theorem 5 of Chapter IV with $E = \{|w| \leq R\}$ and μ Lebesgue measure, and obtain from (5.67)-(5.69) with a bit of care that

$$\begin{aligned} &\text{meas}\{a;\ |a| \leq R,\ N(r,a) < cT(r) \text{ for some } r \in I_n\} \leq \\ &\text{meas}\{a;\ |a| \leq R, N(r_n,a) < c(1+h)T(r_n)\} \\ &\qquad \leq [c(1+h+o(1))T(r_n)]^{-1} \leq A(1+h)^{-n} \end{aligned}$$

Thus

$$\begin{aligned} &\text{meas}\{a;\ |a| \leq R,\ N(r,a) < cT(r) \text{ for some } r \geq r_n\} \\ &\qquad \leq A \sum_{n}^{\infty} (1+h)^{-n} = o(1) \qquad (n \to \infty) \end{aligned} \tag{5.70}$$

which proves that for each $c < 1$, $\{a;\Delta(a) > 1-c\} \cap \{|w| \leq R\}$ h measure 0 and the Theorem follows.

Exercise 7. Prove

$$T(r,f') \leq \{2+o(1)\}T(r,f) \qquad ||$$

and if f is entire

$$T(r,f') \leq \{1+o(1)\}T(r,f) \qquad ||$$

Finally, show $(k \geq 1)$

$$m(r,\frac{f^{(k)}}{f}) = o(\log rT(r)) \qquad ||$$

Exercise 8. Let $f(z)$ be transcendental and meromorphic. Prove that f' assumes every finite value with at most one exception infinitely often (look at the θ's in the defect relation).

VI. APPLICATIONS AND OPEN PROBLEMS

A. APPLICATIONS TO DIFFERENTIAL EQUATIONS

(Good references are [7] and [47], Ch. V).

__Theorem 1__ (Rellich). __Let__ ϕ __and__ f __be entire__, ϕ __not linear__. __Then if__ f __satisfies the differential equation__

$$f'(z) = \phi(f(z))$$

f __is constant__.

__Proof__. According to Theorem 7 of Chapter VI, V_ϕ and V_f have Lebesgue measure 0. Set

$$V = V_\phi \cup \phi(V_f)$$

a set of measure 0 and choose $c \notin V$. Then $c \notin V_\phi$, so $N(r,c,\phi) \sim T(r,\phi)$. Since $T(r,\phi) > (\log r)^2$ we may choose $w_1 \neq w_2$ with $\phi(w_i) = c$ $(i = 1,2)$. Then the w_i are disjoint from V_f, so $N(r,w_i,f) \sim T(r,f)$. Thus

$$2T(r) \sim \sum_1^2 N(r,w_i,f)$$

$$\leq N(r,c,\phi(f)) < T(r,\phi(f))$$

$$\leq T(r,f') \leq \{1+o(1)\}T(r,f) \qquad ||$$

where the last inequality follows from Ex. 7 of Chapter V.

Exercise 1. Let f be a nonconstant entire function and g be ent[illegible] and $\phi = g(f)$. Discuss the behavior of $T(r,\phi)/T(r,f)$ $(r \to \infty)$ (consider the cases ϕ a polynomial, ϕ transcendental).

There are some other pretty theorems on growth of solutions of differential equations. Here is one:

<u>Theorem 2</u> (Frei). <u>In the differential equation</u>

$$(6.1) \qquad f^{(m)} + a_{m-1}f^{(m-1)} + \cdots + a_0 f = 0$$

<u>suppose the</u> a_j <u>are entire with</u> a_μ <u>transcendental and all</u> a_j $(j > \mu)$ <u>polynomials</u>. <u>Then</u> (6.1) <u>has at most</u> μ <u>linear independent solutions of finite order</u>.

<u>Proof</u>. Let g_1 be a non-zero solution of finite order. Then if we set $g_1 u = f$, we obtain a d.e. of the form

$$(6.2) \qquad u^{(m)} + u^{(m-1)}\{\binom{m}{1}\frac{g_1'}{g_1} + a_{m-1}\}$$

$$+ u^{(m-2)}\{\binom{m}{2}\frac{g_1''}{g_1} + \binom{m-1}{1}\frac{g'}{g}a_{m-1}$$

$$+ a_{m-2}\} + \cdots + \cdots u'\{\binom{m}{m-1}\frac{g_1^{(m-1)}}{g_1} + \cdots + a_1\} = 0$$

The important fact is that the coefficients of (6.2) are sums o[f] the a_j $(j \geq 1)$ and logarithmic derivatives of entire function[s] of finite order. Note that a_0 has been eliminated.

If (6.2) has a solution g_2 of finite order, and $f = g_2 u$, we obtain a d.e. of order $(m-2)$ with a_2 eliminated, whose coefficients are sums of logarithmic derivatives of functions of finite order and of the a_j.

After the qth iteration of this step, we have an equation of the form

$$(6.3) \qquad u^{(m)} + b_1 u^{(m-1)} + b_2 u^{(m-2)} + \dots + \{b_{\mu-1} + a_{\mu+1}\} u^{\mu+1} \equiv 0$$

where (according to our hypothesis on the $\{a_j\}$, Ex. 7 of Ch. V and the assumption that the $\{g_j\}$ $(j \leq \mu+1)$ have finite order)

$$m(r, b_j) = 0(\log r) \qquad ||$$

Let $v = u^{(\mu+1)}$. Then after (6.3) with $m = \mu+1$ is divided by v, there results

$$(6.4) \qquad m(r, a_\mu) = 0(\log r) \qquad ||$$

since $m(r, a_\mu)$ increases, it is routine to eliminate the qualification "||" in (6.4). In any case (6.4) yields that a_μ is a polynomial, and this contradiction proves the theorem.

B. SOME OPEN QUESTIONS

(Good collections of problems are in [11] and [26]).

It is very easy to give examples of functions of all order for which equality holds in (5.9), so the questions usually center on the weaker relation

$$(6.5) \qquad \sum \delta(a, f) \leq 2$$

(a) Let f have finite order and

$$(6.6) \qquad \sum \delta(a, f) = 2$$

<u>What can be said about</u> f?

A glance at (5.9) and (5.48) shows that a function f whic satisfies (6.6) must necessarily have

$$(6.7) \qquad N_1(r) = o(T(r))$$

In particular, Theorem 2 of Chapter I implies f cannot be ration

For motivation, suppose f is entire and (6.7) is strengthened to

$$(6.8) \qquad N_1(r) = 0$$

(no multiple values). Then any branch of

$$g(z) = \log f'(z)$$

may be analytically continued in the plane, and the assumption t f has finite order gives

$$(6.9) \qquad \int_0^{2\pi} \mathcal{R}g(re^{i\theta})d\theta = \int_0^{2\pi} \log|f'(re^{i\theta})|d\theta$$

$$\leq 2\pi[T(r,f) + m(r,f'/f)] = O(r^k)$$

for some k. The Borel-Carathéodory inequality ([42], p. 175) gives a similar bound for $\mathcal{I}g$, and hence g is a polynomial $P(z)$ of degree $\leq k$; i.e.

$$g'(z) = e^{P(z)}$$

Thus the plane divides into $2k$ sectors, each of width π/k, on which $P(z)$ tends alternatively to $\pm\infty$. From this it is possible to obtain all one could hope about such f:

(i) f has order $n \in \{1,2,...\}$,

(ii) if $\delta(a) > 0$, then $\delta(a)$ is an integral multiple of $1/n$,

(iii) the deficient values are asymptotic values.

From the work of Pfluger [37] and Edrei-Fuchs [18], [19], we know these conclusions hold even when (6.8) is replaced by (6.7). However, for a <u>meromorphic</u> function f, little is known. Even if (6.8) is assumed, it is nontrivial to determine f. The only method I know is due to F. Nevanlinna ([34]) in which he showed that the order of f must be $\frac{1}{2}(n+1)$ $(n = 1,2,\ldots)$, that each $\delta(a)$ is an integral multiple of $2(n+1)^{-1}$ and the deficient values are asymptotic. This conjecture is still largely unresolved, with the exception of Weitsman's result in [45]: if f is of order $\rho < \infty$ the number of deficient values is at most 2ρ.

(b) <u>The deficiency problem</u> (A. Edrei). For $0 < \rho < \infty$, set

$$\Lambda(\rho) = \sup \sum_a \delta(a,f)$$

where the sup is taken over all meromorphic (entire) functions of order Λ. For $\rho \leq 1$, $\Lambda(\rho)$ is now known ([16],[44] for entire functions; [3], [17] for meromorphic functions). For entire functions, the classical conjecture is

$$(6.10) \qquad \sum_a \delta(a) \leqq \begin{cases} \dfrac{|\rho|}{[\rho]+|\sin \pi\rho|} & 0 < \rho - [\rho] \leq \frac{1}{2} \\[2ex] \dfrac{|\rho|+1-|\sin \pi\rho|}{[\rho]+1} & \frac{1}{2} \leq \rho - [\rho] < 1 \end{cases}$$

Very recently, a conjecture was made on the form of $\Lambda(\rho)$ for general meromorphic functions [15].

(c) <u>Deficiencies and zero-distribution</u>. While the terms $N(r,a)$ are insensitive to the arguments of the points of $f^{-1}(a)$, restrictions on even $f^{-1}(0)$ and $f^{-1}(\infty)$ greatly restrict the behavior of f. The first result of this nature is due to Edrei-Fuchs [18]: let f be entire, with all zeros on the negat[illegible] axis and order $\infty > \rho > 1$; then

$$\delta(0,f) > 0 \tag{6.11}$$

It is conjectured that the restriction $\rho \neq \infty$ is unnecessary a[illegible] if f has infinite order with only negative zeros, then $\delta(0,f) = 1$. In support of this, Hellerstein and Shea [27] have recently shown that, given $\varepsilon > 0$, there exists ρ_0 such that if $\rho_0 \leq \rho \lneqq \infty$ and f is of order ρ with only negative zero[illegible] then

$$\delta(0,f) > 1-\varepsilon$$

If the conjecture were true, it would provide a rare positive result for functions of infinite order.

(d) <u>A connection with differential equations</u>. Let the non-constant meromorphic function f be expressed as g/h where [illegible] and h are entire with no common roots. Then (direct computat[illegible] g and h satisfy the second-order differential equation

$$Ww'' - W'w' + \Lambda w = 0 \tag{6.12}$$

where

$$W(z) = \begin{vmatrix} g & g' \\ h & h' \end{vmatrix} = gh' - g'h$$

is the <u>Wronskian</u> of g and h and

$$\Lambda(z) = g'h'' - g''h'$$

It is also clear that

$$f'(z) = \frac{W(z)}{h(z)^2}$$

and so assumption (6.8) becomes

$$W(z) \neq 0$$

Equation (6.12) suggests study of the equation

(6.13) $$a_0 w'' + a_1 w' + a_2 w = 0$$

which is assumed to have linearly independent solutions g and h. Set

(6.14) $$\rho^* = \max(\rho_g, \rho_h)$$

For example, if the a_j in (6.13) are polynomials, it is classical that $\rho^* = \frac{1}{2}(n+1)$ $(n = 1,2,\ldots)$ (cf. [47], Ch. V).

Suppose instead that the orders of the a_j are less than that of g and h. Is it still the case that if g and h have finite order then $\rho^* = \frac{1}{2}(n+1)$ $(n \geq 1)$? This is suggested by the conjecture in subsection (a).

A more appealing way to state this is:

<u>If</u> W <u>has order</u> $< \max(\rho_g, \rho_h)$,
<u>then</u> $\rho^* = \frac{1}{2}(n+1)$ $(n = 1,2,\ldots)$.

(A comparison of this conjecture with (6.12) and (6.13) shows that W and W' then have smaller order, and since

$$\Lambda = W \frac{w''}{w} - W' \frac{w'}{w}$$

Λ also has order $\leq W$). For $\rho < 1$ the conjecture implies that W has order ρ^*, and for $\rho^* < \frac{1}{3}$, this has been proved by J. G. Clunie (unpublished).

(e) <u>Deficiencies of functions and derivatives</u>. I have recently shown [13] that the deficiency relation (5.9) is sharp in the sense that given sequences a_i, δ_i, θ_i with $0 < \delta_i + \theta_i \leq 1$, $\sum \delta_i + \theta_i \leq 2$, there is a meromorphic function $f(z)$ with $\delta(a_i) = a_i$, $\theta(a_i) = \theta_i$, $\delta(a) = \theta(a) = 0$ for $a \notin \{a_i\}$.

However, our understanding of analogous relations between value distribution of functions and their derivatives is far from complete. For example, the hypothesis that if f is meromorphic and $ff'f^{(n)} \neq 0$ for some $n \geq 2$ implies $f(z) = (Az + B)^{-k}$ or $f(z) = \exp(Az + B)$ for suitable A, B and $k \geq 1$ remains unverified in general (it is known for functions of finite order; cf. the discussion in [11]).

Here is a more modest problem. It is classical that if f is of finite order, then

$$\sum \delta(a,f) \leq \delta(0,f')$$

(if f is entire) and

$$\sum \delta(a,f) \leq \frac{\delta(0,\delta')}{2-\delta(\infty,f)} \tag{6.15}$$

if f is meromorphic (cf. [47]).

It is obvious that these inequalities may be augmented to higher derivatives. The problem is: are these relations all that can be asserted? For example, if f has finite order, $\sum \delta(a,f) = 2$ and $\delta(\infty,f) = 0$, (6.15) implies that f' has 0 as maximal deficiency, and this observation is the key to Weitsman's approach in [45]. However, (6.15) leaves open the possibility that f' might have other deficient values. A study of the examples of F. Nevanlinna [34] leads to the conjecture that this cannot happen: f of finite order, $\sum \delta(a,f) = 2$, $\delta(\infty,f) = 0 \implies \sum \delta(a,f') = \delta(0,f') = 1$. It would be of interest to clarify things here.

Here is a concrete problem: let $f(z)$ be meromorphic with $\delta(\infty,f) = 0$. Can we have

$$\sum\delta(a,f) + \sum\delta(a,f') = 4 \tag{6.16}$$

If not, what is the best bound? (The function e^z shows that (6.16) may be satisfied for entire functions).

(f) <u>Value-distribution in more general settings</u>. The idea of studying value distribution on meromorphic curves and Riemann surfaces is quite classical; cf. Ahlfors's fundamental paper [1] and [46]. Recent developments, however, have been centered on two areas, valid in higher dimensions.

(i) <u>Quasi-regular mappings</u>. These were first introduced by Resetnjak [38] and also well studied by O. Martio, S. Rickman and J. Väisälä (cf. [33]). These mappings go from $\mathbb{R}^n$ to $\mathbb{R}^n$ $(n \geq 2)$ and have the property that infinitesimal spheres are mapped onto regions which are not too far from spherical (ellipsoids, etc.). It is conjectured that Picard's theorem holds for these mappings. In two dimensions, such functions f may be written as $f = g \circ h$ where h is a quasi-conformal homeomorphism of the plane and g is entire (meromorphic). Thus Picard's theorem is trivial in this setting when $n = 2$.

(ii) <u>Analytic mappings between algebraic varieties</u> (good references are [9], [21], [40], [41] and the collection of papers [32]). Now one considers analytic mappings $f: A \to B$ where f is a "nice" manifold (e.g. complex n-space) and B is an algebraic variety of dimension k perhaps with singularities. Much of the classical ideas may be adopted to this setting (at least if $k = n$); for example, our Chapter V is adapted from [9] and [21]. However, there are higher-dimensional phenomena that are still poorly understood.

The problem in this more general setting is to count how often a given subset V of B is covered by f. In the classical setting, the only natural choices of V were points (second fundamental theorem) or lines and circles (cf. Theorem 5 and its corollaries in Ch. IV). Here there are lots of natural choices for V. Principal interest is the case that V be a <u>variety</u> of B; i.e. the zero-set of an algebraic mapping on B. Ahlfors' theory provides very complete answers when n = 1 and the V's are hyperplanes, but results for general n and k are still meagre. Interest in attaining sharp positive results arises from the classic Fatou-Bieberbach example [6] of a mapping from $\mathbb{C}^2$ to $\mathbb{C}^2$ whose range omits an open set (for a simple discussion of this example, cf. [20], p. 173). Nothing of this sort is possible when n = k = 1.

BIBLIOGRAPHY

1. L. Ahlfors, The theory of meromorphic curves, Acta Soc. Sci. Fenn., Ser. A, vol. 3 (1941) 1-31.

2. ———, Complex Analysis, Second Edition, McGraw-Hill, New York, 1966.

3. A. Baernstein, Proof of Edrei's spread conjecture, Proc. London Math. Soc. (3) 26(1973) 418-434.

4. ———, A generalization of the cos $\pi\rho$ theorem, Trans. Amer. Math. Soc., <u>193</u>(1974) 181-197.

5. ———, Integral means, univalent functions and circular symmetrization, Acta Math. <u>133</u>(1974) 139-169.

6. L. Bieberbach, Beispiel zweier ganzer Funktionen zweier komplexer Variablen welche eine schlicht volumentreue Abbildung des R_4 auf einen Teil selbst vermitteln, Preuss. Akad. Wiss. Sitzungsber. (1933) 476-479.

7. ———, Theorie der gewöhnlichen Differentialgleichun auf funktionentheoretischer Grundlage dargestellt, Springer, Berlin, 1953.

8. M. Brelot, Eléments de la théorie classique du potential, Les cours de Sorbonne, Paris, 1959.

9. J. Carlson and Ph. Griffiths, A defect relation for equi-dimensional holomorphic mappings between algebraic varieties, Annals of Math. 95(1972) 557-584.

10. H. Cartan, Sur les zéroes des combinaisions linéares de p functions holomorphes donnés, Mathematica (Cluj) 7(1933) 5-32.

11. J. G. Clunie and W. K. Hayman, Proceedings of the Symposium on Complex Analysis at Canterbury 1973, London Math. Lecture Notes #12, Cambridge, 1974.

12. R. Courant and D. Hilbert, Methods of Mathematical Physics (vol. II), Interscience, New York, 1962.

13. D. Drasin, The inverse problem of the Nevanlinna theory, to appear in Acta Math.

14. D. Drasin and A. Weitsman, The growth of the Nevanlinna proximity function and the logarithmic potential, Indiana Univ. Math. J. 20(1970-71) 699-715.

15. ————————, Meromorphic functions with large sums of deficiencies, Advances in Math. 15(1974) 93-126.

16. A. Edrei, Locally tauberian theorems for meromorphic functions of lower order less than one, Trans. Amer. Math. Soc. 140(1969) 309-332.

17. ————, Solution of the deficiency problem for functions of small lower order, Proc. London Math. Soc. (3) 26(1973) 435-445.

18. ———— and W. H. J. Fuchs, On the growth of meromorphic functions with several deficient values, Trans. Amer. Math. Soc. 93(1959) 292-328.

19. ————, Valeurs déficientes et valeurs asymptotiques des fonctions méromorphes, Comment. Math. Helv. 33(1959)

20. Ph. Griffiths, Two results in the global theory of holomorphic mappings, in Contributions to Analysis: A collection of papers dedicated to Lipman Bers, Academic Press, New York, 1974.

21. ———— with J. King, Nevanlinna theory and holomorphic mappings between algebraic varieties, Acta Math. 130 (1975) 145-220.

22. P. Halmos, Measure Theory, Van Nostrand, Princeton, 1950.

23. W. K. Hayman, On the characteristic of functions meromorphic in the plane and of their integrals, Proc. London Math. Soc. (3) 14a(1965) 143-173.

24. ————————, On the Valiron deficiencies of integral functic of infinite order, Arkiv för Math. 10(1972) 163-172.

25. ————————, Meromorphic Functions, Oxford, 1964.

26. ————————, Research Problems in Function Theory, Athlone Press (Univ. of London), 1967.

27. S. Hellerstein and D. Shea, An extremal problem concerning entire functions with radially distributed zeros, pp. 81-87 of [11].

28. E. Hewitt and K. Stromberg, Real and Abstract Analysis, Springer, New York and Berlin, 1969.

29. E. Hille, Analytic Function Theory (I), Ginn, Boston, 1959.

30. L. Hörmander, Linear Partial Differential Operators, Springe New York and Berlin, 1963.

31. ————————, An Introduction to Complex Analysis in Severa Variables, Van Nostrand, New York, 1964.

32. R. O. Kujala and A. L. Vitter III, Value Distribution Theor (Part A), Marcel Decker, New York, 1974.

33. O. Martio, S. Rickman and J. Väisälä, Definitions of quasi-regular mapping, Ann. Acad. Sci. Fenn. AI 448(1969) 1-

34. F. Nevanlinna, Über eine Klasse meromorpher Funktionen, Septième Congrès Math. Scand., Oslo, 1930, 81-83.

35. R. Nevanlinna, Über Riemannsche Flächen mit endlich vielen Windungspunkten, Acta Math. 58(1932) 295-373.

36. ————————, Analytic Functions, Springer, New York and Berlin, 1970.

37. A. Pfluger, Zur Defektrelation ganzer Funktionen endlicher Ordnung, Comment. Math. Helv. 19(1946) 91-104.

38. Yu. Resetnjak, Estimates of modules of continuity of some mapping (Russian), Sibirsk. Math. Z. 7(1966) 1106-1114

39. L. Schwartz, Théorie des Distributions, Hermann, Paris, 196

40. B. Shiffman, Nevanlinna defect relations for singular divis Inventiones Math. 31 (1975) 155-182.

41. W. Stoll, Value distribution of holomorphic maps into compa complex manifolds, Lecture Notes in Mathematics vol. 13 Springer, New York and Berlin, 1970.

42. E. C. Titchmarsh, Theory of Functions (2nd Ed.) Oxford, 19

43. M. Tsuji, Potential Theory in Modern Function Theory, Maruzen, Tokyo, 1958.

44. A. Weitsman, Asymptotic behavior of meromorphic functions with extremal deficiencies, Trans. Amer. Math. Soc. 140(1969) 333-332).

45. ——————, Meromorphic functions with maximal deficiency sum and a conjecture of F. Nevanlinna, Acta Math. 123(1969) 115-139.

46. H. Weyl and F. J. Weyl, Meromorphic Functions and Analytic Curves, Princeton, 1943.

47. H. Wittich, Neuere Untersuchungen über eindeutige analytische Funktionen, Springer, Berlin, 1955.

48. K. Yosida, Functional Analysis, Springer, New York and Berlin, 1968.

IAEA-SMR-18/7

ELLIPTIC OPERATORS ON MANIFOLDS

J. EELLS
Mathematics Institute,
University of Warwick,
Coventry, Warwickshire,
United Kingdom

Abstract

ELLIPTIC OPERATORS ON MANIFOLDS.

I. An abstract Dirichlet problem: Hilbert space background; Representation of functionals; Compact transformations; The Dirichlet problem. II. Elliptic operators on the torus: Group duality and Fourier analysis; The Hilbert spaces; Dirichlet's problem on T^n; Regularity of solutions; Zero boundary values in $\mathbb{R}^n$; Strongly elliptic systems. III. Differential operators on vector bundles: Sheaves of modules; Vector bundles; Smooth manifolds and vector bundles; Certain operators on vector bundles; Riemannian structures; Differential operators. IV. The existence theorem and applications: The existence theorem; Hodge's theorem.

INTRODUCTION

The primary object of this paper is to present an elementary and self-contained proof of the following fundamental existence theorem. *Let* X *be a compact smooth manifold (without boundary), and let* ξ,η *be smooth vector bundles (finite fibre dimension) over* X. *Let* A *be a smooth elliptic operator from the sections of* ξ *to the sections of* η. *If* ψ *is a smooth section of* η, *then there is a smooth section* ϕ *of* ξ *such that* $A\phi = \psi$ *if and only if* ψ *is orthogonal to the kernel of the adjoint of* A. Furthermore, $\dim \mathrm{Ker}(A) < \infty$, so that deviation from uniqueness is somewhat restricted.

There are several well established approaches to this problem; for example:

(1) **The methods of potential theory**, centering around the properties of singular integral equations and pseudodifferential operators; see Seeley [17] for an exposition in appropriate generality.

(2) **Schwarz's alternating method** (Hildebrandt [7]).

(3) **The heat equation method** of Milgram-Rosenbloom [12], put in a Hilbert space framework by Gaffney [4].

(4) **The theory of coercive quadratic forms** in Hilbert space, based on Gårding's inequality. It is this fourth method that we shall develop. It should be clear that our exposition owes a great deal to Bers-John-Schechter [2] (their chapters on Hilbert space methods are the basis for our Section II), Koszul [9], Singer [18]. A treatment of Hodge's theorem in this direction was given by Morrey-Eells [13].

The material was presented at Cornell University in the Fall of 1964, and at the University of Amsterdam in the Spring of 1966. It is included in the present volume primarily because that fundamental theorem has been so frequently used during our Summer Course.

I. AN ABSTRACT DIRICHLET PROBLEM

1. HILBERT SPACE BACKGROUND

(A) Let us review briefly certain aspects of the theory of Hilbert spaces, over the real number field $\mathbb{R}$; the modifications necessary to produce the analogous theory for the complex field $\mathbb{C}$ will be taken for granted.

Definition. A *pre-Hilbert space* is a vector space E together with an inner product $\langle , \rangle$. We will let $\langle , \rangle_E$ denote $\langle , \rangle$ if we wish to emphasize the space to which the inner product belongs. Thus $\langle , \rangle$ is a symmetric bilinear form $E \times E \to \mathbb{R}$ such that $\langle x, x \rangle \geqslant 0$, and $\langle x, x \rangle = 0$ when and only when $x = 0$. For each $x \in E$ we write $|x| = +\sqrt{\langle x,x \rangle}$; then we have the

Schwarz inequality: For any $x, y \in E$, $|\langle x, y\rangle| \leqslant |x|\,|y|$. It follows at once that the function $x \to |x|$ is a norm on E:

(1) $|x| = 0$ if and only if $x = 0$;
(2) $|ax| = |a|\,|x|$ for all $(a,x) \in \mathbb{R} \times E$, where $|a|$ denotes the absolute value of a;
(3) For any $x,y \in E$ we have $|x+y| \leqslant |x| + |y|$.

In particular, setting $\rho(x,y) = |x-y|$ defines a *metric on* E; and that in turn determines a Hausdorff topology on E relative to which the algebraic operations are continuous.

We have also the

Parallelogram law: $|x+y|^2 + |x-y|^2 = 2(|x|^2 + |y|^2)$

Pythagoras' law: If x and y are orthogonal, then $|x+y|^2 = |x|^2 + |y|^2$.

Definition. A *Hilbert space* is a pre-Hilbert space which is complete in the metric ρ. Every pre-Hilbert space E has a unique completion to a Hilbert space E_1, whose points can be viewed as the totality of Cauchy sequences of E; and E is a dense subspace of E_1.

If $(E, |\ |_1)$ and $(E, |\ |_2)$ are Hilbert spaces, we say that they are *equivalent* if there is a number $c > 0$ such that

$$c^{-1}|x|_1 \leqslant |x|_2 \leqslant c|x|_1 \qquad \text{for } x \in E$$

This is an equivalence relation; the equivalence classes are called *Hilbertian spaces.* Thus a Hilbertian space is a topological vector space whose topology can be given by an inner product whose induced metric is complete.

Example. Let E be the totality of smooth functions $x: I \to \mathbb{R}$, where $I = [0, 1]$. Then

$$\langle x,y \rangle_1 = \int_I (x(t)y(t)\,dt + \int_I x'(t)\,y'(t)\,dt$$

$$\langle x, y \rangle_2 = x(0)\,y(0) + \int_I x'(t)\,y'(t)\,dt$$

are both inner products on E; let E_1 and E_2 be the completions of E in the indicated metric. Then E_1 and E_2 are topologically equivalent Hilbert spaces, and hence are two representations of the same Hilbertian space, really the fundamental entity. The elements are those absolutely continuous functions on I having square integrable first derivatives.

Example. Any two inner products on an n-dimensional vector space determine equivalent Hilbert spaces.

(B) **Theorem.** *A Hilbertian space is locally compact if and only if it is finite dimensional.*

Proof of the sufficiency. Let $(e_i)_{1 \leqslant i \leqslant n}$ be a base for E. Define the isomorphism $\phi : \mathbb{R}^n \to E$ by

$$\phi(x_1, ..., x_n) = \sum_{i=1}^{n} x_i e_i$$

Then ϕ is a homeomorphism; for this is true for n = 1 by the axioms for a norm, and we can apply induction. Thus any n-dimensional vector space E is topologically isomorphic to $\mathbb{R}^n$, whence E is locally compact.

Proof of the necessity. We can suppose $B = \{x \in E : |x| \leqslant 1\}$ is compact without loss of generality. Suppose $\dim E = \infty$; then there is an infinite orthonormal set $(e_i)_{i \geqslant 1}$ in B, which we can suppose is convergent. But that contradicts Pythagoras' law: $2 = |e_i|^2 + |e_j|^2 = |e_i - e_j|^2$.

(C) **Theorem.** *Let* V *be a closed subspace of the Hilbert space* E. *Then*
(1) *for any* $x \in E$, *there is a unique* $y \in V$ *such that* $|x - y| = \rho(x, V)$ $(= \inf[|x - z| : z \in V]$. So y is the point in V nearest to x.
(2) y *is the only point in* V *for which* $x - y \perp V$.

Proof (1). Let $\alpha = \rho(x, V)$, and let $(y_i)_{i \geqslant 1} \subset V$ be a sequence such that $|x - y_i| \to \alpha$. Then (y_i) is Cauchy. By the parallelogram law

$$|y_i - y_j|^2 = |y_i - x + x - y_j|^2 = 2(|x - y_i|^2 + |x - y_j|^2) - 4|x - (y_i + y_j)/2|^2$$

But $(y_i + y_j)/2 \in V$, whence $4|x - (y_i + y_j)/2|^2 \geqslant 4\alpha^2$. Given $\epsilon > 0$, choose i_0 such that $i, j \geqslant i_0$ implies $|x - y_i|^2 \leqslant \alpha^2 + \epsilon$, $|x - y_j|^2 \leqslant \alpha^2 + \epsilon$. Then

$$|y_i - y_j|^2 \leqslant 2(\alpha^2 + \epsilon + \alpha^2 + \epsilon) - 4\alpha^2 = 4\epsilon$$

Since $(y_i) \subset V$ is Cauchy and V is closed, we find that y_i approaches some $y \in V$, whence $\rho(x, y_i) \to \rho(x, y)$ and $\rho(x, y) = \alpha$. Suppose y′ were another such point. Then

$$|y - y'|^2 = |y - x + x - y'|^2 = 2(|y - x|^2 + |y' - x|^2) - 4|x - (y + y')/2|^2$$

$$\leqslant 2(\alpha^2 + \alpha^2) - 4\alpha^2 = 0$$

whence $y = y'$.

Proof (2). For any $z \in V$ such that $z \neq 0$ and any number $\lambda \neq 0$, we have $|x - (y + \lambda z)|^2 > \alpha^2$. Thus $2\lambda\langle z, y - x\rangle < \lambda^2 |z|^2$ for all real $\lambda \neq 0$. But this cannot be (for all $|\lambda|$ small) unless $\langle z, y - x\rangle = 0$.

Suppose now $y' \in V$ is such that $x - y' \perp V$. Then

$$\alpha^2 = |x - y|^2 = |x - y'|^2 + |y - y'|^2$$

But $\alpha \leqslant |x - y'|$, whence $|y - y'|^2 = 0$.

(D) Proposition. *If* E *and* F *are Hilbert spaces and* $\phi : E \to F$ *a linear map, then* ϕ *is continuous if and only if there is a real number* b *such that*

$$|\phi(x)|_F \leqslant b|x|_E \quad \text{for all } x \in E$$

Lemma. *If* $\phi : E \to F$ *is a linear map for which there exists a number* $b > 0$ *such that*

$$b^{-1}|\phi(x)|_F \leqslant |x|_E \leqslant b|\phi(x)|_F \quad \textit{for all } x \in E$$

then ϕ *is continuous, injective, and* $\phi(E)$ *is closed in* F.

Proof. The first two assertions are immediate. To prove that $\phi(E)$ is closed, let $(y_i)_{i \geqslant 1}$ be a Cauchy sequence in $\phi(E)$, and let $x_i \in E$ be the points for which $\phi(x_i) = y_i$. Then $|x_i - x_j|_E \leqslant b|y_i - y_j|_F \to 0$ as $i, j \to \infty$. Let $x \in E$ be the limit of x_i in E, and set $y = \phi(x)$. Then $|y - y_i|_F = |\phi(x - x_i)|_F \leqslant b|x - x_i|_E \to 0$, whence $y \in \phi(E)$ is the limit of the y_i.

(E) Let $\pi : E \to V$ be the nearest-point map $x \to y$ given by (1) in Theorem 1C. Then clearly π is surjective and is linear, and is continuous. (Proof of linearity: $x + y - \pi(x + y)$ is $\perp V$, and so is $x + y - (\pi(x) + \pi(y))$. But there is *only one* vector z such that $(x + y) - z \perp V$, whence $\pi(x + y) = \pi(x) + \pi(y)$. Similarly for $\pi(ax) = a\pi(x)$ for $(a, x) \in \mathbb{R} \times E$.)

Proposition. Kernel $\pi = \{x \in E : \pi(x) = 0\}$ *is the orthogonal complement of* V, *written* $V^\perp$. *We have* $E = V \oplus V^\perp$; i.e. *every* $x \in E$ *can be written uniquely* $x = \pi(x) + (x - \pi(x))$ *with* $\pi(x) \in V$, $x - \pi(x) \in V^\perp$ *and these components are orthogonal.*

Corollary. *If* $V \neq E$, *there exists a* $u \neq 0$ *in* E *such that* $u \perp V$.

Proof. Take any $x \in E$, $x \notin V$. Then we can write $x = \pi(x) + u$, $u \perp V$. $u \neq 0$, for otherwise $x = \pi(x)$, whence $x \in V$.

2. REPRESENTATION OF FUNCTIONALS

(A) Representation theorem. *Let* $f : E \to \mathbb{R}$ *be a continuous linear form. Then there exists a unique* $w \in E$ *such that* $f(x) = \langle x, w \rangle$ *for all* $x \in E$; *if* $|f| = \sup\{|f(x)| : |x| = 1 \text{ in } E\}$, *then* $|f| = |w|$.

Proof. Let V = kernel f; then V is a closed linear subspace, for if $(x_i)_{i \geqslant 1} \subset V$ and $|x_i - x| \to 0$ then $|f(x)| = |f(x - x_i)| \leqslant \text{const}\, |x - x_i| \to 0$; whence $x \in V$. If $V = E$, take $w = 0$. Otherwise there exists a u in E such that $|u| = 1$ and $u \perp V$. Then $f(u) \neq 0$, and for all $x \in E$

$$f\left(x - \frac{f(x)}{f(u)}\, u\right) = f(x) - f(x) = 0$$

Thus $x - \dfrac{f(x)}{f(u)}\, u \in V$, whence $\left\langle x - \dfrac{f(x)}{f(u)}\, u, u \right\rangle = 0$, i.e. $\langle x, u \rangle = \dfrac{f(x)}{f(u)}\, |u|^2$. It follows that $f(x) = \langle x, f(u)u/|u|^2 \rangle$. Taking $w = f(u)u/|u|^2$ satisfies the conditions of the theorem. If w' is another representation of f, then $f(x) = \langle x, w \rangle = \langle x, w' \rangle$, whence $(w' - w) \perp E$, and by Corollary 1E we have $w' - w = 0$.

To prove $|f| = |w|$, we first observe that $|w|^2 = \dfrac{f(u)}{|u|^2} \cdot \dfrac{f(u)}{|u|^2} \langle u, u \rangle$, whence $|w| = \dfrac{|f(u)|}{|u|} \leqslant |f|$. Conversely, for every $\epsilon > 0$ there exists an $x_\epsilon \in E$ such that $|x_\epsilon| = 1$ and $|f| - \epsilon \leqslant f(x_\epsilon) = \langle x_\epsilon, w \rangle \leqslant |x_\epsilon||w| = |w|$. Since this is true for all $\epsilon > 0$, we have $|f| \leqslant |w|$.

(B) Consider now a continuous linear map $\phi: E \to F$. For each $y \in F$ the map $f: x \to \langle \phi(x), y \rangle_F$ is a continuous linear form on E. By Theorem 2A there is a unique element $x_y \in E$ such that

$$f(x) = \langle x, x_y \rangle_E \quad \text{for all } x \in E$$

Thus we have a map $F \to E$ defined by $y \to x_y$, which we will call ϕ^*, *the adjoint of* ϕ. Clearly ϕ^* is a continuous linear map, satisfying $\langle \phi(x), y \rangle_F = \langle x, \phi^* y \rangle_E$ for all $x \in E$, $y \in F$. In particular, $\phi^{**} = \phi$.

(C) **Representation theorem.** *Let* $\beta: E \times E \to \mathbb{R}$ *be a bilinear form such that:*
(1) $|\beta(x, y)| \leqslant \text{const}\, |x|\, |y|$
(2) $|x|^2 \leqslant b\beta(x,x)$
for all $x, y \in E$ *for some strictly positive* $b \in \mathbb{R}$. *Given any continuous linear form* $f: E \to \mathbb{R}$, *there is a unique* $v \in E$ *such that*

$$f(x) = \beta(x, v) \quad \textit{for all } x \in E$$

Suppose first that β is symmetric. Then $|x|^2 \leqslant b\,\beta(x,x) \leqslant \text{const}|x|^2$ shows that β is a topologically equivalent inner product on E, and the theorem follows from Theorem 2A. Thus the emphasis of the present theorem is absence of symmetry of β.

Proof. Take any $y \in E$; then the map $x \to \beta(x, y)$ is a continuous linear form on E, whence by Theorem 2A there is a unique element, which we will call $Sy \in E$, such that

$$\beta(x, y) = \langle x, Sy \rangle$$

It follows that S is a linear endomorphism of E which is continuous: $|Sx|^2 = \langle Sx, Sx \rangle = \beta(Sx, x) \leqslant \text{const}\, |Sx|\, |x|$, whence $|Sx| \leqslant \text{const}\, |x|$ for all $x \in E$.

Furthermore, $|x| \leqslant \text{const}|Sx|$ for all $x \in E$, so that S maps E bijectively onto a closed linear subspace (by Lemma 1D). In fact, S is surjective, for otherwise there is a $w \perp S(E)$ and $w \neq 0$. This would imply that $\langle w, Sw \rangle = 0$ so that $\beta(w, w) = 0$; i.e. $w = 0$, a contradiction.

Now take the form f, and let w be its representative: $f(x) = \langle x, w \rangle$. Then there is a unique $v \in E$ such that $Sv = w$, so that $f(x) = \langle x, Sv \rangle = \beta(x, v)$ for all $x \in E$.

3. COMPACT TRANSFORMATIONS

(A) A linear transformation $\phi: E \to F$ is *compact* (or *completely continuous)* if ϕ maps bounded subsets of E into relatively compact subsets of F. (A subset is relatively compact if its closure is compact.) Such a ϕ is bounded (i.e. is continuous), for otherwise there is a sequence $(x_i)_{i \geqslant 1} \subset E$ with $|x_i|_E = 1$, $|\phi x_i|_F \to \infty$; but $(\phi(x_i))_{i \geqslant 1}$ is relatively compact, so that a subsequence would converge to an element of F.

Proposition. *Let* $\phi: E \to F$ *be a compact linear transformation, and set* $\psi = I - \phi$. *Then* $\psi: E \to E$ *is a continuous linear map whose kernel* $K(\psi)$ *has finite dimension.*

Proof. Clearly ψ is continuous and linear. Let $(x_i)_{i \geqslant 1} \subset K(\psi)$ be a bounded sequence. Then $x_i = \phi(x_i)$ for all $i \geqslant 1$, whence $(x_i)_{i \geqslant 1}$ is a relatively compact subset of E. Thus every bounded sequence in $K(\psi)$ has a convergent sequence, which implies that Kernel ψ is locally compact.

Example. Let E be the completion of the smooth functions $x: I \to \mathbb{R}$ with the inner product

$$\langle x, y \rangle_E = \int_I x(t)y(t)\, dt + \int_I x'(t)\, y'(t)\, dt$$

Let F be the completion of the same space in

$$\langle x, y\rangle_F = \int_I x(t)\, y(t)\, dt$$

Then the inclusion map $\phi: E \to F$ is compact, for if (x_i) is a bounded sequence in E, then it can be proved that (x_i) is equicontinuous, whence by Ascoli's theorem (x_i) has a convergent subsequence

Let $K^\perp(\psi) = \{x \in E: \langle x, y\rangle = 0$ for all $y \in \operatorname{Ker}\psi\}$. $K^\perp(\psi)$ is called the orthogonal complemen $K(\psi)$; it is clearly a closed linear subspace of E.

Lemma. *There is a number* $c > 0$ *such that* $c^{-1}|x| \leqslant |\psi(x)| \leqslant c|x|$ *for all* $x \in K^\perp(\psi)$.

Proof. The second inequality is clear. If the first were false, there would exist a sequence $(x_i)_{i\geqslant 1} \subset K^\perp(\psi)$ such that $|x_i| = 1$ and $|\psi(x_i)| \to 0$. But $x_i = \phi(x_i) + \psi(x_i)$ and $(\phi(x_i))_{i\geqslant 1}$ has a convergent subsequence, whence $(x_i)_{i\geqslant 1}$ has a convergent subsequence (still called $(x_i)_{i\geqslant 1}$) converging to a point $x \in K^\perp(\psi)$. The continuity of ϕ implies that $\phi(x) = x$, so that $x \in K(\psi)$; i.e. $x = 0$. On the other hand, continuity of the norm shows that $|x| = 1$, contradiction.

(B) **Lemma.** *If* $\phi: E \to E$ *is compact, then so is its adjoint* ϕ^*.

Proof. Suppose (x_i) is a bounded sequence in E. Since ϕ^* is continuous, $(\phi^*(x_i))$ is also bounde so that $(\phi\phi^*(x_i))$ has a convergent subsequence, still called $(\phi\phi^*(x_i))$. Then

$$|\phi^*(x_i) - \phi^*(x_j)|^2 = \langle \phi^*(x_i - x_j), \phi^*(x_i - x_j)\rangle = \langle x_i - x_j, \phi\phi^*(x_i - x_j)\rangle \leqslant \text{const}\,|\phi\phi^*(x_i - x_j)| \to$$

i.e. $(\phi^*(x_i))$ is convergent.

Proposition. *Given* $y \in E$, *there is a* $v \in E$ *such that* $\psi^*(v) = v - \phi^*(v) = y$ *if and only if* $y \in K^\perp(\psi$

Proof. The necessity is clear. To prove the sufficiency, we first remark that by Lemma 3A the functions $x \to |x|$ and $x \to |\psi(x)|$ are equivalent norms on the Hilbertian space $K^\perp(\psi)$. Since the form $f(x) = \langle x, y\rangle$ is continuous on $K^\perp(\psi)$, it follows from Theorem 2A that there is some $u \in K^\perp(\psi)$ for which

$$\langle x, y\rangle = f(x) = \langle \psi(x), \psi(u)\rangle \quad \text{for all } x \in K^\perp(\psi)$$

In fact, this holds for all $x \in E$. For $K(\psi)$ has finite dimension and is therefore closed; by Proposition 1E we can write every $x \in E$ uniquely in the form $x = x' + x''$ with $x' \in K^\perp(\psi)$, $x'' \in K(\psi)$. Then $\langle \psi(x), \psi(u)\rangle = \langle \psi(x'), \psi(u)\rangle = \langle x', y\rangle = \langle x, y\rangle$, the last equality because y is orthogonal to every $x'' \in K(\psi)$. If we now set $v = \psi(u)$, we find that $\langle x, \psi^*(v)\rangle = \langle \psi(x), v\rangle = \langle x, y$ for all $x \in E$, whence $\psi^*(v) = y$.

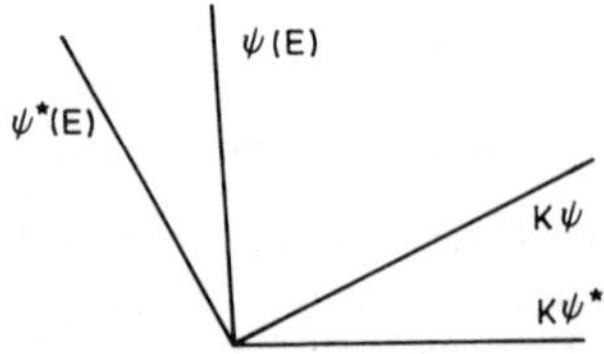

Corollary. *Given* $y \in E$ *there is a* $v \in E$ *such that* $\psi(v) = v - \phi(v) = y$ *if and only if* $y \in K^\perp(\psi^*)$; i.e. $K^\perp(\psi) = \psi^*(E)$ and $K^\perp(\psi^*) = \psi(E)$. In particular, ψ and ψ^* have closed ranges and finite-dimensional kernels and cokernels.

(C) It is an easy matter to check that the composition and linear combinations of compact linear endomorphisms of E are again compact. Thus the nth iterate:

$$\psi^n = (I-\phi)^n = I + \sum_{k=1}^{n} \binom{n}{k} (-1)^k \phi^k = I - T$$

where T is a compact linear endomorphism of E. In particular, setting $K^p = K(\psi^p)$, we find that (K^p) is a non-decreasing sequence of finite-dimensional subspaces of E.

Lemma. *There is a positive integer* n *such that*

$$K^p \subsetneqq K^{p+1} \quad \text{for } p < n$$

$$K^p = K^n \quad \text{for } p > n$$

Proof. First of all, that $K^p = K^{p+1}$ implies $K^p = K^{p+2} = \ldots = K^{p+k}$ follows at once. If the lemma were false, we could find a sequence $(x_p)_{p\geqslant 1} \subset E$ such that $|x_p| = 1$, $x_p \in K^{p+1}$ and $\langle x_p, K^p\rangle = 0$. Then $(\phi x_p)_{p\geqslant 1}$ has no convergent subsequence, because $|\phi x_p - x_{p-q}|^2 \geqslant 1$. Namely, $\phi(x_p - x_{p-q}) = x_p - (\psi x_p + \phi x_{p-q})$, and $\psi^p(\psi x_p + \phi x_{p-q}) = \psi^{p+1} x_p + \phi\psi^p x_{p-q} = 0$, whence x_p and $\psi x_p + \phi x_{p-q}$ are orthogonal. By the law of Pythagoras, $|\phi x_p - \phi x_{p-q}|^2 =$ $= |x_p|^2 + |\psi x_p + \phi x_{p-q}|^2 \geqslant 1$. But this contradicts the compactness of ϕ.

Theorem. *If $\phi: E \to E$ is compact, then $\psi = I - \phi$ is injective if and only if ψ is surjective.*

Proof of the sufficiency. Suppose there is a non-zero element $x_0 \in K(\psi)$. Let $(x_i)_{i\geqslant 1}$ be chosen inductively so that $\psi x_i = x_{i-1}$. Then $\psi^p x_p = x_0 \neq 0$, $\psi^{p+1}(x_p) = \psi x_0 = 0$; i.e. $K^p \subsetneqq K^{p+1}$ for all p, contradicting the lemma. Thus $K(\psi) = 0$.

Proof of the necessity. The hypothesis $K(\psi) = 0$ implies that ψ^* is surjective by Proposition 3B. We then apply the preceding argument to ψ^* to conclude that $K(\psi^*) = 0$. Corollary 3B shows that ψ is surjective.

(D) **Theorem.** *$K(\psi)$ and $K(\psi^*)$ have the same finite dimension.*

Proof. Let these spaces have dimensions n and n*. Without loss of generality we can suppose $n^* \geqslant n$; and we shall show that strict inequality leads to a contradiction.

Let (u_i) and (u_i^*) be orthonormal bases for $K(\psi)$ and $K(\psi^*)$ respectively. The operator

$$x \to \phi(x) - \sum_{i=1}^{n} \langle u_i, x\rangle u_i^*$$

is compact, being the sum of two compact operators. We define

$$\theta(x) = \psi(x) + \sum_{i=1}^{n} \langle u_i, x\rangle u_i^*$$

Now if $\theta(x) = 0$, then $x = 0$. For

$$0 = \langle u_j^*, \theta(x)\rangle = \langle u_j^*, \psi(x)\rangle + \sum_{i=1}^{n} \langle u_i, x\rangle\langle u_i^*, u_j^*\rangle = \langle \psi^* u_j^*, x\rangle + \langle u_j, x\rangle = \langle u_j, x\rangle$$

for all $1 \leqslant j \leqslant n$, whence $\psi(x) = 0$, and therefore $x = 0$. Thus θ is an injective map of the type considered in Theorem 3C, from which we conclude that there is an element v of E such that $\theta(v) = u_{n+1}^*$. But

$$1 = |u_{n+1}^*|^2 = \langle u_{n+1}^*, \theta v\rangle = \langle u_{n+1}^*, \psi v\rangle + \sum_{i=1}^{n} \langle u_i, v\rangle\langle u_{n+1}^*, u_i^*\rangle = 0$$

contradicting the assumption $n^* > n$.

(E) For any $\lambda \in \mathbb{R}$, the endomorphism $\lambda\phi$ is compact if ϕ is.

Proposition. *Let* $\psi_\lambda = I - \lambda\phi$. *Then* $\dim K(\psi_\lambda) > 0$ *for at most countably many* λ *having no finite accumulation point.*

Proof. Suppose there were a sequence $(\lambda_i)_{i\geqslant 1}$ of distinct bounded non-zero numbers such that each $\dim K(\psi_{\lambda_i}) > 0$. Choose $0 \neq x_i \in K(\psi_{\lambda_i})$ for each i; then for each n, the elements $x_1, \ldots, x_n$ are linearly independent. For, suppose $x_1, \ldots, x_{n-1}$ are linearly independent and

$$\sum_{j=1}^{n} c_j x_j = 0$$

Then

$$0 = \psi_{\lambda_n}\Big(\sum_{j=1}^{n} c_j x_j\Big) = \sum_{j=1}^{n} c_j x_j - \sum_{j=1}^{n} \lambda_n c_j \phi(x_j)$$

and because $\phi(x_j) = x_j/\lambda_j$, we find

$$\sum_{j=1}^{n-1}\Big(1 - \frac{\lambda_n}{\lambda_j}\Big) c_j x_j = 0$$

so that $c_1 = \ldots = c_{n-1} = 0$; it follows that $c_n = 0$ too; i.e. $x_1, \ldots, x_n$ are linearly independent.

Let V_n be the subspace spanned by $x_1, \ldots, x_n$; then there are elements $v_n \in V_n$ such that $|v_n| = 1$ and $v_n \perp V_{n-1}$. Further, if $w \in V_n$, then $w - \lambda_n\phi(w) \in V_{n-1}$.

$$w = \sum_{j=1}^{n} c_j x_j$$

whence

$$w - \lambda_n \phi(w) = \sum_{j=1}^{n} c_j x_j - \lambda_n \sum_{k=1}^{n} \frac{c_j}{\lambda_j} x_j = \sum_{j=1}^{n-1} \left(1 - \frac{\lambda_n}{\lambda_j}\right) c_j x_j$$

But $(\phi(\lambda_n v_n))_{n \geq 1}$ has no convergent subsequence since $\phi(\lambda_n v_n - \lambda_m v_m) = v_n - (v_n - \lambda_n \phi(v_n) + \lambda_m \phi(v_m))$, and for $n > m$ we can apply Pythagoras' law:

$$|\phi(\lambda_n v_n - \lambda_m v_m)|^2 = |v_n|^2 + |v_n - \lambda_n \phi(v_n) + \lambda_m \phi(v_m)|^2 \geq 1$$

But because ϕ is compact, this contradicts the boundedness of $(\lambda_i)_{i \geq 1}$. Hence the numbers $\lambda: \dim K(\psi_\lambda) > 0$ can be counted.

Let us collect several of the preceding results as a theorem, often referred to as the

Fredholm Alternative Theorem. *Let ϕ be a compact endomorphism of the Hilbertian space* E, *and form the endomorphism $\psi_\lambda = I - \lambda\varphi$. Then there is:*

(1) at most a countable sequence of real numbers λ for which $\dim K(\psi_\lambda) > 0$. *For such values there is a solution of $\psi_\lambda(x) = x - \lambda\phi(x) = y$ if and only if* $y \perp K(\psi_\lambda^*)$.

(2) If λ is a value for which $\dim K(\psi_\lambda) = 0$, *then for all* $y \in E$, *there is a unique* $x \in E$ *for which*

$$\psi_\lambda(x) = x - \lambda\phi(x) = y$$

4. THE DIRICHLET PROBLEM

(A) Let V and E be Hilbert spaces, and suppose we have a compact injection of V into E; in particular, considering V as a vector subspace of E (from the algebraic viewpoint only), there is a constant such that $|x|_E \leq \text{const}\,|x|_V$ for all $x \in V$. Suppose furthermore that V is dense in E.

Let $\alpha: V \times V \to \mathbb{R}$ be a bilinear form for which:

(1) there exists $a \in \mathbb{R}$ such that $|\alpha(x, y)| \leq a|x|_V|y|_V$ for all $x, y \in V$:

(2) there are numbers $c > 0$, $\lambda_0 > 0$, such that $\alpha(x, x) + \lambda_0 |x|_E^2 \geq c|x|_V^2$ for all $x \in V$. This will be called the *coercivity condition* on α. In physical terms $\alpha(x, x)$ has an interpretation as a sort of energy.

For any positive $\lambda \in \mathbb{R}$, set $\alpha_\lambda(x, y) = \alpha(x, y) + \lambda\langle x, y\rangle_E$. Then $|\alpha_\lambda(x, y)| \leq a|x|_V|y|_V + \lambda|x|_E|y|_E \leq \text{const}|x|_V|y|_V$, where the constant depends on λ. Also, for $\lambda \geq \lambda_0$ we have $\alpha_\lambda(x, x) = \alpha(x, x) + \lambda_0|x|_E^2 + (\lambda - \lambda_0)|x|_E^2 \geq c|x|_V^2$ for all $x \in V$.

(B) Suppose now we are given $u \in E$; then the linear form $f: V \to \mathbb{R}$ given by $f(x) = \langle x, u\rangle_E$ is continuous on V, for $|\langle x, u\rangle_E| \leq |x|_E|u|_E \leq \text{const}|x|_V$ for all $x \in V$. Thus by Theorem 2C for any $\lambda \geq \lambda_0$ there is a unique $v \in V$ such that $f(y) = \alpha_\lambda(v, y)$ for all $y \in V$; i.e. for $u \in E$, there is a unique $v \in V$ such that

$$\alpha(v, y) + \lambda\langle v, y\rangle_E = \langle u, y\rangle_E \quad \text{for all } y \in V$$

Thus for $\lambda \geq \lambda_0$ we have a linear map $G_\lambda: E \to V$ given by $u \to v$ (G_λ plays the role of Green's function in potential theory). We have

$$c|v|_V^2 \leq \alpha_\lambda(v, v) = \langle u, v\rangle_E \leq |u|_E|v|_E$$

whence $|G_\lambda(u)|_V^2 \leqslant$ const$|u|_E|G_\lambda(u)|_E$ so that $|G_\lambda(u)|_V \leqslant$ const $|u|_E$; i.e. G_λ is a continuous linear map. Because the injection $V \to E$ is compact, we obtain the

Proposition. *Take $\lambda \geqslant \lambda_0$. The composition $G_\lambda : E \to E$ is a compact endomorphism such that for any $u \in E$, $G_\lambda(u)$ is the unique element such that*

$$\alpha(G_\lambda(u), y) + \lambda\langle G_\lambda(u), y\rangle_E = \langle u, y\rangle_E$$

for all $y \in V$.

This result leads us to consideration of the operator $I - \lambda G_\lambda$, to which we can apply the methods of Section 3.

(C) **Definition.** The *first null space* of α is $K_1(\alpha) = \{x \in V : \alpha(x, y) = 0 \text{ for all } y \in V\}$. Similarly for the *second null space* $K_2(\alpha)$. If we define $\alpha^*(x, y) = \alpha(y, x)$, then clearly $K_1(\alpha^*) = K_2(\alpha)$.

Proposition. $x \in K_1(\alpha)$ *if and only if for* $\lambda \geqslant \lambda_0$, $x - \lambda G_\lambda(x) = 0$. *In particular,* $\dim K_1(\alpha) < \infty$.

Proof. For any $x \in V$, $\alpha(G_\lambda(x), y) + \lambda\langle G_\lambda(x), y\rangle_E = \langle x, y\rangle_E$, so that $\alpha(G_\lambda(x), y) = \langle x - \lambda G_\lambda(x), y$ If $x = \lambda G_\lambda(x)$, then $0 = \alpha(G_\lambda(x), y) = \lambda^{-1}\alpha(x, y)$ for all $y \in V$.

Conversely, we compute

$$\begin{aligned}\alpha_\lambda(x - \lambda G_\lambda(x), y) &= \alpha_\lambda(x, y) - \lambda\alpha_\lambda(G_\lambda(x), y)\\ &= \alpha_\lambda(x, y) - \lambda\langle x, y\rangle_E\\ &= \alpha(x, y) = 0 \quad \text{for all } y \in V\end{aligned}$$

The coercivity condition now shows that $x - \lambda G_\lambda(x) = 0$.

Lemma. *If G_λ^* is Green's function associated with α^*, then G_λ and G_λ^* are adjoints relative to the inner product of E for $\lambda \geqslant \lambda_0$.*

Proof.

$$\begin{aligned}\langle x, G_\lambda^*(y)\rangle_E &= \alpha_\lambda(G_\lambda(x), G_\lambda^*(y))\\ &= \alpha(G_\lambda(x), G_\lambda^*(y)) + \lambda\langle G_\lambda(x), G_\lambda^*(y)\rangle_E\\ &= \alpha^*(G_\lambda^*(y), G_\lambda(x)) = \langle y, G_\lambda(x)\rangle_E\end{aligned}$$

It follows that G_λ^* is a compact endomorphism of E, and the corresponding results above apply to G_λ^*. For instance, $\dim K_2(\alpha) = \dim K_1(\alpha^*) < \infty$.

(D) **Proposition.** Given $u \in E$, *there is a solution* $v \in V$ *of* $\alpha(v, y) = \langle u, y\rangle_E$ *for all* $y \in V$ *if and only if* $u \perp_E K_2(\alpha) = K_1(\alpha^*)$.

Proof. First of all, $\alpha(v, y) = \langle u, y\rangle_E$ for all $y \in V$ if and only if $(I - \lambda G_\lambda)v = G_\lambda u (\lambda \geqslant \lambda_0)$; for $\alpha(G_\lambda(u), y) + \lambda\langle G_\lambda(u), y\rangle_E = \langle u, y\rangle_E$, and to prove the sufficiency we substitute for $G_\lambda u$ to obtai

$$\alpha(v - \lambda G_\lambda(v), y) + \lambda\langle v - \lambda G_\lambda(v), y\rangle_E = \langle u, y\rangle_E$$

$$\alpha(v, y) - \lambda[\alpha(G_\lambda(v), y) + \lambda\langle G_\lambda v, y\rangle_E] + \lambda\langle v, y\rangle_E = \langle u, y\rangle_E$$

whence $\alpha(v, y) = \langle u, y\rangle_E$. The necessity is obtained by reversing the steps, using the uniqueness of $G_\lambda(u)$.

By Corollary 3B there is a $v \in E$ with $v - \lambda G_\lambda(v) = G_\lambda(u)$ (or $G_\lambda(u) \in \text{image } I - \lambda G_\lambda$) if and only if $G_\lambda(u) \perp_E K(I - \lambda G_\lambda^*)$. But then $v \in V$. By Proposition 4C, $K(I - \lambda G_\lambda^*) = K_1(\alpha^*)$, where $\lambda \geqslant \lambda_0$. But $G_\lambda(u) \perp K(I - \lambda G_\lambda^*)$ when and only when $u \perp K(I - \lambda G_\lambda^*)$, for $\langle u, y\rangle_E = \lambda\langle u, G_\lambda^* y\rangle_E = \lambda\langle G_\lambda u, y\rangle_E$ for all $y \in K(I - \lambda G_\lambda^*)$.

For future reference let us formulate our results as follows:

Theorem. *Let* V *and* E *be Hilbert spaces and* $V \to E$ *a compact dense injection. Suppose that* $\alpha: V \times V \to \mathbb{R}$ *is a* V*-continuous coercive bilinear form. Then*

(1) *For all* $u \in E$ *and* $\lambda \geqslant \lambda_0$, *there is a unique solution* v *of* $\alpha(v, y) + \lambda\langle v, y\rangle_E = \alpha_\lambda(v, y) = \langle u, y\rangle_E$ *for all* $y \in V$.

(2) $\alpha(v, y) = \langle u, y\rangle_E$ *has a solution if and only if* $u \perp_E K_2(\alpha)$.

(3) $\dim K_1(\alpha) = \dim K_2(\alpha) < \infty$.

(E) Consider $N = \{x \in V: \text{the form } y \to \alpha(x, y) \text{ is E continuous on } V\}$; because V is dense in E it follows that we can extend this form to be defined on E. By Theorem 2A, for each $x \in N$ there is an element $A(x) \in E$ such that $\alpha(x, y) = \langle A(x), y\rangle_E$ for all $y \in V$. Then A is linear on N, called *the domain of* A and written henceforth Dom(A), but not necessarily continuous.

Lemma. Dom(A) *is dense in* E.

Proof. Take $u \in E$ such that $\langle x, u\rangle_E = 0$ for all $x \in \text{Dom}(A)$. For $\lambda \geqslant \lambda_0$, there is a unique $v \in \text{Dom}(A^*)$ for which $A_\lambda^*(v) = u$, whence $\langle A_\lambda(x), v\rangle_E = \langle x, A_\lambda^*(v)\rangle_E \langle x, u\rangle = 0$ for all $x \in \text{Dom}(A)$. But A_λ is surjective, whence v is E-orthogonal to all E. It follows that $v = 0$, whence $u = 0$.

Remark. Let V and E be as above, and let $A: V \to E$ be a continuous linear map with closed range and $\dim K(A) < \infty$. Then the bilinear form $\alpha(x, y) = \langle Ax, Ay\rangle_E$ on V is a special case of the preceding situation, and is the object of the study [6].

In Part II we shall be interested in densely defined operators $A: \text{Dom}(A) \to E$ (not necessarily satisfying the strong conditions of the preceding Remark). These seem to be most easily studied – as we have done in this section – through the associated bilinear form $\alpha(x, y) = \langle Ax, y\rangle_E$, with solutions given by Theorem 4D; if we think of A as a differential operator, then these solutions are thus given, so to speak, in integrated form.

II. ELLIPTIC OPERATORS ON THE TORUS

1. GROUP DUALITY AND FOURIER ANALYSIS

(A) In the next section we shall study elliptic differential operators on the torus. There are certain special features of this case which provide simplification in the analytic theory. Basically this happens because the Fourier transform is a tool used essentially in the study of differential operators, and on the torus it is especially simple. (Here we have an example of a principle prevalent in modern mathematics: in certain problems a dualization is possible, and the dual problem is sometimes easier to handle. For instance, cohomology is richer in structure than homology; the dual space of a Banach space has special features not always present in the given space.) In our case dualization is provided by the Fourier transform. We illustrate the idea briefly now before specializing to the torus.

Let G be a locally compact abelian group; its dual group $\hat{G}$ is the totality of continuous homomorphisms of G into the unit circle S^1. With natural (weak) topology $\hat{G}$ also has the structure of a locally compact abelian group. The Pontriagin duality theorem asserts that the natural map $G \times \hat{G} \to S^1$ is a dual pairing, written $x, y \to (x, y)$ and called the character function; i.e. the canonical map $G \to \hat{\hat{G}}$ is a topological isomorphism.

With every such G we have an (essentially unique) invariant Radon measure, called *Haar measure,* relative to which we can form the complex Hilbert space H(G) of square integrable complex functions on G. Given $f \in H(G)$ we can construct its *Fourier transform* $\hat{f} = F(f) \in H(\hat{G})$ by

$$\hat{f}(y) = \int_G f(x)\,\overline{(x, y)}\,dx$$

When the Haar measures on G and $\hat{G}$ are suitably normalized, the Plancherel theorem states that $F : H(G) \to H(\hat{G})$ is a bijective isometry.

Example. $G = \mathbb{R}^n$. Then $\hat{G} = \mathbb{R}^n$ too, and the character function is $(x, y) = \exp(i\langle x, y\rangle)$ for all $x, y \in \mathbb{R}^n$, where $\langle x, y\rangle = x_1 y_1 + \ldots + x_n y_n$. The Haar measure is (except for a multiplicative factor) Lebesgue measure. The Fourier transformation of $u \in H(\mathbb{R}^n)$ is

$$\hat{u}(y) = \int_{\mathbb{R}^n} u(x) e^{-i\langle x, y\rangle} dx$$

The Plancherel theorem asserts that

$$\int_{\mathbb{R}^n} |u(x)|^2\,dx = (2\pi)^n \int_{\mathbb{R}^n} |\hat{u}(y)|^2\,dy$$

(B) *Example.* The following example is basic motivation for all that follows. Let $\mathbb{Z}^n$ be the lattice subgroup of $\mathbb{R}^n$ consisting of the n-tuples of integers. The quotient group $G = T^n = \mathbb{R}^n/2\pi\mathbb{Z}^n$ is called the n-dimensional torus; its points can be thought of as n-tuples of real numbers modulo 2π. Then clearly $\hat{G} = \mathbb{Z}^n$ with character function $T^n \times \mathbb{Z}^n \to S^1$ defined by $(x, k) \to \exp(i\langle k, x\rangle)$. We form the Hilbert space $H(T^n)$ of square integrable (relative to the natural invariant measure on T^n) complex functions on T^n with inner product

$$\langle u, v\rangle_{H(T^n)} = \int_{T^n} u(x)\bar{v}(x)\,dx$$

The functions $x \to \exp(i\langle k, x\rangle)/(2\pi)^{n/2}$ for $k \in \mathbb{Z}^n$ form an orthonormal base for $H(T^n)$; i.e. if $e_k = \exp(i\langle k, x\rangle)/(2\pi)^{n/2}$, then $\langle e_p, e_q\rangle = \delta_{pq}$ for all $p, q \in \mathbb{Z}^n$, and their finite linear combinations (called *trigonometric polynomials)* are dense in $H(T^n)$, by the Riesz-Fischer theorem.

Now $H(\mathbb{Z}^n)$, the totality of maps $\mathbb{Z}^n \to \mathbb{C}$ of the form $k \to a_k$ with

$$\sum_{k \in \mathbb{Z}^n} |a_k|^2 < \infty$$

is also a Hilbert space, relative to the inner product

$$\langle a, b \rangle_{H(\mathbb{Z}^n)} = (2\pi)^n \sum_{k \in \mathbb{Z}^n} a_k \bar{b}_k$$

The *Fourier transform* $F : H(T^n) \to H(\mathbb{Z}^n)$ is given by $F(u) = \hat{u}$, where

$$\hat{u}(k) = \int_{T^n} u(x) e^{-i\langle k, x \rangle} dx$$

Thus we transfer the study of $H(T^n)$ to $H(\mathbb{Z}^n)$, thereby obtaining a definite simplification. Furthermore, if $H^s(T^n)$ denotes the completion of the smooth functions in the inner product (with s a positive integer)

$$\langle u, v \rangle_{H^s(T^n)} = \sum_{|\alpha| \leqslant s} \langle D^\alpha u, D^\alpha v \rangle_{H(T^n)}$$

then the Fourier transform defines a topological isomorphism of $H^s(T^n)$ onto a Hilbert space $H^s(\mathbb{Z}^n)$; in fact, $H^s(\mathbb{Z}^n)$ is the completion of the trigonometric polynomials in the inner product

$$\langle a, b \rangle_{H^s(\mathbb{Z}^n)} = (2\pi)^n \sum_{k \in \mathbb{Z}^n} (1 + |k|^2)^s a_k \bar{b}_k$$

where $|k|^2 = k_1^2 + \ldots + k_n^2$. We shall not prove this, but mention it now only as motivation for the constructions to follow.

A powerful property of the Fourier transform is that it changes differentiation into multiplication. This fact lies behind the equivalence of the norms on $H^s(T^n)$ expressed in Proposition 2A below.

2. THE HILBERT SPACES

(A) We wish to consider real-valued *trigonometric polynomials on* T^n, i.e. functions u expressible in the form $u(x) = \sum u_k \exp(i\langle k, x \rangle)$, where the sum is extended over all $k \in \mathbb{Z}^n$, where the $u_k \in \mathbb{C}$ satisfy $u_{-k} = \bar{u}_k$, and where only finitely many $u_k \neq 0$. For any real number s let H^s denote the completion of these trigonometric polynomials in the inner product

$$\langle u, v \rangle_s = (2\pi)^n \sum (1 + |k|^2)^s u_k v_{-k}$$

Note that

$$|u|_s \leqslant |u|_t \quad \text{if} \quad s \leqslant t$$

whence we have a continuous injection $H^t \to H^s$ if $s \leqslant t$.

Define $H^\infty = \cap\{H^t : t \in \mathbb{R}\}$ and $H^{-\infty} = \cup\{H^t : t \in \mathbb{R}\}$:

$$H^\infty \subset \ldots \subset H^s \subset \ldots \subset H^0 \subset \ldots \subset H^{-t} \subset \ldots \subset H^{-\infty}$$

Lemma **(Schwar(t)z inequality).** *For any trigonometric polynomials* u, v *we have*

$$|\langle u, v\rangle_s| \leqslant |u|_{s+t} |v|_{s-t}$$

In particular, $|u|_s^2 \leqslant |u|_{s+t} |u|_{s-t}$.

This follows from

$$\left[\sum (1+|k|^2)^{\frac{s+t}{2}} u_k (1+|k|^2)^{\frac{s-t}{2}} v_{-k}\right]^2 \leqslant \left[\sum (1+|k|^2)^{s+t} u_k u_{-k}\right]\left[\sum (1+|k|^2)^{s-t} v_k v_{-k}\right]$$

Lemma. *For any* $0 < \epsilon < 1$ *and* $t_2 < s < t_1$ *we have*

$$|u|_s^2 \leqslant \epsilon |u|_{t_1}^2 + \epsilon^{-(s-t_2)/(t_1-s)} |u|_{t_2}^2$$

for all trigonometric polynomials. Also,

$$|u|_s \leqslant \epsilon |u|_{t_1} + \epsilon^{-(s-t_2)/(t_1-s)} |u|_{t_2}$$

Proof. We first note the inequality $a^s \leqslant \epsilon a^{t_1} + \epsilon^{-(s-t_2)/(t_1-s)} a^{t_2}$ for $a \geqslant 1$. Namely,

$$\epsilon a^{t_1-s} + \epsilon^{-(s-t_2)/(t_1-s)} a^{t_2-s} = \epsilon a^{t_1-s} + (\epsilon a^{t_1-s})^{-(s-t_2)/(t_1-s)}$$

But this is always $\geqslant 1$, as we see by considering the cases $\epsilon a^{t_1-s} \geqslant 1$ and $\epsilon a^{t_1-s} \leqslant 1$; in either case one of the terms in question is $\geqslant 1$.

Applying this to $a = 1 + |k|^2$ for $k \in \mathbb{Z}^n$, multiplying by $u_k u_{-k}$ and summing gives the desired inequality for all trigonometric polynomials $x \to u(x) = \sum u_k \exp(i\langle k, x\rangle)$. The second inequality follows, replacing ϵ by ϵ^2.

Lemma. *For all* ϵ *such that* $0 < \epsilon < 1$ *we have*

$$|u|_s |u|_{s-1} \leqslant \epsilon |u|_s^2 + (\epsilon^{-2s+1}/2)\, |u|_0^2$$

for all trigonometric polynomials u, *if* $s \geqslant 1$. (This lemma is for use in Section 3.)

Proof. Assume first that $s > 1$.

$$|u|_s^2 |u|_{s-1}^2 \leqslant |u|_s^2 [\epsilon^2 |u|_s^2 + \epsilon^{-2(s-1)} |u|_0^2]$$

using the first inequality of the preceding lemma with ϵ replaced by ϵ^2 and taking $0 < s - 1 < s$.

By completion of the square

$$|u|_s^2|u|_{s-1}^2 \leqslant \epsilon^2[|u|_s^4 + \epsilon^{-2s}|u|_s^2|u|_0^2 + (\epsilon^{-2s}/2)^2|u|_0^4] \leqslant \epsilon^2[|u|_s^2 + (\epsilon^{-2s}/2)|u|_0^2]^2$$

Now observe that the same estimates hold for $s = 1$, even without the preceding lemma.

Lemma. *The differential operator* $D^\alpha = \partial^{|\alpha|}/\partial x_1^{\alpha_1} \ldots x_n^{\alpha_n}$ (*where* $|\alpha| = \alpha_1 + \ldots + \alpha_n$) *is a continuous map* $D^\alpha: H^{s+|\alpha|} \to H^s$, *with* $|D^\alpha u|_s \leqslant |u|_{s+|\alpha|}$ *for all* $u \in H^{s+|\alpha|}$. *There is a constant depending only on* s *such that*

$$|u|_s \leqslant \text{const} \sum_{|\alpha| \leqslant s} |D^\alpha u|_0$$

Proof. First of all, $D^\alpha u$ is only defined for trigonometric polynomials u. $D^\alpha u(x) = \sum (ik)^\alpha u_k \times \exp(i\langle k, x\rangle)$, where $(ik)^\alpha = (ik_1)^{\alpha_1} \ldots (ik_n)^{\alpha_n}$. Then

$$|D^\alpha u|_s^2 = (2\pi)^n \sum (1 + |k|^2)^s (ik)^\alpha u_k u_{-k} (-ik)^\alpha = (2\pi)^n \sum (1 + |k|^2)^s k^{2\alpha} u_k u_{-k}$$

But $k^{2\alpha} \leqslant (|k|^2)^{|\alpha|} \leqslant (1 + |k|^2)^{|\alpha|}$, so that

$$|D^\alpha u|_s^2 \leqslant (2\pi)^n \sum (1 + |k|^2)^{s+|\alpha|} u_k u_{-k} = |u|_{s+|\alpha|}^2$$

It now follows that D^α has a unique extension to a continuous linear map $H^{s+|\alpha|} \to H^s$.

On the other hand,

$$(1 + |k|^2)^s \leqslant \text{const} \sum_{|\alpha| \leqslant s} k^{2\alpha}$$

hence

$$|u|_s^2 = (2\pi)^n \sum (1 + |k|^2)^s u_k u_{-k} \leqslant \text{const} \sum_{|\alpha| \leqslant s} k^{2\alpha} u_k u_{-k} \leqslant \text{const} \sum_{|\alpha| \leqslant s} |D^\alpha u|_0^2$$

Putting these together, we obtain the

Proposition. *Suppose* s *is a positive integer. There are numbers* $a > 0$ *and* $c > 0$ *such that*

$$a^{-1}|u|_s^2 \leqslant \sum_{|\alpha| \leqslant s} |D^\alpha u|_0^2 \leqslant a|u|_s^2$$

$$c^{-1} \sum_{|\alpha| \leqslant s} |D^\alpha u|_0^2 \leqslant \left[|u|_0^2 + \sum_{|\alpha| = s} |D^\alpha u|_0^2\right] \leqslant c \sum_{|\alpha| \leqslant s} |D^\alpha u|_0^2$$

for all $u \in H^s$.

Proof. The first equivalence of inner products follows at once from the preceding lemma. To prove the first inequality of the second equivalence we note that for every $\epsilon > 0$ there is a number t such that (taking $t_2 = 0$, $t_1 = t$, in a lemma above)

$$|D^\alpha u|_0 \leqslant |u|_{|\alpha|} \leqslant b|u|_0 + \epsilon |u|_t$$

for all α for which $|\alpha| < t$.

Thus we regard H^t as a Hilbertian space and are free to choose any one of the above three inner products to describe its topology.

(B) **Lemma.** *Let*

$$\Delta = \sum_{i=1}^{n} \frac{\partial^2}{\partial x_i^2}$$

be the Laplacian on $\mathbb{R}^n$, *and define* $K = I - \Delta$. *Then for any* t *we have*

$$\langle K^t u, v\rangle_s = \langle u, K^t v\rangle_s = \langle u, v\rangle_{s+t}$$

for any trigonometric polynomial u. *Furthermore,* $K^{t/2}: H^{s+t} \to H^s$ *is a bijective isometry; in particular,* $|K^{t/2}u|_s = |u|_{s+t}$.

This is immediate, because

$$Ku(x) = \sum (1 + |k|^2) u_k \exp(i\langle k, x\rangle) \quad \text{if} \quad u = \sum u_k \exp(i\langle k, x\rangle)$$

Proposition.[1] *Let* $s \geqslant 0$. *The* H^t*-inner product dually pairs* $H^{t-s} \times H^{t+s} \to \mathbb{R}$; *i.e. the continuous linear forms* f *on* H^{t+s} *are uniquely representable as the elements* $v \in H^{t-s}$, *with* $f(u) = \langle u, v\rangle_t$ *for all* $u \in H^{t+s}$; *and every* $v \in H^{t-s}$ *thus represents a continuous linear form on* H^{t+s}. *Furthermo*

$$|v|_{t-s} = \sup\{\langle u, v\rangle_t : |u|_{t+s} = 1\}$$

The Schwar(t)z inequality shows that $\langle u, v\rangle_t$ is defined for $u \in H^{t+s}$ and $v \in H^{t-s}$.

Proof. First of all, clearly every $v \in H^{t-s}$ defines such a form. Given such a form $f: H^{t+s} \to \mathbb{R}$, there is a unique $\hat{v} \in H^{t+s}$ such that $f(u) = \langle u, \hat{v}\rangle_{t+s}$ for all $u \in H^{t+s}$, and $|\hat{v}|_{t+s} = \sup\{|f(u)| : |u|_{t+s}$ Let $v = K^s\hat{v} \in H^{t-s}$, whence $|v|_{t-s} = |\hat{v}|_{t+s}$. Then $\langle u, v\rangle_t = \langle u, K^s\hat{v}\rangle_t = \langle u, \hat{v}\rangle_{t+s} = f(u)$.

If there were another such $v' \in H^{t-s}$, then $\langle u, v' - v\rangle_t = 0$ for all $u \in H^{t+s}$. Taking $u = K^{-s}(v'$ gives

$$0 = \langle K^{-s}(v' - v), v' - v\rangle_t = |v' - v|^2_{t-s}$$

whence $v' = v$.

(C) **Lemma.** *If* ϕ *is a smooth function on* T^n, $u \in H^s$, *then* $\phi u \in H^s$ *and* $|\phi u|_s \leqslant \text{const}|u|_s$. *The constant depends on* ϕ.

---

[1] We have not yet defined K^t for t not a positive integer. Note that we can define K^t for t not a positive integer with the relation $\langle K^t u, v\rangle_s = \langle u, v\rangle_{s+t}$.

Proof. If $s \geqslant 0$, then by Proposition 2A

$$|\phi u|_s^2 \leqslant \text{const} \sum_{|\alpha| \leqslant s} |D^\alpha(\phi u)|_0^2 < \text{const}_\phi \sum_{|\alpha| \leqslant s} |D^\alpha u|_0^2 \leqslant \text{const}_\phi |u|_s^2$$

If $-t = s < 0$ then by Proposition 2B and the Schwar(t)z inequality,

$$|\phi u|_{-t} = \sup\{\langle \phi u, v\rangle_0 : |v|_t = 1\} = \sup\{\langle u, \phi v\rangle_0 : |v|_t = 1\} \leqslant |u|_{-t}|\phi v|_t \leqslant \text{const}_\phi |u|_{-t}$$

As an application we have the

Proposition. *Let* A *be a smooth differential operator* T^n *of order* r; *i.e.*

$$A = \sum_{|\alpha| \leqslant r} a_\alpha D^\alpha$$

where $a_\alpha : T^n \to \mathbb{R}$ *are smooth functions. Then* $A : H^s \to H^{s-r}$ *is continuous for all* s.

(D) **Theorem (Rellich).** *If* $s < t$, *then the injection* $H^t \to H^s$ *is compact and dense.*

Proof. The image is dense because it contains the trigonometric polynomials. If

$$u^j(x) = \sum u_k^j \exp(i\langle k, x\rangle)$$

is a sequence ($j \geqslant 1$) such that $|u^j|_t \leqslant M$ for all $j \geqslant 1$, we shall find a sequence which is H^s-Cauchy.

First of all, we have $|u_k^j| \leqslant M(1 + |k|^2)^{-t/2}(2\pi)^{-n/2}$. We order $\mathbb{Z}^n$ (denoted by $k_1, k_2, \ldots$) and then select inductively and successively on $k_1, k_2, \ldots$ convergent subsequences of the rows of

$$u_{k_1}^1, u_{k_1}^2, \ldots$$

$$u_{k_2}^1, u_{k_2}^2, \ldots$$

$$u_{k_3}^1, \ldots$$

The diagonal sequence, still called (u_k^j), will be convergent for every $k \in \mathbb{Z}^n$.

Now fix a positive integer N and write $u^j = u_N^j + v_N^j$, where

$$u_N^j(x) = \sum_{|k| \leqslant N} u_k^j \exp(i\langle k, x\rangle)$$

Then

$$|v_N^i - v_N^j|_s = (2\pi)^n \sum_{|k| \geqslant N} (1 + |k|^2)^s (u_k^i - u_k^j)(u_{-k}^i - u_{-k}^j)$$

$$= \sum_{|k| \geqslant N} \frac{1}{(1 + |k|^2)^{t-s}} (1 + |k|^2)^t |u_k^i - u_k^j|^2 \leqslant \frac{1}{(1 + N^2)^{t-s}} |u^i - u^j|_t^2$$

is arbitrarily small for N sufficiently large. Take such N. Then

$$|u^i_N - u^j_N|^2_s = \left| \sum_{|k| \leqslant N} (u^i_k - u^j_k) \exp \sqrt{-1} \langle k, x \rangle \right|^2_s$$

is arbitrarily small for i, j sufficiently large, since $(u^j_k)_{j=1}$ converge uniformly for $|k| \leqslant N$. It follows that the subsequence (u^j) is H^s convergent.

3. DIRICHLET'S PROBLEM ON T^n

(A) Let

$$A = \sum_{|\alpha| \leqslant r} a_\alpha D^\alpha$$

be an elliptic differential operator of order r with smooth real periodic coefficients on $\mathbb{R}^n$. That means that its *symbol* $\sigma_A : T^n \times \mathbb{R}^n \to \mathbb{R}$ defined by

$$\sigma_A(x, \xi) = (-1)^s \sum_{|\alpha| = r} a_\alpha(x) \xi^\alpha$$

is positive definite, where we have written $r = 2s$ and $\xi^\alpha = \xi_1^{\alpha_1} \ldots \xi_n^{\alpha_n}$. Viewed as a differential operator on T^n, it follows that there is a number $\lambda_0 > 0$ such that $\sigma_A(x, \xi) \geqslant \lambda_0 |\xi|^r$ for all $x \in T^n$, $\xi \in \mathbb{R}^n$. The following result is immediate:

Lemma. *The composition of elliptic operators on* T^n *is elliptic. If* A *is elliptic, then so is its formal adjoint* A^*, *given by*

$$A^*u = \sum_{|\alpha| \leqslant r} (-1)^{|\alpha|} D^\alpha (a_\alpha u)$$

Furthermore, $\langle Au, v\rangle_0 = \langle u, A^*v\rangle_0$ *for all trigonometric polynomials* u, v.

Form $\alpha(u, v) = \langle Au, v\rangle_0$ for all trigonometric polynomials u, v. It is our aim in this section to show that the hypotheses of Theorem 4D of Part I are satisfied, with $V = H^s$, $E = H^0$. Theorem 2D shows that $H^s \to H^0$ is a compact dense injection. Since $A : H^s \to H^{-s}$ is continuous by Proposition 2C, we have $|\alpha(u, v)| \leqslant |\langle Au, v\rangle_0| \leqslant |Au|_{-s} |v|_s \leqslant \text{const } |u|_s |v|_s$ for all $u, v \in H^s$.

Proposition (Gårding's inequality). *If* A *is an* r^{th}*-order smooth elliptic differential operator on* T^n, *then there are numbers* $c > 0$, $\lambda_0 > 0$ *such that*

$$\langle Au, u\rangle_0 + \lambda_0 |u|^2_0 \geqslant c|u|^2_s$$

for all $u \in H^s$. *Again* $r = 2s$. This will be proved in Section 3C.

Example: The operator $K^s = (I - \Delta)^s$ is elliptic of order 2s. Lemma 2B shows that $\langle K^s u, u\rangle_0 = $ so that Gårding's inequality is verified in a very strong sense.

(B) Before proving Gårding's inequality, we need to know that in $\mathbb{R}^n$ there are sufficiently many smooth functions to separate closed sets.

Lemma. *Let* $0 \leqslant a \leqslant b$. *Then there is a smooth function* $\phi: \mathbb{R} \to \mathbb{R}$ *such that* $\phi(t) = 0$ *if* $t \leqslant a$, $\phi(t) = 1$ *if* $t \geqslant b$ *and* $0 < \phi(t) < 1$ *otherwise.*

In fact, define $\phi_0(t) = \exp(-1/(t-a)(b-t))$ for $a < t < b$, and 0 elsewhere. Then set

$$\phi(t) = \int_{-\infty}^{t} \phi_0(s)\, ds \Big/ \int_{-\infty}^{+\infty} \phi_0(s)\, ds$$

Proposition. *If* C_0, C_1 *are disjoint non-void, closed subsets of* $\mathbb{R}^n$, *then there is a smooth function* $\phi: \mathbb{R}^n \to \mathbb{R}$ *such that* $\phi(x) = 0$ *if* $x \in C_0$; $\phi(x) = 1$ *if* $x \in C_1$; *and* $0 \leqslant \phi(x) \leqslant 1$ *for all* $x \in \mathbb{R}^n$.

Proof. If C_1 is the closed disc centred at $0 \in \mathbb{R}^n$ and of radius $a > 0$, and C_0 is the complement of the open disc centred at 0 and of radius $b > a$, then set $\phi(x) = 1 - \psi(|x|)$, where ψ is defined as in the Lemma.

In the general case we start by taking an open covering $(U_k)_{k \geqslant 1}$ of C_1 by discs such that the family $(\bar{U}_k)_{k \geqslant 1}$ is locally finite and each $\bar{U}_k \cap C_0 = \emptyset$. Let $\phi_{1k}: \mathbb{R}^n \to \mathbb{R}$ be a smooth function such that $\phi_{1k}(x) > 0$ if $x \in U_k$, and $\phi_{1k}(x) = 0$ if $x \in \mathbb{R}^n - \bar{U}_k$. Define $\phi_1: \mathbb{R}^n \to \mathbb{R}$ by

$$\phi_1(x) = \sum_{k \geqslant 1} \phi_{1k}(x)$$

Thus $\phi_1(x) > 0$ if $x \in C_1$, whence there is a neighbourhood U of C_1 in which $\phi_1 > 0$. Set $C_1^* = \mathbb{R}^n - U$, $C_0^* = C_1$. These are disjoint closed subsets of $\mathbb{R}^n$, and we repeat the construction, giving a smooth function $\phi_2: \mathbb{R}^n \to \mathbb{R}$ such that $\phi_2(x) > 0$ for $x \in \mathbb{R}^n - U$ and $\phi_2(x) = 0$ if $x \in C_0^*$.

Finally, the function

$$x \to \phi(x) = \phi_1(x)/[\phi_1(x) + \phi_2(x)]$$

satisfies the requirements of the proposition.

Definition. The *support of a function* ϕ is the closure $\{x : \phi(x) \neq 0\}$; denote it by $\mathrm{spt}(\phi)$.

The following result is a special case of a theorem proved later.

Lemma. *Let* $(U_k)_{1 \leqslant k \leqslant m}$ *be a finite cover of* T^n *by open sets. Then there are smooth functions* $\phi_k: T^n \to I = [0, 1]$ *for* $1 \leqslant k \leqslant m$ *such that*

$$\sum_{k=1}^{m} \phi_k^2 = 1$$

and $\mathrm{spt}(\phi) \subset U_k$.

Proof. First of all, construct by induction an open covering $(V_k)_{1 \leqslant k \leqslant m}$ of T^n such that each $\bar{V}_k \subset U_k$. Use it and the preceding proposition to define a smooth function $\psi_k: T^n \to \mathbb{R}$ such that

$\psi_k(x) > 0$ if $x \in V_k$ and $\operatorname{spt}(\psi_k) \subset U_k$. Then

$$x \to \sum_{k=1}^{m} \psi_k^2(x)$$

is a smooth positive function on T^n, whence the function

$$x \to \phi_k(x) = \psi_k(x) \Bigg/ \left[\sum_{k=1}^{m} \psi_k^2(x) \right]^{1/2}$$

satisfies the requirements of the lemma.

Similarly for the following result, needed in Section 5C.

Lemma. *Let* $(U_j)_{j \geqslant 0}$ *be a locally finite open cover of the open set* U *in* $\mathbb{R}^n$. *Then there are smo functions* $(\phi_j)_{j \geqslant 0}$ *on* U *such that* $\operatorname{spt}(\phi_j) \subset U_j$ *for all* $j \geqslant 0$, $\sum_{j \geqslant 0} \phi_j(x) = 1$ *for all* $x \in U$, *and each* $0 \leqslant \phi_j(x) \leqslant 1$.

(C) We now proceed to the proof of Proposition 3A for all trigonometric polynomials $u(x) = \sum u_k \exp(i\langle k, x\rangle)$.

Case 1. All coefficients a_α are constant functions, and $a_\alpha = 0$ for $|\alpha| < r$. Then

$$\langle Au, u\rangle_0 = \left\langle \sum_k \sum_{|\alpha| = r} a_\alpha (ik)^\alpha u_k \exp(i\langle k, x\rangle)\ ,\ \sum_k u_k \exp(i\langle k, x\rangle) \right\rangle_0$$

But $a_\alpha (ik)^\alpha = (-1)^s a_\alpha k$, whence

$$\langle Au, u\rangle_0 = (-1)^s \sum_k \sum_{|\alpha| = r} a_\alpha k^\alpha |u_k|^2 = \sum_k \sigma_A(x, k)|u_k|^2 \geqslant \sum_k \lambda_0 |k|^r |u_k|^2$$

$$= \lambda_0 \sum_k [1 + |k|^r]|u_k|^2 - \lambda_0 |u|_0^2$$

Consider the function $f(x) = \lambda_0 (1 + x^{2s})/(1 + x^2)^s$. Then there is a number such that $0 < c \leqslant f(x)$ for all positive $x \in \mathbb{R}$; in particular, $\lambda_0(1 + |k|^r) \geqslant c(1 + |k|^2)^s$. Therefore,

$$\langle Au, u\rangle_0 \geqslant c \sum_k (1 + |k|^2)^s |u_k|^2 - \lambda_0 |u|_0^2$$

$$\geqslant c|u|_s^2 - \lambda_0 |u|_0^2$$

Case 2. The coefficients $a_\alpha = 0$ for $|\alpha| < r$, and $A = A_0 + A_1$, where A_0 has constant coefficients,

$$A_1 = \sum_{|\alpha|=r} b_\alpha D^\alpha$$

with all $|b_\alpha(x)| \leqslant \eta$ for some sufficiently small $\eta > 0$; the size of η will be determined in Cases 3, 4.

We apply integration by parts over T^n to obtain $\langle A_1 u, u\rangle_0 = I_1 + I_2$, where

$$I_1 = \sum \int b_{\beta\gamma} D^\beta u D^\gamma u$$

is the collection of all terms with $|\beta| = |\gamma| = s$;

$$I_2 = \sum \int b_{\beta\gamma} D^\beta u D^\gamma u$$

where $|\beta| + |\gamma| < r$ and $|\beta| \leqslant s$, $|\gamma| \leqslant s$. Now in I_1 the $b_{\beta\gamma}$ are just relabellings of the b_α, whence by Proposition 2A we have

$$|I_1| \leqslant \eta \left| \sum \int D^\beta u D^\gamma u \right| \leqslant \eta \,\text{const}\, |u|_s^2$$

Similarly, by the second and third lemmas in Section 2A, for every $\epsilon > 0$,

$$|I_2| \leqslant \text{const}\, |u|_s |u|_{s-1} \leqslant \epsilon \,\text{const}\, |u|_s^2 + \epsilon^{-2s+1} \text{const}\, |u|_0^2$$

Suppose $c > 0$ and $\lambda_0 > 0$ are constants for Gårding's inequality for A_0. Because the preceding estimates for $|I_1|$ and $|I_2|$ are valid for any $\eta > 0$, $\epsilon > 0$, we can redefine $c > 0$ and $\lambda_0 > 0$:

$$\begin{aligned}\langle A_0 u, u\rangle_0 &\geqslant c|u|_s^2 - \lambda_0 |u|_0^2 + I_1 + I_2 \\ &\geqslant c|u|_s^2 - \lambda_0 |u|_0^2 - |I_1| - |I_2| \geqslant c'|u|_s^2 - \lambda_0' |u|_0^2\end{aligned}$$

and the proposition follows for this case.

Case 3. $A = A_0 + A_1 + A_2$, with A_0, A_1 as in Case 2, and

$$A_2 = \sum_{|\alpha|<r} c_\alpha D^\alpha$$

Then arguing as in Case 2,

$$\langle A_2 u, u\rangle_0 = \sum \int c_{\beta\gamma} D^\beta u D^\gamma u$$

summed over all β, γ such that $|\beta| \leqslant s$, $|\gamma| \leqslant s$, $|\beta| + |\gamma| < r$. Again, for all $\epsilon > 0$

$$
\begin{aligned}
|\langle A_2 u, u\rangle_0| &\leqslant \text{const}\, |u|_s |u|_{s-1} \\
&\leqslant \epsilon\, \text{const}\, |u|_s^2 + \epsilon^{-2s+1}\, \text{const}\, |u|_0^2
\end{aligned}
$$

whence we can also absorb this contribution into Gårding's inequality for $A_0 + A_1$.

Case 4. Thus far we have proved the proposition for those elliptic operators A with nearly const coefficients. We now prove the general case by cutting up T^n into small domains on which A is nearly constant. Let $(U_i)_{1 \leqslant i \leqslant m}$ be a finite open disc cover of T^n such that in each U_i every a_α with $|\alpha| = r$ has oscillation $< \eta$, with η chosen as in Cases 2, 3. Let $(\phi_i)_{1 \leqslant i \leqslant m}$ be functions on T chosen in the second lemma of Section 3B. Now write

$$
\langle Au, u\rangle_0 = \sum_{j=1}^{m} \phi_j^2 \langle Au, u\rangle_0 = \sum_{j=1}^{m} \langle \phi_j Au, \phi_j u\rangle_0 = \sum_{j=1}^{m} \langle A(\phi_j u), \phi_j u\rangle_0 + R
$$

We can apply Case 3 to the summands in the right member, to obtain

$$
\langle Au, u\rangle_0 \geqslant c \sum_{j=1}^{m} |\phi_j u|_s^2 - \lambda_0 \sum_{j=1}^{m} |\phi_j u|_0^2 + R
$$

But $|\phi_j u|_0 \leqslant |u|_0$, and using Proposition 2A we can find $a > 0$, $b > 0$, such that

$$
\sum_{j=1}^{m} |\phi_j u|_s^2 \geqslant a|u|_s^2 - b|u|_0^2
$$

Thus by adjusting c and λ_0 if necessary, $\langle Au, u\rangle_0 \geqslant c|u|_s^2 - \lambda_0 |u|_0^2 + R$. The term R may involve derivatives of u through order $< r$, and through integration by parts we can express it in the form

$$
R = \sum \int c_{\beta\gamma} D^\beta u D^\gamma u
$$

summed over all $|\beta| \leqslant s$, $|\gamma| \leqslant s$, $|\beta| + |\gamma| < r = 2s$. Once again, for any $\epsilon > 0$ we have

$$
|R| \leqslant \epsilon\, \text{const}\, |u|_s^2 + \epsilon^{-2s+1}\, \text{const}\, |u|_0^2
$$

By choosing ϵ sufficiently small we can absorb R to obtain $\langle Au, u\rangle_0 \geqslant c|u|_s^2 - \lambda_0 |u|_0^2$. This comp the proof.

4. REGULARITY OF SOLUTIONS

(A) Let $C^s(T^n)$ be the Banach space of functions on T^n having continuous derivatives of order $\leqslant s$, with norm

$$
|u|_{C^s(T^n)} = \sup\{|D^\alpha u(x)| : x \in T^n \quad \text{and} \quad |\alpha| \leqslant s\}
$$

Clearly $C^s(T^n) \subset H^s(T^n)$ for all integers $s \geqslant 0$. Let $C^\infty(T^n) = \cap\{C^s(T^n) : s \geqslant 0\}$, the vector space of smooth functions on T^n.

Theorem (Sobolev). *If* $t > n/2 + s$, *then we have a continuous injection* $H^t(T^n) \to C^s(T^n)$. *In particular, considering this as an inclusion, we have* $H^\infty(T^n) = C^\infty(T^n)$.

Proof. In case $s = 0$, if $u(x) = \sum u_k \exp(i\langle k, x\rangle)$ is a trigonometric polynomial, then for all $x \in T^n$

$$|u(x)|^2 \leqslant \left(\sum |u_k|^2\right)^2 \leqslant \left\{\sum (1 + |k|^2)^t |u_k|^2\right\} \left\{\sum (1 + |k|^2)^{-t}\right\} = |u|_t^2 \sum (1 + |k|^2)^{-t}$$

But $t > n/2$ implies that the sum is convergent, whence $|u(x)| \leqslant \text{const}\,|u|_t$. Thus the Fourier series representation of any $u \in H^t$ converges uniformly, whence its limit is continuous. In case s is any positive integer such that $t > n/2 + s$ and $|\alpha| \leqslant s$, then $|D^\alpha u(x)| \leqslant \text{const}\,|D^\alpha u|_{t-s} \leqslant \text{const}\,|u|_t$, which proves the theorem in general.

(B) Let A be an elliptic operator of order $r = 2s$ on T^n.

Lemma. *Define* $A_\lambda = A + \lambda$. *Let* t *be an integer. Then there are numbers* $c > 0$ *and* $\lambda_0 > 0$ *such that* $\langle A_\lambda u, u\rangle_t \geqslant c|u|_{t+s}^2$ *for all* $u \in H^{t+s}$ *and* $\lambda \geqslant \lambda_0$.

Proof. If t is positive and $K = I - \Delta$, then $K^t A$ and AK^t are elliptic of order $2(t+s)$. Gårding's inequality shows the existence of c and λ_0 such that

$$\begin{aligned} \langle A_\lambda u, u\rangle_t = \langle K^t A_\lambda u, u\rangle_0 &\geqslant c|u|_{t+s}^2 - \lambda_0 |u|_0^2 + \lambda|u|_t^2 \\ &\geqslant c|u|_{t+s}^2 + (\lambda - \lambda_0)|u|_0^2 \\ &\geqslant c|u|_{t+s}^2 \quad \text{if} \quad \lambda \geqslant \lambda_0 \end{aligned}$$

If t is negative, then

$$\begin{aligned} \langle A_\lambda u, u\rangle_t = \langle A_\lambda K^{-t}(K^t u), K^t u\rangle_0 &\geqslant c|K^t u|_{-t+s}^2 - \lambda_0 |K^t u|_0^2 + \lambda\langle u, K^t u\rangle_0 \\ &\geqslant c|u|_{t+s}^2 - \lambda_0 |u|_{2t}^2 + \lambda|u|_t^2 \\ &\geqslant c|u|_{t+s}^2 \quad \text{if} \quad \lambda \geqslant \lambda_0 \end{aligned}$$

Lemma. *There is a unique continuous linear map* $A^* : H^{t+s} \to H^{t-s}$ *such that for all* $u, v \in H^{t+s}$ *we have* $\langle Au, v\rangle_t = \langle u, A^*v\rangle_t$.

Proof. Fix v and consider $f : H^{t+s} \to \mathbb{R}$ defined by $f(u) = \langle Au, v\rangle_t$. Then $|f(u)| \leqslant |Au|_{t-s}|v|_{t+s} \leqslant \text{const}|u|_{t+s}|v|_{t+s}$. By Proposition 2B there is a unique $A^*v \in H^{t-s}$ for which $f(u) = \langle u, A^*v\rangle_t$ for all $u \in H^{t+s}$; and A^* is bounded on H^{t+s}.

Proposition. *There is a number* $\lambda_0 > 0$ *such that for all* $\lambda \geqslant \lambda_0$ *the map* $A_\lambda : H^{t+s} \to H^{t-s}$ *is bijective, with universal bound independent of* $\lambda \geqslant \lambda_0$.

Proof. First of all, $|A_\lambda u|_{t-s}|u|_{t+s} \geqslant |\langle A_\lambda u, u\rangle_t| \geqslant c|u|_{t+s}^2$, whence $|A_\lambda u|_{t-s} \geqslant c|u|_{t+s}$ for all $u \in H^{t+s}$ and $\lambda \geqslant \lambda_0$. By Part I, Lemma 1D, $R_\lambda = A_\lambda H^{t+s}$ is closed in H^{t-s} and A_λ is injective.

If R_λ is a proper closed subspace, there is a non-zero $w \in H^{t+s}$ for which $\langle A_\lambda u, w\rangle_t = \langle u, A_\lambda^* w\rangle$ for all $u \in H^{t+s}$. Therefore $A_\lambda^* w = 0$; but our choice of λ shows that $\langle v, A_\lambda^* u\rangle_t = \langle A_\lambda u, u\rangle_t = c|u|_{t+}^2$ so that A_λ^* is injective; i.e. $v = 0$, a contradiction.

Theorem. *If* $v \in H^{-\infty}(T^n)$ *and* $Av \in H^t(T^n)$, *then* $v \in H^{t+r}(T^n)$. *In particular, if* $Av \in H^\infty(T^n)$ *then* v *is smooth.*

Proof. Assume $v \in H^p = H^p(T^n)$ and set $u = Av$. Then $u + \lambda v \in H^{\min(t,p)}$. But $v = A_\lambda^{-1}(u + \lambda v) \in H^{\min(t+r, p+r)}$ if λ is suitably large. Repeating the process, $u + \lambda v \in H^{\min(t, p+r)}$ whence $v \in H^{\min(t+r, p+2r)}$ for suitably large λ. The process can be continued until we find $v \in H^{t+r}$.

Remark. It can be shown using this result that if the coefficients of A are analytic on T^n and if Av is analytic, then so is v.

(C) Finally we are in the position to apply the results of Part I, in particular Theorem 4D.

Theorem. *Let* A *be a smooth* r^{th}*-order elliptic operator on* T^n $(r = 2s)$. *Then*

(1) A *maps* $H^{t+s}(T^n)$ *onto* $H^{t-s}(T^n) \cap K(A^*)^\perp$, *with kernel* $K(A)$ *for all* t.

(2) $K(A)$ *and* $K(A^*)$ *are subspaces of* $C^\infty(T^n)$ *of the same finite dimension.*

(3) *For every* t *there is a number* $\lambda_t > 0$ *such that if* $\lambda \geqslant \lambda_t$ *then* $A_\lambda = A + \lambda : H^{t+s}(T^n) \to H^{t-s}(T^n)$ *is a continuous bijection. In particular,* $A_\lambda : C^\infty(T^n) \to C^\infty(T^n)$ *is a bijection.*

(4) $\dim K(A_\lambda) > 0$ *for at most a countable number of* λ *with no finite accumulation point.*

Proof. We define $\alpha : H^s \times H^s \to \mathbb{R}$ by $\alpha(u, v) = \langle Au, v\rangle_0$. Then α is a continuous coercive bilinear form on H^s, whence α satisfies the hypotheses of Part I, Theorem 4D; if $\alpha_\lambda(v, y) = \langle A_\lambda v, y\rangle_0$, then $\alpha_\lambda(v, y) = \langle u, y\rangle_0$ for all $y \in H^s$ implies that $A_\lambda v = u$ by Proposition 2B. That the solutions actua lie in the designated spaces H^{t+s} follows from Theorem 4B. The statement (4) is a consequence the observation $K(A_\lambda) = K(I - (\lambda_0 - \lambda)G_{\lambda_0})$; for $G_{\lambda_0} : H^0 \to H^0$ is compact, so that we can apply Part I, Proposition 3E.

5. ZERO BOUNDARY VALUES IN $\mathbb{R}^n$

Our next step is to make certain minor modifications to obtain corresponding results in Euclidean space. The Dirichlet problem in its general form is a theory paying special attention to boundary value assignments; our applications go in another direction, and we shall consider only the elementary case of zero boundary values (taken in the Hilbert space sense of belonging to the space $H_0^t(U)$ below).

It would be possible to imitate rather closely the existence and regularity theory of Sections 2–4, starting with Hilbert spaces $H_0^t(U)$ defined through the Fourier transform in $\mathbb{R}^n$ for $t \geqslant 0$ by the norm:

$$|u|_t^2 = \int_{\mathbb{R}^n} (1 + |\xi|^2)^t \, |\hat{u}(\xi)|^2 \, d\xi \qquad \text{for all} \quad t \in \mathbb{R}^n$$

We shall not do that; rather, we shall use the results just obtained to derive the analogous results.

(A) Let U be a bounded open set in $\mathbb{R}^n$. For smooth function $u, v : \overline{U} \to \mathbb{R}$ we define the scalar product for positive integers s:

$$\langle u, v\rangle_{H^s(U)} = \sum_{|\alpha| \leqslant s} \int_U D^\alpha u(x) D^\alpha v(x)\, dx$$

We let $H^s(U)$ be the completion of the vector space of smooth function on $\overline{U}$ in that inner product.

Definition. A *test function in* U is a smooth function on $\mathbb{R}^n$ whose support is contained in U. Let $H_0^s(U)$ be the closure in $H^s(U)$ of the space of test functions in U. We observe that $H_0^0(U) = H^0(U)$; however, $H_0^s(U)$ is a proper subspace of $H^s(U)$ for $s > 0$.

Proposition. (Poincaré's inequality). *Consider on* $H_0^s(U)$ *the inner product*

$$\langle\langle u, v\rangle\rangle_{H_0^s(U)} = \sum_{|\alpha| = s} \binom{s}{\alpha} \langle D^\alpha u, D^\alpha v\rangle_0$$

where $\binom{s}{\alpha} = s!/\alpha_1! \ldots \alpha_n!$ *and* $\langle D^\alpha u, D^\alpha v\rangle_0 = \langle D^\alpha u, D^\alpha v\rangle_{H^0(U)}$. *This is equivalent to that given on* $H_0^s(U)$ *considered as a subspace of* $H^s(U)$.

Proof. Since the test functions are dense in $H_0^s(U)$ it suffices to work with these. First of all, it is clear that

$$||u||^2_{H_0^s(U)} \leqslant \text{const} |u|^2_{H^s(U)}$$

To prove an opposite inequality we start with

$$|u|^2_{H^s(U)} = \sum_{t=0}^{s} \sum_{|\alpha| = t} |D^\alpha u|_0^2 \leqslant \text{const} \sum_{t=0}^{s} ||u||^2_{H_0^t(U)}$$

It therefore suffices to prove

$$||u||^2_{H_0^t(U)} \leqslant \text{const}\, ||u||^2_{H_0^s(U)} \quad \text{if} \quad t \leqslant s$$

By iteration we reduce the problem to that of proving

$$|u|_0^2 \leqslant \text{const} \left|\frac{\partial u}{\partial x_i}\right|_0^2$$

For that we observe that

$$u(x) = \int_{-\infty}^{x_i} \frac{\partial u(x)}{\partial x_i}\, dx_i$$

since u has compact support; it follows that

$$|u(x)|^2 \leqslant \text{const} \int_{-\infty}^{x_i} \left|\frac{\partial u}{\partial x_i}\right|^2 dx_i$$

and the desired inequality follows by integration.

(B) The spaces $H_0^s(U)$ have many of the properties of $H^s(T^n)$, by virtue of the following lemma.

Let C be a closed cube in $\mathbb{R}^n$ whose interior contains $\bar{U}$. Without loss of generality we can suppose that C is the period domain defining T^n.

Lemma. *For each* $u \in H_0^s(U)$ *define* $\tilde{u} = u$ *in* U, 0 *in* C−U *and extend* $\tilde{u}$ *over* $\mathbb{R}^n$ *by periodicit*
Then $u \to \tilde{u}$ *defines an inclusion* $H_0^s(U) \to H^s(T^n)$, *and* $|u|_{H_0^s(U)} = |\tilde{u}|_{H^s(T^n)}$ (= $|\tilde{u}|_s$ in the notatio of Section 3).

Proof. Given $u \in H_0^s(U)$, let $(\phi_j)_{j\geq 1}$ be a sequence of test functions such that $|\phi_j - u|_{H_0^s(U)} \to 0$. View each ϕ_j as defined on T^n; then

$$|\phi_j - \tilde{u}|_{H^s(T^n)} = |\phi_j - u|_{H^s(C)} \to 0$$

and the lemma follows.

We can now transfer some of our results from T^n to U.

Theorem (Rellich). *If* $s < t$, *the injection* $H_0^t(U) \to H_0^s(U)$ *is compact and dense.*

Let $C^s(U)$ denote the vector space of C^s-functions on U.

Theorem (Sobolev). *If* $t > n/2 + s$ *then we have an injection* $H_0^t(U) \to C^s(U)$. *In particular,* $H_0^\infty(U)$ *consists of smooth functions in* U.

(C) We shall write $U \subset\subset W$ to mean that $\bar{U} \subset W$.

Lemma. *Given disjoint closed sets* C_0 *and* C_1 *in an open set* U *of* $\mathbb{R}^n$ *and* r^{th}*-order elliptic operators*

$$A_j = \sum_{|\alpha| \leq r} a_{j\alpha} D^\alpha$$

defined in two disjoint open subsets U_j *of* U *with* $C_j \subset U_j$ $(j = 0, 1)$. *Then there exists an* r^{th}*-or elliptic operator in* U *extending these in* C_j.

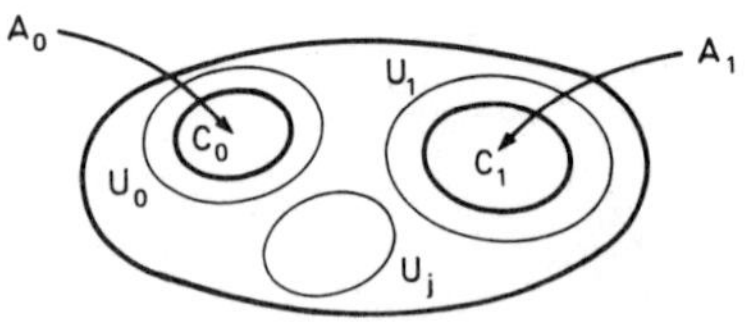

Proof. Construct a locally finite open covering $(U_j)_{j\geq 0}$ of U such that $U_j \cap C_i = \emptyset$ for $j \neq i$. Let $(\phi_j)_{j\geq 0}$ be a smooth partition of unity with $0 \leq \phi_j(x) \leq 1$ and each spt $\phi_j \subset U_j$ as in the third ler of Section 3B; furthermore, note that $\phi_j | C_j \equiv 1$ for $j = 0, 1$. For each $j \geq 2$ let

$$A_j = \sum_{|\alpha| \leq r} a_{j\alpha} D^\alpha$$

be a smooth r^{th}-order elliptic operator in U_j; e.g. we can take for A_j the iterated Laplacian $\Delta^{r/2}$. Then define A in U by

$$A = \sum_{j \geqslant 0} \sum_{|\alpha| \leqslant r} \phi_j a_{j\alpha} D^\alpha$$

Clearly A is well defined, and has the symbol

$$\sigma_A(x, \xi) = (-1)^{r/2} \sum_{|\alpha| = r} \sum_{j \geqslant 0} \phi_j(x) a_{j\alpha}(x) \xi$$

In particular, for each $x \in U$ all $\phi_j(x) \geqslant 0$ and some $\phi_k(x) > 0$, whence A is elliptic throughout U. Also, $A = A_j$ in C_j by our choice of covering.

Corollary. *If* A *is an elliptic operator in* W *with* $U \subset\subset W \subset\subset \operatorname{Int} C$, *then there is an extension* $\tilde{A}$ *of* A *in* U *to a periodic elliptic operator with* C *as the period domain.*

Proof. Take $C_0 = \bar{U}$ and C_1 = a closed neighbourhood of bdy C which does not meet W. Then use the lemma to extend A in C_0 and $\Delta^{r/2}$ in C_1.

(D) If A is an elliptic operator in $W \subset\subset U$ of order $r = 2s$, then we define $\alpha(u,v)$ for test functions in U by

$$\alpha(u, v) = \langle Au, v\rangle_0 = \sum \int_U b_{\beta\gamma} D^\beta u D^\gamma v \, dx$$

where the sum extends over all β, γ with $|\beta| \leqslant s$, $|\gamma| \leqslant s$. By using the right member to define α we have a unique extension of α as a continuous bilinear form on $H_0^s(U)$; thus there is a number $a > 0$ such that

$$|\alpha(u, v)| \leqslant a |u|_{H_0^s(U)} |v|_{H_0^s(U)}$$

for all $u, v \in H_0^s(U)$.

Theorem (Gårding's inequality). *If* A *is an elliptic operator of order* $r = 2s$ *in an open set* W *such that* $U \subset\subset W \subset\subset \operatorname{Int} C$, *then there are numbers* $c > 0$ *and* $\lambda_0 > 0$ *such that*

$$\alpha(u, u) + \lambda_0 |u|_0^2 \geqslant c|u|^2_{H_0^s(U)} \quad \textit{for all} \quad u \in H_0^s(U)$$

First of all, it suffices to verify this for test functions in U. We extend A to a periodic operator $\tilde{A}$ in C by Corollary 5C, and then apply Proposition 3A.

(E) Next we come to the problem of regularity.

Theorem. *Let* $v \in H^0(U)$, *and suppose that for some integer* $t \geqslant 0$ *we have* $Av \in H^t(U)$. *Then for all open sets* U_1 *with* $U_1 \subset\subset U$ *we have* $v \in H^{t+r}(U_1)$. *If* $Av \in H^\infty(U)$, *then* v *is smooth in* U.

Proof. Let U_2, U_3 be open sets for which $U_1 \subset\subset U_2 \subset\subset U_3 \subset\subset U$. Choose a test function ζ in U_3 which is 1 on U_2. Set $Av = u$. Then

$$\langle \zeta u, w\rangle_{H^0(U)} = \langle Av, \zeta w\rangle_{H^0(U)} = \langle v, A^*(\zeta w)\rangle_{H^0(U)} = \langle \zeta v, A^* w\rangle_{H^0(U)} + \langle v, A_1^* w\rangle_{H^0(U)}$$

for all periodic smooth functions w in $\mathbb{R}^n$ with period domain C, where A_1^* is a smooth differential operator of order $\leqslant r-1$ whose coefficients have supports in U_3. It follows that

$$\langle \zeta u, w\rangle_{H^0(C)} = \langle v, A^* w\rangle_{H^0(C)} + \langle v, A_1^* w\rangle_{H^0(C)}$$

Let η be a test function in U which is 1 on U_3, and define $\widetilde{v}$

$$\widetilde{v} = \begin{cases} v \text{ in } U \\ 0 \text{ in the rest of } C \end{cases}$$

similarly for the definition of $\widetilde{u} = \eta u$. Extend ζ and A_1^* to $\widetilde{\zeta}$ and $\widetilde{A}_1^*$ in C by defining them to be zero in $C - U_3$.

Let $\widetilde{A}$ be an extension of A to a periodic elliptic operator with period domain C, by Corollary 5C.

If $\widetilde{A}_1$ denotes the $H^0(C)$-adjoint of $\widetilde{A}_1^*$,

$$\langle \zeta\widetilde{u}, w\rangle_{H^0(C)} = \langle \widetilde{A}(\zeta\,\widetilde{v}), w\rangle_{H^0(C)} + \langle \widetilde{A}_1\widetilde{v}, w\rangle_{H^0(C)}$$

for all w, so that $\zeta\widetilde{u} = \widetilde{A}(\zeta\,\widetilde{v}) + \widetilde{A}_1(\widetilde{v})$. But $\zeta\widetilde{u} - \widetilde{A}_1(\widetilde{v}) \in H^{1-r}(T^n)$ and $\widetilde{A}$ is elliptic on T^n, whenc $\zeta\widetilde{v} \in H^1(T^n)$ by Theorem 5B. In particular, $v \in H^1(U_2)$.

We now repeat the process, starting with an open set U_2' for which $U_1 \subset\subset U_2' \subset\subset U_2$; we find that $\zeta\widetilde{u} - \widetilde{A}_1(\widetilde{v}) \in H^{2-r}(T^n)$, so that $v \in H^2(U_2')$. We can continue in this way until $v \in H^{t+r}(U_2'')$ for some open set U_2'' containing $\overline{U}_1$. The final statement of the theorem follows from Sobolev's theorem.

(F) Taken together, these results yield a conclusion similar to that of Theorem 4C, with T^n replaced by U.

First of all, let $H^t_{loc}(U)$ be the vector space of functions in U which have all derivatives of orders $\leqslant t$ square integrable on every open $U_1 \subset\subset U$.

Theorem. *Let* A *be an elliptic operator in an open set* $W \supset U$ *of order* $r = 2s$. *Then*

(1) A *maps* $H^s_0(U) \cap H^{t+r}_{loc}(U)$ *onto* $H^t_{loc}(U) \cap K^{\perp}(A^*)$ *with kernel* $K(A) \subset C^\infty(U)$.

(2) $\dim K(A) = \dim K(A^*) < \infty$.

(3) *There is a* $\lambda_0 > 0$ *such that* $A_\lambda : H^s_0(U) \cap H^{t+r}_{loc}(U) \to H^t_{loc}(U)$ *is a bijection for all* $\lambda \leqslant \lambda_0$.

(4) $\dim K(A_\lambda) > 0$ *for at most countably many* λ *having no finite accumulation point.*

Remark. It is quite appropriate that these statements should involve the spaces $H^t_{loc}(U)$ in expressing higher order differentiability; for differentiability is a local concept, whereas belonging to $H^t(U)$ is a global one.

It is possible to obtain more precise interpretations of the assumption of boundary values in case U is a bounded region with smooth boundary. Roughly speaking, solutions $v \in H^s_0(U)$ can then be shown to have all derivatives of order $< s$ vanishing on bdy U; see Bers-John-Schechter and Nirenberg [11]. We shall not pursue that aspect of the theory now.

Remark. Suppose we are given $u \in H^t_{loc}(U)$ and $\phi \in H^{t+r}_{loc}(U)$ and are asked to find $v \in H^{t+r}_{loc}(U)$ such that $Av = u$ in U, with v assuming the boundary values ϕ in the sense that $u - \phi \in H_0^{t+r}(U)$. Setting $u' = u - A\phi$, we reduce the problem to solving $Av' = u'$ for $v' \in H_0^s(U) \cap H^{t+r}_{loc}(U)$; for then we can take $v = v' + \phi$ to solve the given problem.

Example. Let $U = \{x \in \mathbb{R}^2 : x_1^2 + x_2^2 < 1\}$. Then the negative of the Laplace operator $-\Delta$ is elliptic, with symbol $\sigma_{-\Delta}(x, \zeta) = \zeta_1^2 + \zeta_2^2$. Suppose we set $u \equiv 0$ in U, but require that v equals (in the above sense) a given function ϕ on bdy U. The solution describes in $\mathbb{R}^3$ a surface spanning ϕ which is a minimal surface, in the sense that the mean normal curvature is zero.

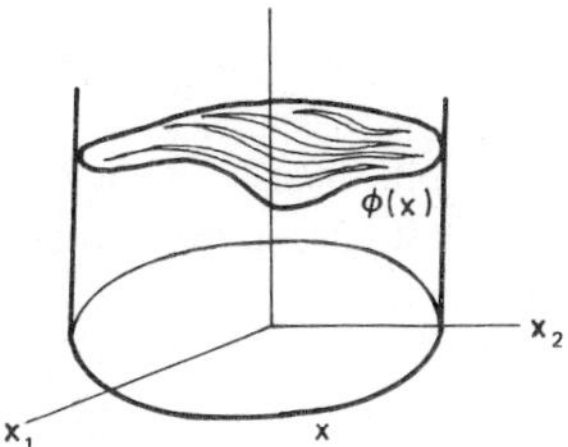

Remark. Given a subspace H such that $H_0^s(U) \subset H \subset H^s(U)$, we can consider membership in H as expressing boundary conditions on the problem. If we have a form of coerciveness (such as Gårding's inequality) for functions in H, then we can apply the above theory. We have considered the simplest case, with $H = H_0^s(U)$; the opposite extreme $H = H^s(U)$ has been developed by Aronszajn [1], and intermediate cases by Lions [10], Schechter [15], and others.

6. STRONGLY ELLIPTIC SYSTEMS

On T^n let us suppose given for each n-tuple α of natural numbers with $|\alpha| \leq r$ a smooth map $a_\alpha : T^n \to M(p, p)$, the vector space of real $p \times p$ matrices. If $C^\infty(T^n, \mathbb{R}^p)$ denotes the vector space of smooth maps $T^n \to \mathbb{R}^p$, then the differential operator

$$A = \sum_{|\beta| \leq r} a_\beta D^\beta$$

is an endomorphism of $C^\infty(T^n, \mathbb{R}^p)$. We define its symbol

$$\sigma_A : T^n \times \mathbb{R}^n \to M(p, p)$$

by

$$\sigma_A(x, \xi) = (-1)^{r/2} \sum_{|\beta| = r} a_\beta(x) \xi^\beta$$

and say that A *is strongly elliptic* if σ_A is positive definite at every point $x \in T^n$. Note that there is a number $\lambda_0 > 0$ such that for all $(x, \xi, \eta) \in T^n \times \mathbb{R}^n \times \mathbb{R}^p$ we have $(r = 2s)$

$$\langle \sigma_A(x, \xi)\eta, \eta \rangle_{\mathbb{R}^p} \geq \lambda_0 |\xi|^{2s} |\eta|^2$$

Let $E = H^0 \times \ldots \times H^0$ (p copies), with inner product

$$\langle u, v\rangle_E = \sum_{i=1}^{p} \langle u_i, v_i\rangle_0$$

where $u = (u_1, \ldots, u_p)$, $v = (v_1, \ldots, v_p)$. Similarly for $V = H^s \times \ldots \times H^s$. Then we have the bilinear form $\alpha: V \times V \to \mathbb{R}$ defined by

$$\alpha(u, v) = \langle Au, v\rangle_E$$

As in the case $p = 1$, we again find that α is continuous and coercive. Consequently, *if* $A: H^r(T^n, \mathbb{R}^p) \to H^0(T^n, \mathbb{R}^p)$ *is a strongly elliptic operator, then Theorem 4C is valid with* $H^t = H^t(T^n, \mathbb{R}^p)$. *Similarly, Theorem 5F is also valid with* $H^t_{loc}(U)$ *replaced by* $H^t_{loc}(U, \mathbb{R}^p)$.

III. DIFFERENTIAL OPERATORS ON VECTOR BUNDLES

We present in Part III an abbreviated treatment of smooth manifolds, vector bundles, and of certain differential operators associated with them. The theory is primarily local, and thus it is most appropriate to formulate the concepts in terms of sheaves. The basic reference for sheaf theory is Godement [5]; the viewpoint in the theory of connections is that of Koszul [9].

1. SHEAVES OF MODULES

(A) **Definitions.** Let R be a commutative ring with unit. A *sheaf of* R-*modules* is a triple $(\mathscr{S}, \pi, X)$ where $\mathscr{S}$ and X are topological spaces and $\pi: \mathscr{S} \to X$ is a continuous surjective map, subject to the following conditions:

(1) π is a local homeomorphism; i.e. each $s \in \mathscr{S}$ has an open neighbourhood S such that the restriction $\pi|S$ of π is a homeomorphism of S onto an open neighbourhood of $\pi(s)$.

(2) For each $x \in X$ the *stalk* $\mathscr{S}_x = \pi^{-1}(x)$ is an R-module, and the algebraic operations $(s, t) \to s+$ and $(r, s) \to rs$ are continuous where they are defined, for $r \in R$, $s, t \in \mathscr{S}$.

The totality $\mathscr{S}(X)$ of sections of $(\mathscr{S}, \pi, X)$, – i.e. of continuous maps $\phi: X \to \mathscr{S}$ such that $\pi\phi(x) = x$ for all $x \in X$ – is an R-module, with algebraic operations defined pointwise.

If U is an open subspace of X, then the *restriction* $\mathscr{S}|U$ of $(\mathscr{S}, \pi, X)$ is the sheaf $(\pi^{-1}(U), \pi|\pi^{-1}(U), U)$.

A triple $(\mathscr{S}', \pi', X)$ *is a subsheaf* of $(\mathscr{S}, \pi, X)$ if (1) $\mathscr{S}'$ is an open subspace of $\mathscr{S}$; (2) $\pi' = \pi|\mathscr{S}'$; (3) each stalk $\mathscr{S}'_x$ is a submodule of the stalk $\mathscr{S}_x$.

Given sheaves $(\mathscr{S}, \pi, X)$ and $(\mathscr{T}, \rho, Y)$ and a continuous map $f: X \to Y$, an f-homomorphism is a continuous map $\bar{f}: \mathscr{S} \to \mathscr{T}$ for which the following diagram is commutative, and for which

$$\begin{array}{ccc} \mathscr{S} & \xrightarrow{\bar{f}} & \mathscr{T} \\ \downarrow \pi & & \downarrow \rho \\ X & \xrightarrow{f} & Y \end{array}$$

each restriction $\bar{f}:\mathscr{S}_x \to \mathscr{T}_{f(x)}$ is an R-homomorphism. In particular, for X = Y and f the identity map we have the notion of R-homomorphism $\lambda:\mathscr{S} \to \mathscr{T}$ of sheaves of R-modules over X. Every λ induces a homomorphism $\Lambda:\mathscr{S}(X) \to \mathscr{T}(X)$ of their modules of sections; and Λ *is local*, in the sense that for all $\phi \in \mathscr{S}(X)$ we have $\text{spt}(\Lambda\phi) \subset \text{spt}(\phi)$. With mild conditions on $\mathscr{S}$ we can prove that every local homomorphism $\mathscr{S}(X) \to \mathscr{T}(X)$ induces a sheaf homomorphism; see Section 2D.

We say that $(\mathscr{S}, \pi, X)$ *is modelled on* $(\mathscr{T}, \rho, Y)$ if every $x \in X$ has a neighbourhood U and homeomorphisms $f:U \to Y$ and $\bar{f}:\mathscr{S}|U \to \mathscr{T}$ such that $\bar{f}$ is an f-isomorphism.

Example. Let $\mathscr{T}$ be the *simple sheaf* $\mathscr{T} = X \times R$, where R is given the discrete topology, and $\rho:\mathscr{T} \to X$ is the projection on the first factor. We shall sometimes write $\mathscr{T} = R$. If $\mathscr{S} \to X$ is modelled on $\mathscr{T}$, with $f:X \to X$ the identity map, then $\mathscr{S}$ is said to be a *locally simple sheaf*. Its components are covering spaces of X.

(B) Let $(\mathscr{S}, \pi, X)$ be a sheaf of R-modules, and $U \supset V$ open subsets of X. Then we have the R-homomorphism

$$r_V^U:\mathscr{S}(U) \to \mathscr{S}(V)$$

defined by restriction. Clearly, if $W \subset V \subset U$ are open, then $r_W^U = r_W^V \cdot r_V^U$, and $r_U^U = \text{identity}$.

Conversely, if for all pairs of open sets $U \supset V$ of a base for the topology of X we have R-modules $\mathscr{S}(U)$ and R-homomorphisms r_V^U defined with the above properties, then this assignment (called a presheaf) defines a sheaf of R-modules over X. In fact, for each point $x \in X$ let $\mathscr{S}_x = \varinjlim \mathscr{S}(U)$, the direct limit taken over the open neighbourhoods of x; define $r_x^U:\mathscr{S}(U) \to \mathscr{S}_x$ as the natural map. Set $\mathscr{S} = \cup\{\mathscr{S}_x : x \in X\}$, and topologize $\mathscr{S}$ by constructing the following base for this topology: for each open U in X and $\phi \in \mathscr{S}(U)$ let $V(U, \phi) = \{r_y^U(\phi) : y \in U\}$. The map $\pi:\mathscr{S} \to X$ is defined by $\mathscr{S}_x \to x$ for all $x \in X$. It is immediate that $(\mathscr{S}, \pi, X)$ satisfies the conditions of a sheaf.

Example 1. Let $\mathscr{R} \to X$ be a sheaf of commutative rings with unit. An *$\mathscr{R}$-module over* X is a sheaf $\mathscr{S} \to X$ of abelian groups such that (1) for every open U in X, $\mathscr{S}(U)$ is an $\mathscr{R}(U)$-module; (2) for every open $V \subset U$ the restriction $\mathscr{S}(U) \to \mathscr{S}(V)$ is a homomorphism of modules compatible with the restrictions $\mathscr{R}(U) \to \mathscr{R}(V)$ of rings of operators.

Example 2. If $\mathscr{S}$ and $\mathscr{T}$ are $\mathscr{R}$-modules over X, their *tensor product* $\mathscr{S} \otimes_{\mathscr{R}} \mathscr{T} \to X$ is the $\mathscr{R}$-module characterized by the assignment $U \to \mathscr{S}(U) \otimes_{\mathscr{R}(U)} \mathscr{T}(U)$ for every open U in X. Note in particular that $\mathscr{R} \otimes_{\mathscr{R}} \mathscr{T} \approx \mathscr{T}$.

Example 3. Let X be a topological space and R the real number field. For each open U in X let $\mathscr{C}(U, R^n)$ be the R-module of continuous maps $U \to R^n$. The resulting sheaf $\mathscr{C}^n = \mathscr{C}_{(X, R^n)} \to X$ is the *sheaf of continuous maps of* X *into* R^n; it is a $\mathscr{C}^1$-module. Then $\mathscr{C}^n$ is isomorphic to the direct sum $\mathscr{C}^1 \oplus \ldots \oplus \mathscr{C}^1$ (n copies; definition evident), and $\mathscr{C}^n$ is said to be *free of rank* n. A sheaf of $\mathscr{C}^1$-modules is *locally free of rank* n if it is locally isomorphic to $\mathscr{C}^n$.

Example 4. If $\phi:X \to Y$ is a continuous map of toplogical spaces, for each open U in X let $\mathscr{S}(U) = \{\sigma:U \to \mathscr{C}_{(Y,R)} : \eta\sigma(x) = \phi(x) \text{ for all } x \in U\}$, where $\eta:\mathscr{C}_{(Y,R)} \to Y$ is the sheaf of continuous functions on Y. Thus $\mathscr{S}(U)$ is an R-algebra, and if V is open in U, we have the restriction homomorphism $\mathscr{S}(U) \to \mathscr{S}(V)$. The resulting sheaf is denoted by $\phi^{-1}\mathscr{C}_{(Y,R)} \to X$. For each open U in X we have $\phi^{-1}\mathscr{C}_{(Y,R)}(U)$ identified with a subalgebra of $\mathscr{C}_{(X,R)}(U)$, and thus $\phi^{-1}\mathscr{C}_{(Y,R)}$ as a subsheaf of $\mathscr{C}_{(X,R)}$.

2. VECTOR BUNDLES

For our purposes vector bundles over a space X can be viewed from three standpoints:

(1) As a fibre bundle whose fibres are vector spaces; this is convenient for introducing related structure (e.g. G-structures);

(2) As a locally free sheaf of finite rank; this is appropriate for utilizing local structures such as the differential structure of a manifold;

(3) As a special sort of $\mathscr{C}(R)$-module, where $\mathscr{C}(R)$ is the algebra of continuous functions on X; this viewpoint stresses the formal manipulative aspect of the theory.

(A) **Definition.** Let X be a topological space. A (real) *vector bundle over* X *of fibre dimension* m is a topological space E together with a continuous surjective map $\xi: E \to X$ such that:

(1) each *fibre* $E_x = \xi^{-1}(x)$ has a structure of a (real) vector space;

(2) each point of X has a neighbourhood U and a homeomorphism (a trivialization) $\rho: U \times R^m \to \xi^{-1}(U)$ such that $\xi\rho(x, v) = x$ for all $(x, v) \in U \times R^m$, and the map $v \to \rho(x, v)$ is linear for all $x \in U$.

If E and F are vector bundles over X, a *homomorphism* $\lambda: E \to F$ is a continuous map such that for each $x \in X$ the restriction $\lambda: E_x \to F_x$ is linear.

(B) Let $\xi: E \to X$ be a vector bundle. Then for any open set U in X the totality of continuous sections $U \to E$ forms a vector space $\underline{E}(U)$; for V open in U the restriction map $r^U_V: \underline{E}(U) \to \underline{E}(V)$ defines a presheaf of vector spaces, whose sheaf (a $\mathscr{C}^1$-module over X) we shall denote by $\underline{E} \to X$. From the definition of vector bundle E it follows that each point of X has a neighbourhood U and m continuous sections $\sigma_i: U \to E$ with $\xi\sigma_i(x) = x$ for all $x \in U$, and each $(\sigma_1(x), ..., \sigma_m(x))$ is a base for E_x. It follows that $\underline{E}$ is locally free of rank m.

Conversely, let $\mathscr{E} \to X$ be a locally free sheaf of rank m. Let $M_x = \{\gamma \in \mathscr{C}^1_x : \gamma(x) = 0\}$. Then the sequence

$$0 \to M_x \mathscr{E}_x \to \mathscr{E}_x \to \mathscr{E}_x / M_x \mathscr{E}_x \to 0$$

of vector spaces is exact. Set $E_x = \mathscr{E}_x / M_x \mathscr{E}_x$, and $E = \cup\{E_x : x \in X\}$, and define the map $\xi: E \to X$ by the condition $E_x \to x$.

If U is an open set in X over which $\mathscr{E}|U$ is isomorphic to $\mathscr{C}_{(U, R^m)}$, then we have sections $\sigma_1, ..., \sigma_m \in \mathscr{E}(U)$ defined so that for each $x \in U$, $r^U_x \sigma_1, ..., r^U_x \sigma_m$ is a base for $\mathscr{E}_x$ over $\mathscr{C}^1_x$. Their image $\sigma_1(x), ..., \sigma_m(x)$ under the coset mapping form a base for E_x over R. Furthermore, the map $\rho: U \times R^m \to \xi^{-1}(U)$ defined by $\rho(x; a_1, ..., a_m) = \sum a_i \sigma_i(x)$ is a bijection; we topologize E by requiring that each ρ be a homeomorphism.

The following consequence is immediate:

Proposition (1). *If* $\xi: E \to X$ *is a vector bundle, then there is a canonical isomorphism of* E *onto the vector bundle constructed from its sheaf* $\underline{E}$ *of sections.*

Proposition (2). *If* $\mathscr{E} \to X$ *is a locally free sheaf of rank* m, *then there is a canonical isomorphism of* $\mathscr{E}$ *onto the sheaf of sections of the vector bundle just constructed from* $\mathscr{E}$.

(C) If E and F are finite-dimensional vector spaces, then we can form their *space,* Hom(E, F) *of linear maps,* their *direct sum* $E \oplus F$, *their tensor product* $E \otimes F$, the pth *tensor power* $\otimes^p E = E \otimes ... \otimes E$ (p copies), the pth *exterior power* $\wedge^p E$, the pth *symmetric power* $\odot^p E$. We agree that $\otimes^p E = R$ if $p = 0$. Let

$$\otimes E = \sum_{p \geqslant 0} \otimes^p E, \quad \wedge E = \sum_{p \geqslant 0} \wedge^p E, \quad \odot E = \sum_{p \geqslant 0} \odot^p E,$$

direct sums in each case. These all have structures of graded algebras over R, whose multiplications pair p- and q-summands to the (p+q)-summand. In particular, viewing $\wedge^p E$ as the totality of alternating p-linear forms on the *dual space* $E^* = \mathrm{Hom}(E, R)$, we define the product of $\alpha \in \wedge^p E$ and $\beta \in \wedge^q E$ by

$$\alpha \wedge \beta(x_1, \ldots, x_{p+q}) = \sum \epsilon_\sigma \alpha(x_{\sigma(1)}, \ldots, x_{\sigma(p)})\, \beta(x_{\sigma(p+1)}, \ldots, x_{\sigma(p+q)})$$

with summation taken over all permutations σ of $(1, \ldots, p+q)$ and ϵ_σ the sign of the permutation, with $\sigma(1) < \ldots < \sigma(p)$ and $\sigma(p+1) < \ldots < \sigma(p+q)$. Thus $\alpha \wedge \beta = (-1)^{pq} \beta \wedge \alpha$, and $\wedge E$ is an associative, commutative (in the graded sense) graded algebra of $\dim E = 2^n$, where $n = \dim E$.

These constructions are all immediately transferable to sheaves $\underline{E}$ of sections of vector bundles E over X. Thus, for instance, we define the $\mathscr{C}^1$-module $\underline{E} \otimes_{\mathscr{C}^1} \underline{F} \to X$ as in Section 1B; it is an easy matter to verify that it is locally free and corresponds canonically to a vector bundle whose fibre over x is the vector space $E_x \otimes F_x$.

(D) Let X be a normal space, and $\xi : E \to X$ a vector bundle over X with $\mathscr{C}(E) = \underline{E}(X)$ as its $\mathscr{C}^1$-module of sections.

Then $\mathscr{C}(R) = \mathscr{C}(X \times R)$ is the R-algebra of continuous functions on X.

Proposition. *If* E *and* F *are vector bundles over* X, *then there is a natural isomorphism:*

$$\mathscr{C}(\mathrm{Hom}_R(E, F)) \to \mathrm{Hom}_{\mathscr{C}(R)}(\mathscr{C}(E), \mathscr{C}(F))$$

Proof. Given $\lambda \in \mathscr{C}(\mathrm{Hom}_R(E, F))$ and $\phi \in \mathscr{C}(E)$, define $\lambda\phi \in \mathscr{C}(F)$ by $(\lambda\phi)x = \lambda(x)(\phi(x))$ for all $x \in X$. Then $\phi \to \lambda\phi$ is clearly a $\mathscr{C}(R)$-homomorphism. The inverse map is defined as follows. Given a $\mathscr{C}(R)$-homomorphism $\Lambda : \mathscr{C}(E) \to \mathscr{C}(F)$ and $x \in X$, $v \in E_x$, we construct (using the normality of X) $\phi \in \mathscr{C}(E)$ whose support is in some open U over which E is trivial and $\phi(x) = v$. Define $\lambda(x)(v) = (\Lambda\phi)x$. If ϕ' were another such section, then $(\Lambda\phi) = (\Lambda\phi')x$. For if $\sigma_1, \ldots, \sigma_m \in \underline{E}(U)$ are a set of sections linearly independent at every point $x \in U$, then we can write $\phi = \sum \rho_i \sigma_i$ and $\phi' = \sum \rho_i' \sigma_i$ for suitable $\rho_i, \rho_i' \in \mathscr{C}(R)$ with $\rho_i(x) = \rho_i'(x)$ $(1 \leqslant i \leqslant m)$. Then $\Lambda(\phi') = \sum \rho_i' \Lambda(\sigma_i)$, and $(\Lambda\phi')x = \sum \rho_i'(x)\Lambda(\sigma_i) = \sum \rho_i(x)\Lambda(\sigma_i) = (\Lambda\phi)x$. The proposition follows.

Example. If $E^* = \mathrm{Hom}_R(E, R)$ is the dual bundle of E, then $\mathscr{C}(E^*) = \mathrm{Hom}_{\mathscr{C}(R)}(\mathscr{C}(E), \mathscr{C}(R))$. Also, $\mathscr{C}(E) = \mathrm{Hom}_{\mathscr{C}(R)}(\mathscr{C}(E^*), \mathscr{C}(R))$. More generally, if $L^p_R(E, R)$ is the bundle of p-linear maps of $E \times \ldots \times E \to R$, then $\mathscr{C}(L^p_R(E, R)) = L^p_{\mathscr{C}(R)}(\mathscr{C}(E), \mathscr{C}(R))$.

(E) *Remark.* Any local homomorphism (i.e. support decreasing) $\Lambda : \mathscr{C}(E) \to \mathscr{C}(F)$ defines a sheaf homomorphism $\lambda : \underline{E} \to \underline{F}$.

From the viewpoint of differential operators the following result is useful.

Proposition. *Let* E *and* F *be vector bundles over* X *and* $\Lambda : \mathscr{C}(E) \to \mathscr{C}(F)$ *be a* $\mathscr{C}(R)$*-homomorphism. Let* D *be an* R*-endomorphism of* $\mathscr{C}(R)$, *and suppose* $\Lambda(\gamma\phi) = D(\gamma)\phi + \gamma\Lambda(\phi)$ *for all* $\gamma \in \mathscr{C}(R)$, $\phi \in \mathscr{C}(E)$. *Then* Λ *is local.*

Proof. Let $\phi \in \mathscr{C}(E)$, and let U be an open neighbourhood of a point $x \in X$ such that $\phi | U = 0$. Take a function $\gamma \in \mathscr{C}(R)$ for which $\gamma(x) = 0$ and $\gamma = 1$ in a neighbourhood of $X - U$. Then $\phi = \gamma\phi$, whence $(\Lambda\phi)x = D(\gamma)(x)\phi(x) + \gamma(x)\Lambda\phi(x) = 0$; i.e. $\Lambda(\phi)|U = 0$.

Taking D as the zero homomorphism, we obtain the

Corollary. *Every* $\mathscr{C}(R)$*-homomorphism is local.*

3. SMOOTH MANIFOLDS AND VECTOR BUNDLES

(A) A Hausdorff space X is a *topological* n-*manifold* if it is locally homeomorphic to $\mathbb{R}^n$; thus with every $x \in X$ we have a *chart* (θ, U), consisting of an open neighbourhood U of x and a homeomorphism θ of U onto an open subset of $\mathbb{R}^n$. For such a space the following properties are equivalent:

(1) X is paracompact;
(2) X is metrizable;
(3) X is expressible as a countable union of compact sets;
(4) X has a countable base for its topology.

We assume henceforth that X *has these properties.*

Definitions. Let X be a topological n-manifold. A *differential structure* on X is a subsheaf $\mathcal{D} = \mathcal{D}_{(X,\mathbb{R})}$ of the sheaf $\mathcal{C}_{(X,\mathbb{R})}$ of continuous functions, which is modelled on $\mathcal{D}_{(\mathbb{R}^n,\mathbb{R})}$. We shall call the pair $(X, \mathcal{D})$ *a smooth manifold,* and sometimes denote it by X alone. We shall let $\mathcal{D}(\mathbb{R}) = \mathcal{D}_{(X,\mathbb{R})}(X)$ denote the algebra of all sections of $\mathcal{D}$, called the smooth functions on X. A *smooth map* $f: X \to Y$ of smooth manifolds is a continuous map such that $f^{-1}\mathcal{D}_{(Y,R)}$ is a subsheaf of $\mathcal{D}_{(X,R)}$; this means that for every open V in Y, composition with f defines a homomorphism $\mathcal{D}_{(Y,R)}(V) \to \mathcal{D}_{(X,R)}(f^{-1}(V))$. A *diffeomorphism* is a smooth homeomorphism f whose inverse f^{-1} is also smooth. Note that each chart $\theta: U \to R^n$ is a diffeomorphism.

Remark. It is possible to define a smooth manifold by giving (in place of $\mathcal{D}$) a sheaf of sets on the space X which corresponds to the sheaf of sections of the principal bundle of X. That approach would have the advantage of emphasizing the fibre bundle theoretic aspects of manifolds. See Section 4F.

Remark. Not every compact topological n-manifold admits a differential structure $(n \geqslant 8)$; nor is a differential structure unique $(n \geqslant 7)$.

The following result follows easily from the constructions in Section 3B of Part II:

Proposition. *If* $\mathcal{U} = (U_i)$ *is a locally finite open covering of* X, *then there is a smooth partition of unity subordinate to* $\mathcal{U}$, *i.e. a family* (λ_i) *of smooth functions on* X *such that* $\mathrm{spt}(\lambda_i) \subset U_i$ *for all* i, $\sum_i \lambda_i(x) = 1$ *for all* $x \in X$, *and each* $0 \leqslant \lambda_i(x) \leqslant 1$.

(B) *A smooth vector bundle* $\xi: E \to X$ *of fibre dimension* m is a vector bundle in which E is a smooth manifold, ξ is a smooth map, and all trivializations ρ are smooth. The correspondence between smooth vector bundles and locally free $\mathcal{D}$-modules (i.e. sheaves of $\mathcal{D}$-modules locally isomorphic to $\mathcal{D}^m$) proceeds as in Proposition 2B. In particular, the space $\mathcal{D}(E)$ of smooth sections of E is a $\mathcal{D}(R)$-module.

For any section $\phi \in \mathcal{D}(E)$ we have the notion of its *support:* $\mathrm{spt}(\phi) = \mathrm{Closure}\{x \in X: \phi(x)$ Then $\mathrm{spt}(\phi + \psi) \subset \mathrm{spt}(\phi) \cup \mathrm{spt}(\psi)$, $\mathrm{spt}(\gamma\phi) \subset \mathrm{spt}(\gamma) \cap \mathrm{spt}(\phi)$ for all $\varphi, \psi \in \mathcal{D}(E)$, $\gamma \in \mathcal{D}(R)$. We shall say that a section ϕ *is compact* if $\mathrm{spt}(\phi)$ is compact. The totality of compact sections is a vector subspace $\mathcal{D}_0(E)$ of $\mathcal{D}(E)$; of course, $\mathcal{D}_0(E) = \mathcal{D}(E)$ if X is compact.

Henceforth we assume that all manifolds, maps and vector bundles are smooth.

Remark. A sheaf is *soft* if every section defined over a closed subset $A \subset X$ can be extended to section over all X. The sheaf of sections of a smooth vector bundle is always soft. (For proof see Kobayashi-Nomizu: Foundations of Differential Geometry Vol. I, Interscience, p.58.)

(C) We now construct certain sheaves on X defined in terms of its differential structure. For each open set U of X let $\mathscr{D}_1(U)$ be the algebra of all smooth functions φ in U which admit a smooth extension to an open $U_\varphi \supset U$, which domain is allowed to vary with φ. Let $\mathscr{T}(U)$ be the totality of derivations on $\mathscr{D}_1(U)$, i.e. R-endomorphisms $v:\mathscr{D}_1(U) \to \mathscr{D}_1(U)$ such that $v(\varphi\psi) = v(\varphi)\psi + \varphi v(\psi)$ Thus v is a local homomorphism (it decreases supports, as in Proposition 2E). If V is an open subset of U, we define the restriction map $r_V^U : \mathscr{T}(U) \to \mathscr{T}(V)$ as follows.

Take any $v \in \mathscr{T}(U)$. Then given a $\varphi \in \mathscr{D}_1(V)$, there is (by Part II, Section 3B) a $\varphi_1 \in \mathscr{D}_1(U)$ such that $\varphi_1 | V = \varphi$.

Define $w(\varphi) = v(\varphi_1)$ restricted to V. If φ_1' were another such extension, then $(\varphi_1' - \varphi_1)|V = 0$, and because v is local, we find $v(\varphi_1') - v(\varphi_1) = 0$ in V. Thus $w(\varphi)$ is independent of the choice of extension of φ, and we define $r_V^U(v) = w$.

It is clear that we have just defined vector spaces and homomorphisms of a presheaf. The associated sheaf $\mathscr{T} \to X$ is called the *tangent sheaf* of X. This sheaf is locally free of rank n: in fact, for each point of X let (θ, U) be a chart containing it, and define the derivation v_i in $\mathscr{D}(U)$ by

$$(v_i(\phi))x = \left(\frac{\partial}{\partial x_i}(\phi \circ \theta^{-1})\right)\theta(x) \qquad (1 \leqslant i \leqslant n)$$

for $\phi \in \mathscr{D}(U)$. These are easily seen to be a base over $\mathscr{D}(R)$ for $\mathscr{T}(U)$. The vector bundle $\pi : T(X) \to X$ of fibre dimension n associated with $\mathscr{T} \to X$ by Proposition 2B is called the *tangent vector bundle of* X. The fibre $\pi^{-1}(x) = X(x)$ is called the *tangent space of* X at x, and its elements are called the *tangent vectors at* x. These can be identified with those mappings $v : \mathscr{D}(R) \to R$ which are R-linear, vanish on the constant functions, and satisfy $v(\phi\psi) = v(\phi)\psi(x) + \phi(x)v(\psi)$ for all $\phi, \psi \in \mathscr{D}(R)$. Furthermore, the space $\mathscr{D}(T(X)) = \mathscr{T}(X)$ is precisely the Lie algebra over R of all derivations of $\mathscr{D}(R)$ with bracket $[u, v] = u \cdot v - v \cdot u$. These derivations are called the *tangent fields* of X.

By the constructions in Section 2C we can form the vector bundles

$$\otimes^p T(X), \quad \wedge^p T(X), \quad \odot^p T(X),$$

the dual bundle $T^*(X) = \mathrm{Hom}(T(X), R)$ and its powers. The elements in the $\mathscr{D}(R)$-modules $\mathscr{D}(\otimes^p T(X))$, $\mathscr{D}(\otimes^p T^*(X))$ are called p-*contravariant,* p-*covariant tensor fields* on X. Those of $\mathscr{D}(\wedge^p T^*(X))$ are called *exterior* p-*forms* on X. Note that

$$\mathscr{D}(\wedge T^*(X)) = \sum_{p=0}^{n} \mathscr{D}(\wedge^p T^*(X))$$

is a graded commutative algebra. Similarly for the spaces $\mathscr{D}_0(\otimes^p T(X))$, etc., of compact tensor fields. Observe that if X is not compact, then $\mathscr{D}_0(\wedge T^*(X))$ does not have a unit.

If $f : X \to Y$ is a map of manifolds, then f determines an f-homomorphism $T(f) : T(X) \to T(Y)$ of the tangent vector bundles, which at each point $x \in X$ is the *differential* $f_*(x) : X(x) \to Y(f(x))$, given for each $v \in X(x)$ by the function which assigns to each $\psi \in \mathscr{D}_{(Y,R)}(Y)$ the number $(w(\psi))x = v(\psi \circ f)x$. It is clear that w has the properties of a tangent vector at f(x), and we define $f_*(x)v = w$. The map f therefore induces f-homomorphisms:

$$\otimes^p T(f) : \otimes^p T(X) \to \otimes^p T(Y), \quad \otimes^p T^*(f) : \otimes^p T^*(Y) \to \otimes^p T^*(X)$$

(with evident interpretation of the term f-homomorphism in the second case). f induces homomorphisms of the spaces of sections of the covariant bundles; not of the contravariant bundles.

In particular, we have the algebra homomorphism

$$\wedge(f): \mathscr{D}(\wedge T^*(Y)) \to \mathscr{D}(\wedge T^*(X))$$

which on the p-components is given by

$$((\wedge^p f)\varphi)\, x(u_1, ..., u_p) = \varphi(f(x))(f_*(x)u_1, ..., f_*(x)u_p)$$

for all $\varphi \in \mathscr{D}(\wedge^p T^*(Y))$; $u_1, ..., u_p \in X(x)$.

(D) For each open set U in X let $\mathscr{T}_p(U)$ be the submodule of $\mathrm{Hom}_R(\mathscr{D}_1(U), \mathscr{D}_1(U))$ generated by the monomials $v_1 \circ ... \circ v_k$ $(1 \leqslant k \leqslant p)$, where each v_i is a derivation on $\mathscr{D}_1(U)$. Just as in the construction in (C), we have a sheaf $\mathscr{T}_p \to X$ defined, called the *sheaf of* p^{th}*-order contact vectors* c X (or *sheaf of differential operators of order* p). Again, $\mathscr{T}_p \to X$ is locally free of rank

$$\nu(n, p) = \sum_{k=1}^{p} \binom{n+k-1}{k}$$

because in a chart (θ, U) the iterated partial derivatives $D^\alpha (0 < |\alpha| \leqslant p)$ form a base over $\mathscr{D}(R)$. Thus we have a vector bundle $\pi: T_p(X) \to X$ canonically associated.

Its dual bundle $J^p(X) = \mathrm{Hom}_R(T_p(X), R) \to X$ is called the *bundle of* p*-jets of functions on* X Note that the fibre of $J^p(X)$ over $x \in X$ is $\mathscr{D}(R)/Z_x^p$, where $Z_x^p = \{\phi \in \mathscr{D}(R): \text{all derivatives of orders} \leqslant p \text{ of } \phi \text{ are } 0 \text{ at } x\}$. The natural map $j^p: \mathscr{D}(R) \to \mathscr{D}(J^p(X))$ such that $j^p(\phi)$ assigns to each $x \in X$ its coset in $\mathscr{D}(R)/Z_x^p$ is called the pth *jet extension*.

Again, if $f: X \to Y$ is a map of manifolds, we have induced f-homomorphisms $T_p(f): T_p(X) \to T_p$ and $J^p(f): J^p(Y) \to J^p(X)$.

(E) *Exercise.* Throughout the past three sections we have constructed many functors on various categories; e.g.:

(1) $\mathscr{C}$ mapping the category of sheaves of R-modules over a space to the category of R-modules over a space to the category of R-modules. Also $\mathscr{D}$.
(2) The functor mapping vector bundles over a space X to sheaves of R-modules over X; its inve functor (which is exact, in a sense easily made precise).
(3) The various products $\otimes, \wedge, \odot$, in the category of vector bundles over X.
(4) T_p mapping the category of smooth manifolds to the category of smooth vector bundles.

Organize these sections categorically, and discuss the relations between these functors.

4. CERTAIN OPERATORS ON VECTOR BUNDLES

(A) **Definitions.** Let $\xi: E \to X$ be a vector bundle over X. A *linear connection* on E is a map

$$\nabla: \mathscr{D}(T(X)) \times \mathscr{D}(E) \to \mathscr{D}(E)$$

written $\nabla(u, \phi) = \nabla_u \phi$, such that:

(1) the map $u \to \nabla_u \phi$ is $\mathscr{D}(R)$-linear for each $\phi \in \mathscr{D}(E)$;
(2) the map $\phi \to \nabla_u \phi$ is R-linear for each $u \in \mathscr{D}(T(X))$;
(3) $\nabla_u(\gamma\phi) = (\nabla_u \gamma)\phi + \gamma \nabla_u \phi$ for each $\phi \in \mathscr{D}(E)$, $\gamma \in \mathscr{D}(R)$ where $\nabla_u \gamma = u(\gamma)$. We shall refer to $\nabla_u \phi$ as the *covariant derivative of* ϕ *with respect to* u.

The *curvature* of ∇ is the map

$$R : \Lambda^2 \mathscr{D}(T(X)) \times \mathscr{D}(E) \to \mathscr{D}(E)$$

defined by

$$R(u_1, u_2)\phi = \nabla_{u_1} \nabla_{u_2} \phi - \nabla_{u_2} \nabla_{u_1} \phi - \nabla_{[u_1, u_2]} \phi$$

$$= -R(u_2, u_1)\phi$$

The curvature is $\mathscr{D}(R)$-linear in each variable u_1, u_2, ϕ. The *covariant differential* is the map

$$\nabla : \mathscr{D}(E) \to \mathscr{D}(T^*(X) \otimes E) = \mathscr{D}(\mathrm{Hom}(T(X), E))$$

given by $(\nabla\phi)u = \nabla_u\phi$. If ∇^E and ∇^F are connections on the vector bundles E and F, we define the connection ∇ on $E \oplus F$ by

$$\nabla_u(\phi \oplus \psi) = \nabla^E_u(\phi) \oplus \nabla^F_u(\psi) \qquad \text{for } u \in \mathscr{D}(T(X)),\ \ \phi \in \mathscr{D}(E),\ \ \psi \in \mathscr{D}(F)$$

Also the connection ∇ on $E \otimes F$, characterized by

$$\nabla_u(\phi \otimes \psi) = \nabla^E_u(\phi) \otimes \psi + \phi \otimes \nabla^F_u(\psi)$$

In particular, we have induced connections on $\otimes^p E$, $\Lambda^p E$, $\odot^p E$. Also the connection ∇^* on the dual bundle E^*, given by

$$\nabla^*_u(\psi)v = \nabla_u(\psi v) - \nabla_u(\psi v) - \psi\nabla_u(v) \qquad \text{for } \psi \in \mathscr{D}(E^*),\ v \in \mathscr{D}(E)$$

where again $\nabla_u(\psi v) = u(\psi v)$. More generally, define the connection ∇ on the space $L^p(E, F)$ $= \mathrm{Hom}(\otimes^p E, F)$ of p-linear maps $E \to F$ by

$$(\nabla_u\phi)(v_1, \ldots, v_p) = \nabla^F_u(\phi(v_1, \ldots, v_p)) - \sum_{i=1}^{p} \phi(v_1, \ldots, \nabla^E_u(v_i), \ldots, v_p)$$

Similarly for $\mathrm{Hom}(\Lambda^p E, F)$, $\mathrm{Hom}(\odot^p E, F)$.

Remark. Throughout this section it would be possible to generalize the notions and results, replacing $\mathscr{D}(T(X))$ by $\mathscr{D}(S)$, where S is a vector sub-bundle of T(X) which is completely integrable, i.e. such that $\mathscr{D}(S)$ is a Lie subalgebra of $\mathscr{D}(T(X))$. By Frobenius' theorem these are just the leaved structures on X.

B) **Definition.** If $\xi : E \to X$ is a vector bundle, the *Lie derivative* is the map

$$\theta : \mathscr{D}(T(X)) \times L^p(\mathscr{D}(T^*(X)), \mathscr{D}(E)) \to L^p(\mathscr{D}(T^*(X)), \mathscr{D}(E))$$

defined by

$$(\theta_u\alpha)(\phi_1, \ldots, \phi_p) = \sum_{i=1}^{p} \alpha(\phi_1, \ldots, \theta_u(\phi_i), \ldots, \phi_p) \qquad \text{for } \phi_i \in \mathscr{D}(T^*(X))$$

where $(\theta_u \phi)v = \phi([u, v])$ for $v \in \mathscr{D}(T(X))$. Similarly for alternating and symmetric p-linear α. The Lie derivative is R-linear in each variable and satisfies $\theta_u(\gamma\alpha) = (\nabla_u \gamma)\alpha + \gamma\theta_u \alpha$ for $\gamma \in \mathscr{D}(R)$, $\alpha \in L^p(\mathscr{D}(T^*(X)), \mathscr{D}(E))$; however, $\theta_{\gamma u}(\alpha) \neq \gamma\theta_u(\alpha)$ in general, so that θ is not usually a connection.

Now suppose ∇ is a connection on $\xi: E \to X$. Then we can define the *dual Lie derivative (relative to ∇):*

$$\theta: \mathscr{D}(T(X)) \times L^p(\mathscr{D}(T(X)), \mathscr{D}(E)) \to L^p(\mathscr{D}(T(X)), \mathscr{D}(E))$$

by

$$\theta_u(\alpha)(u_1, \ldots, u_p) = \nabla_u \alpha(u_1, \ldots, u_p) - \sum_{i=1}^{p} \alpha(u_1, \ldots, [u, u_i], \ldots, u_p)$$

Similarly for alternating and symmetric p-linear α.

(C) If $\xi: E \to X$ is a vector bundle, let $\mathscr{A}^p(E) = AL^p(\mathscr{D}(T(X)), \mathscr{D}(E))$ be the space of alternating p-linear maps from $\mathscr{D}(T(X)) \to \mathscr{D}(E)$; we agree that $\mathscr{A}^0(E) = \mathscr{D}(E)$, and that $\mathscr{A}^p(E) = 0$ for $p < 0$ and for $p > n$. Thus $\mathscr{A}(E) = \sum \mathscr{A}^p(E)$ is a graded commutative vector space over R with only finitely many non-zero summands. We call the elements of $\mathscr{A}^p(E)$ the E-*valued differential* p-*forms on* X.

Definition. Let ∇ be a connection on $\xi: E \to X$. The *exterior differential (relative to ∇)* is the map $d: \mathscr{A}^p(E) \to \mathscr{A}^{p+1}(E)$ defined by

$$(d\phi)(u_1, \ldots, u_{p+1}) = \sum_{i=1}^{p+1} (-1)^{i+1} \nabla_{u_i} \phi(u_1, \ldots, \hat{u}_i, \ldots, u_{p+1})$$

$$+ \sum_{i<j} (-1)^{i+j} \phi([u_i, u_j], u_1, \ldots, \hat{u}_i, \ldots, \hat{u}_j, \ldots, u_{p+1})$$

where $u_i \in T(X)$.

Given $u \in \mathscr{D}(T(X))$, let us define the *interior product* $i_u: \mathscr{A}^p(E) \to \mathscr{A}^{p-1}(E)$ by $(i_u \phi)(u_1, \ldots, u_{p-1}) = \phi(u, u_1, \ldots, u_{p-1})$.

The following formal computations are left as an exercise.

Proposition.

$$\theta_u i_v - i_v \theta_u = i_{[u,v]}$$

$$i_u i_u = 0$$

$$\theta_u = d i_u + i_u d$$

$$\theta_v \theta_v - \theta_v \theta_u = \theta_{[u,v]} + R(u, v) \quad \textit{for all} \quad u, v \in \mathscr{D}(T(X)).$$

In particular,

$$\theta_u i_u = i_u \theta_u$$

(D) Let E_1, E_2, F be vector bundles over X with linear connections $\nabla^1, \nabla^2, \nabla$ respectively. Suppose we are given a bilinear pairing $E_1 \times E_2 \to F$, written $(v_1, v_2) \to v_1 \cdot v_2$, such that $\nabla_u(\phi_1 \cdot \phi_2) = (\nabla^1_u \phi_1)\cdot \phi_2 + \phi_1 \cdot (\nabla^2_u \phi_2)$ for all $u \in \mathscr{D}(T(X))$, $\phi_i \in \mathscr{D}(E_i)$. Then we have the induced bilinear pairing

$$\wedge : \mathscr{A}^p(E_1) \times \mathscr{A}^q(E_2) \to \mathscr{A}^{p+q}(F)$$

relative to which

$$\nabla_u(\phi \wedge \psi) = (\nabla^1_u \phi) \wedge \psi + \phi \wedge (\nabla^2_u \psi)$$

$$\theta_u(\phi \wedge \psi) = (\theta_u \phi) \wedge \psi + \phi \wedge (\theta_u \psi)$$

$$i_u(\phi \wedge \psi) = (i_u \phi) \wedge \psi + (-1)^p \phi \wedge (i_u \psi)$$

$$d(\phi \wedge \psi) = (d\phi) \wedge \psi + (-1)^p \phi \wedge (d\psi) \qquad \text{(proofs by induction).}$$

For instance, taking $E_1 = \mathrm{Hom}_R(E, E)$, $E_2 = E = F$ and the natural pairing $\mathrm{Hom}_R(E, E) \times E \to E$, we have the

Proposition. *For any connection* ∇ *on* E,

$$\theta_u d - d\theta_u = i_u R \qquad \text{i.e.} \quad \theta_u d(\phi) - d\theta_u(\phi) = (i_u R)\phi$$

$$d^2 = R \qquad \text{i.e.} \quad d^2\phi = R \wedge \phi$$

$$dR = 0$$

(E) *Example 1.* $E = X \times R$. Then $\nabla_u \phi = u(\phi) = d\phi \cdot u$ is a connection. Its curvature $R(u, v) = 0$ for all $u, v \in \mathscr{D}(T(X))$, by definition of the bracket $[u, v]$. The Lie derivative $\theta_u(v) = [u, v]$; for $\phi \in \mathscr{D}(T^*(X))$ we have $(\theta_u \phi)v = d(\phi \cdot v)u - \phi([u, v])$. The product formula for the exterior differential in $\mathscr{A}^p(R)$ is obtained through the exterior product $\wedge : \Lambda^p T^*(X) \times \Lambda^q T^*(X) \to \Lambda^{p+q} T^*(X)$. In particular,

(1) $d\phi$ is the differential of $\phi \in \mathscr{D}(R)$;
(2) $d(\phi \wedge \psi) = d\phi \wedge \psi + (-1)^p \phi \wedge d\psi$ for all $\phi \in \mathscr{A}^p(R)$, $\psi \in \mathscr{A}^q(R)$;
(3) $d^2\phi = 0$ for all $\phi \in \mathscr{A}^p(R)$. In fact, these properties characterize d.

Note also that

$$\theta_u d = d\theta_u$$

$$\theta_u \theta_v - \theta_v \theta_u = \theta_{[u, v]}$$

Example 2. Let $E_1 = E$, $E_2 = E^*$, $F = X \times R$ with the natural pairing. Then given any connection ∇ on E, the dual connection and the differential (on functions) are related in the desired way – by definition of dual connection:

$$d(\phi \cdot \psi)u = (\nabla_u \phi)\cdot \psi + \phi \cdot (\nabla_u \psi) \qquad \text{for} \quad \phi \in \mathscr{D}(E), \quad \psi \in \mathscr{D}(E^*)$$

Example 3. Take $E = T(X)$ (or more generally any completely integrable sub-bundle). A connection ∇ on $T(X)$ is called a *connection on* X. A special phenomenon occurs because of the existence of the identity map $\alpha:\mathscr{D}(T(X)) \to \mathscr{D}(T(X))$. Its exterior differential

$$d\alpha(u_1, u_2) = \nabla_{u_1}\alpha(u_2) - \nabla_{u_2}\alpha(u_1) - \alpha([u_1, u_2])$$

The map $T:\mathscr{D}(T(X)) \times \mathscr{D}(T(X)) \to \mathscr{D}(T(X))$ given by

$$T(u_1, u_2) = \nabla_{u_1}(u_2) - \nabla_{u_2}(u_1) - [u_1, u_2] = -T(u_2, u_1)$$

is called *the torsion of the connection* on X, and ∇ is said to be *symmetric* if $T = 0$.

Bianchi identities:

(1) $(dT)(u, v, w) = R(u, v)w + R(v, w)u + R(w, u)v;$

(2) $(\nabla_u R)(v, w) + (\nabla_v R)(w, u) + (\nabla_w R)(u, v) = 0$ for all $u, v, w \in \mathscr{D}(T(X))$, if $T = 0$.

The connection ∇ induces a connection ∇^p on $\otimes^p T(X)$; the tensor pairing $\otimes^p T(X) \times \otimes^q T(X$ $\to \otimes^{p+q} T(X)$ has the required derivation property $\nabla^{p+q}_u(\phi \otimes \psi) = (\nabla^p_u \phi) \otimes \psi + \phi \otimes (\nabla^q_u \psi)$.

The *covariant differential* on X is the map $\mathscr{D}(\otimes^p T^*(X)) \to \mathscr{D}(\otimes^{p+1} T^*(X))$ characterized by

$$(\nabla\phi)(v_0, \ldots, v_p) = \sum_{i=0}^{p} \nabla_{v_i}\phi(v_0, \ldots, \hat{v}_i, \ldots, v_p)$$

Proposition. *Let ∇ be a symmetric connection on X and ∇^E a connection on the vector bundle $\xi: E \to X$. Then we have the following formula for the exterior differential* $d:\mathscr{A}^p(E) \to \mathscr{A}^{p+1}(E$

$$(d\phi)(u_1, \ldots, u_{p+1})$$

$$= \sum_{i=1}^{p+1} (-1)^{i+1} \left\{ \nabla^E_{u_i}\phi(u_1, \ldots, \hat{u}_i, \ldots, u_{p+1}) - \sum_{j=1}^{p+1} \phi(u_1, \ldots, \hat{u}_j, \ldots, \nabla_{u_i}(u_j), \ldots, u_{p+1}) \right\}$$

(If the right member is denoted by $(d'p)(u_1, \ldots, u_{p+1})$, then $i_u d' + d' i_u = \theta_u$. But the exterior differential is the unique operator with that property.)

Example 4. A pth-order linear connection on X is a connection ∇ on $T_p(X)$ such that $\nabla_u \phi = u$ for all $u \in \mathscr{D}(T(X))$, $\phi \in T_{p-1}(X)$. The operators constructed above can be viewed in this conte

(F) For each open set U of the n-manifold X let $\mathscr{S}(U)$ be the totality of charts with domain U. If V is open in U we define $\mathscr{S}(U) \to \mathscr{S}(V)$ by restriction, whence we have the sheaf (of sets) $\mathscr{S} \to X$.

Let G be a subgroup of the general linear group L_n of linear automorphisms of R^n. Say tha $\theta, \theta' \in \mathscr{S}(U)$ are G-*equivalent* if $\theta' \circ \theta^{-1}$ has its differential belonging to G for every point of U. Let $\mathscr{S}(U)/G$ denote these equivalence classes and $\mathscr{S}/G \to X$ the corresponding sheaf.

Definition. An *(integrable)* G-*structure on* X is a section of $\mathscr{S}/G$.

For instance, *an orientation of* X is an L_n^+-structure, where $L_n^+ = \{\lambda \in L_n : \det(\lambda) > 0\}$; *an oriented manifold* is a manifold with a particular orientation. A *Riemannian structure on* X is an O_n-structure, where O_n is the subgroup of orthogonal matrices (λ^{-1} = transpose λ); a *Rieman manifold* is a manifold with a particular Riemannian structure. Most of differential geometry centres around the theory of G-structures.

Remark. The viewpoint in the preceding definition admits a far-reaching generalization, as emphasized by Spencer [19]. Let $\mathscr{M}$ be a category of topological spaces and continuous maps, closed under unions and restrictions to open sets. Given any topological space X, we define the sheaf $\mathscr{S} \to X$ whose space $\mathscr{S}(U)$ of sections over an open set U of X is the totality of homeomorphisms θ of U onto an object $\theta(U)$ of $\mathscr{M}$. Say $\theta, \theta' \in \mathscr{S}(U)$ are equivalent (written $\theta \sim \theta'$) if $\theta' \circ \theta^{-1}$ is a map of $\mathscr{M}$; let $\mathscr{S}(U)/(\sim)$ denote the space of these equivalence classes. An *$\mathscr{M}$-structure on* X is a section of the sheaf

$$\mathscr{S}/(\sim) \to X$$

For instance, taking for $\mathscr{M}$ the totality of domains of R^n, whose maps are the diffeomorphisms of one domain to another, an $\mathscr{M}$-structure is a differential structure.

Remark. A large class of vector bundles can be constructed as follows: Let $\Theta \in \mathscr{S}/G(X)$ be a G-structure on X, and define the sheaf $\mathscr{P} \to X$ whose space of sections over U is $\mathscr{P}(U) = \{(\theta' \circ \theta^{-1})_*: U \to G \text{ with } \theta, \theta' \in \Theta(U)\}$.

If F is a finite-dimensional vector space on which G operates on the left (i.e. F is a left G-module), then G operates on each $\mathscr{P}(U) \times F$ by $((\theta' \cdot \theta^{-1})_* f) \cdot g = ((\theta' \cdot \theta^{-1})_* g, g^{-1} f)$; clearly $(\theta \cdot \theta^{-1})_* \cdot g = (\theta \cdot \theta^{-1} \cdot g)_*$, so that the operation is well defined. Let $\mathscr{P}(U) \times_G F$ be the orbit space, and $\mathscr{P} \times_G F \to X$ the resulting sheaf. This is easily verified to be a sheaf of vector spaces, locally free of rank = dim F.

5. RIEMANNIAN STRUCTURES

In this section we present briefly the few ideas that we shall need from Riemannian geometry: Riemannian bundles, dual differential forms, integration of scalar densities, the Levi-Cività connection.

(A) Let E be a Euclidean n-dimensional vector space with inner product $\langle\,\rangle$ or $\langle\,\rangle_E$. We have the canonical isometric isomorphism $P: E \to E^*$ given by $P(x)y = \langle x, y\rangle = P(y)x$.

If E and F are both Euclidean spaces, then we define inner products on the direct sum $E \oplus F$ and the tensor product $E \otimes F$, characterized by

$$\langle x_1 \oplus y_1, x_2 \oplus y_2\rangle = \langle x_1, x_2\rangle_E + \langle y_1, y_2\rangle_F$$

$$\langle x_1 \otimes y_1, x_2 \otimes y_2\rangle = \langle x_1, x_2\rangle_E \times \langle y_1, y_2\rangle_F$$

respectively. In particular, we have an induced inner product in each $\otimes^p E$, $\otimes^p E^*$.

Introduce the projection map $A: \otimes^p E \to \wedge^p E$ by

$$A(x_1 \otimes \dots \otimes x_p) = \sum_\sigma e_\sigma x_{\sigma(1)} \otimes \dots \otimes x_{\sigma(p)}/p!$$

summed over all permutations σ of $(1, \dots, p)$, where e_σ is the sign of σ. The elements $x_1 \wedge \dots \wedge x_p = A(x_1 \otimes \dots \otimes x_p)$ generate $\wedge^p E$. We define the inner product on $\wedge^p E$ by setting

$$\begin{aligned}\langle x_1 \wedge \dots \wedge x_p, y_1 \wedge \dots \wedge y_p\rangle &= \langle A(x_1 \otimes \dots \otimes x_p), A(y_1 \otimes \dots \otimes y_p)\rangle \\ &= \langle x_1 \otimes \dots \otimes x_p, A(y_1 \otimes \dots \otimes y_p)\rangle \\ &= \det\langle x_i, y_i\rangle\end{aligned}$$

Similarly for the inner product on $\odot^p E$.

(B) **Definition.** An *orientation* in an n-dimensional vector space E is a choice of ray in the one-dimensional space $\Lambda^n E$; if E is Euclidean, this is equivalent to a choice of $\alpha_0 \in \Lambda^n E$ with $|\alpha_0| = 1$, using the norm in $\Lambda^n E$ induced from the inner product in E.
The duality isomorphism $\mathscr{D}: \Lambda^p E^* \to \Lambda^{n-p} E$ is defined by $\mathscr{D}(\varphi)\psi = (\varphi \wedge \psi)\alpha_0$ for all $\psi \in \Lambda^{n-p}$ E.
The *star isomorphism* $*: \Lambda^p E^* \to \Lambda^{n-p} E^*$ is the composition $* = P\mathscr{D}$, where $P: \Lambda^{n-p} E \to \Lambda^{n-p} E^*$ is the isomorphism of Euclidean spaces induced from $P: E \to E^*$.

Then $*$ *is self adjoint:* $\langle *\varphi, \psi\rangle = (-1)^{p(n-p)}\langle \varphi, *\psi\rangle$ for all $\varphi \in \Lambda^p E^*$, $\psi \in \Lambda^{n-p} E^*$; namely

$$\langle P\mathscr{D}(\varphi), \psi\rangle = \mathscr{D}(\varphi)\psi = (\varphi \wedge \psi)\alpha_0 = (-1)^{p(n-p)}\mathscr{D}(\psi)\varphi = (-1)^{p(n-p)}\langle P\mathscr{D}(\psi), \varphi\rangle$$

Next, $*$ *is an isometry:* $\langle *\varphi, *\psi\rangle = \langle \varphi, \psi\rangle$; namely, P is an isometry, and so is $\mathscr{D}$ (as we can see by introducing a base in E).

It follows that $*$ *is an involution:* $**\varphi = (-1)^{p(n-p)}\varphi$ for all $\varphi \in \Lambda^p E^*$. The number $1 \in \Lambda^0 E^*$ is mapped to $*1 \in \Lambda^n E^*$, characterized by $(*1)\alpha_0 = 1$. The $*$ operation does not preserve the algebra structure in ΛE^*; however, for all $\varphi, \psi \in \Lambda^p E^*$ we have

$$\varphi \wedge (*\psi) = \langle \phi, \psi\rangle * 1 = \psi \wedge (*\varphi)$$

because

$$(\varphi \wedge (*\psi))\,\alpha_0 = \mathscr{D}(\varphi)P\mathscr{D}(\psi) = \langle \mathscr{D}\varphi, \mathscr{D}\psi\rangle = \langle \varphi, \psi\rangle$$

If E and F are both Euclidean spaces and E is oriented, we extend the definition of the star isomorphism to $*: \Lambda^p E^* \otimes F \to \Lambda^{n-p} E^* \otimes F^*$ by setting $*(\varphi \otimes y) = (*\varphi) \otimes Py$. The above properties of $*$ continue to hold; for instance, using the natural pairing $F \times F^* \to R$ to define the exterior product,

$$(\varphi \otimes y) \wedge (*(\psi \otimes z)) = (\varphi \wedge (*\psi))\,(yP(z)) = \langle \varphi, \psi\rangle\langle y, z\rangle * 1 = \langle \varphi \otimes y, \psi \otimes z\rangle * 1$$

(C) **Definition.** Let $\xi: E \to X$ be a vector bundle. A *Riemannian metric* in E is an element $g \in \mathscr{D}(\odot^2 E^*)$ such that each g(x) is positive definite; i.e. each g(x) is a symmetric bilinear form on the fibre E_x, and we require that g(x) be a Euclidean structure on E_x. Write $g(x)(u, v) = \langle u, v\rangle$. A Riemannian metric in E determines the bundle isomorphism $P: E \to E^*$.

A *Riemannian structure* on the manifold X is a Riemannian metric in the tangent bundle $T(X) \to X$; we shall call X a *Riemannian manifold* if it has a Riemannian structure. That this definition is equivalent to the one given in Section 4F (in fact, the assertion that there is a natural bijective correspondence between the Riemannian metrics on T(X) and the O_n-structures on X) is left as an exercise (of which we shall not make use).

An application of a partition of unity on X yields the

Lemma. *Every vector bundle admits a Riemannian metric.*

Remark. Suppose that X is a Riemannian n-manifold. Then the n-covectors of length one form an orientable submanifold $\widetilde{X}$ of the total space of $\pi: \Lambda^n T^*(X) \to X$, and $\pi: \widetilde{X} \to X$ is a two-leaved covering map. Over each orientable component of X there are precisely two components of $\widetilde{X}$, which (viewed as sections over X) determine the orientations of X.

(D) We now extend the definition of the star isomorphism to vector bundles. That concept requires some sort of orientation – which can be achieved either by: (1) restricting attention to oriented manifolds; or (2) twisting the differential forms. We introduce the machinery for (2) which should not obscure the path of a reader wishing to follow (1).

Definition. Let X be an n-manifold. The *bundle of twisted real numbers of* X is the vector bundle of fibre dimension 1 constructed as in Section 4F through the action $L_n \times R \to R$ given by $(\lambda, r) \to \text{sign}(\det \lambda)r$. To conform with our convention of writing R for the trivial line bundle $X \times R$ we let $\widetilde{R} \to X$ denote the bundle of twisted real numbers. If X is orientable, then a choice of orientation determines a bundle isomorphism $\widetilde{R} \to R$; X is non-orientable if and only if $\widetilde{R} \to X$ is a non-trivial bundle. (The twisted real numbers are a special case of what in topology is called a system of local coefficients.)

Given a vector bundle $\xi : E \to X$ we shall speak of $E \otimes \widetilde{R} \to X$ as the *twisted bundle* of E and of its sections $\varphi \in \widetilde{\mathscr{D}}(E) = \mathscr{D}(E \otimes \widetilde{R})$ as *twisted sections* of E. In particular, we have the notion of a twisted p-form (called a p-form of odd kind by de Rham). A twisted n-form is usually called a *scalar density* on X.

Exercise. Define the notion of an *oriented map* $f : X \to Y$ from one manifold to another. Describe the behaviour of twisted forms under an oriented map.

Definition. Let X be Riemannian n-manifold, and $\xi : E \to X$ a vector bundle with Riemannian metric. We define the *star isomorphism*

$$* : \Lambda^p T^*(X) \otimes E \to \Lambda^{n-p} T^*(X) \otimes \widetilde{R} \otimes E^*$$

as follows. If X is oriented then we define $*$ by using the definition in (B) on the fibres. If X is not oriented then we take any chart (θ, U) at x and define

$$* : \Lambda^p X^*(x) \otimes E_x \to \Lambda^{n-p} X^*(x) \otimes E_x^*$$

using the orientation in U provided by θ. Another chart (θ', U') at x defines another $*' = \text{sign}(\theta' \cdot \theta^{-1})*$, where $\text{sign}(\theta' \cdot \theta^{-1})$ is the sign of the Jacobian of $\theta' \cdot \theta^{-1}$ at x; it follows that the star of an E-valued p-form is well defined as an E^*-valued twisted (n-p)-form.

From the natural pairings of bundles

$$R \times \widetilde{R} \to \widetilde{R},\ \widetilde{R} \times \widetilde{R} \to R,\ E \times E^* \to R$$

we obtain pairings of the type

$$\wedge : (\Lambda^p T^*(X) \otimes E) \times (\Lambda^q T^*(X) \otimes \widetilde{R} \otimes E^*) \to \Lambda^{p+q} T^*(X) \otimes \widetilde{R}$$

In particular, we have

$$(\Lambda^p T^*(X) \otimes E) \times (\Lambda^p T^*(X) \otimes E) \to \Lambda^n T^*(X) \otimes \widetilde{R}$$

given by

$$(\varphi, \psi) \to \varphi \wedge * \psi = \varphi * \psi$$

Definition. The *volume density* of the Riemannian n-manifold X is the twisted n-form $*1$. If X is oriented then we view $*1$ as an n-form. Note that if (θ, U) is a chart at $x \in X$ and $e^{1 \cdots n} = e^1 \wedge \ldots \wedge e^n$ is the unit n-covector of R^n defining its orientation, then $*1(x)$ has the representation $|\Lambda^n \theta(x) e^{1 \cdots n}|$.

(E) Let X be any n-manifold.

Definition. Given any compact twisted n-form $w \in \mathscr{D}_0(\Lambda^n T^*(X) \otimes \widetilde{R})$, we define its *integral* $\int_X w$ as follows.

Suppose first that spt(w) is contained in a chart (θ, U) of X. Then θ orients U, and $(\Lambda^n \theta^{-1})w$ is a compact n-form on R^n. We define

$$\int_X w = \int_{R^n} (\Lambda^n \theta^{-1})w$$

If (θ', U') is another chart containing spt(w), then $\theta' \cdot \theta^{-1}$ is a diffeomorphism $\theta(U \cap U') \to \theta'(U \cap U')$. By the transformation of the integral formula

$$\int_{R^n} (\Lambda^n \theta^{-1})w = \text{sign}(\theta' \cdot \theta^{-1}) \int_{R^n} (\Lambda^n \theta'^{-1})w$$

where $\text{sign}(\theta' \cdot \theta^{-1})$ is the sign of the Jacobian of $\theta' \cdot \theta^{-1}$. It follows that the value of the integral is independent of the choice of chart.

For any compact w let (γ_i) be a finite partition of unity whose supports are contained in charts whose union covers spt(w). We define

$$\int_X w = \sum_i \int_X \gamma_i w$$

If (β_j) were another such partition of unity, then we mix the partitions:

$$\sum_i \int_X \gamma_i w = \sum_i \int_X \sum_j \beta_j \gamma_i w = \sum_j \int_X \sum_i \gamma_i \beta_j w = \sum_j \int_X \beta_j w$$

Thus the value of the integral is independent of the choice of partition.

Remark. If X is oriented, then we can restrict attention to those charts preserving orientation, so that every $\text{sign}(\theta' \cdot \theta^{-1}) = 1$ in the transformation of integral formula. Thus we obtain the definition of the integral over X if $w \in \mathcal{D}_0(\Lambda^n T^*(X))$.

Theorem. *If* w *is a compact twisted* (n−1) *form on* X, *then* dw *is also twisted, and* $\int_X dw = 0$.

Proof. It suffices to suppose that spt(w) is contained in a chart; for if (γ_i) is a partition of unity on X whose supports are contained in charts whose union covers spt(w), then

$$\sum d(\gamma_i w) = \left(\sum_i d\gamma_i\right) \wedge w + \sum_i \gamma_i \wedge dw = dw$$

Thus we have reduced the theorem to that of a compact (n − 1) form on R^n, an elementary case.

(F) Suppose now that ∇ is a connection in the bundle $\xi : E \to X$; in particular, ∇ induces a connection in E^* and in $\odot^2 E^*$.

Definition. *A Riemannian bundle* $\xi: E \to X$ consists of a pair (∇, g) on E, where ∇ is a connection, g is a Riemannian metric, and $\nabla_u g = 0$ for all $u \in \mathscr{D}(T(X))$, i.e.

$$\nabla_u \langle \phi, \psi \rangle = \langle \nabla_u \phi, \psi \rangle + \langle \phi, \nabla_u \psi \rangle$$

where once again we have written

$$\nabla_u \langle \phi, \psi \rangle = u(\langle \phi, \psi \rangle)$$

It is natural to determine when a Riemannian metric on $\xi: E \to X$ induces a Riemannian bundle structure. There are two standard situations: the fundamental theorem of Riemannian geometry which follows, and a class of examples of holomorphic vector bundles.

Theorem (Levi-Cività). *If* g *is a Riemannian metric on* $T(X) \to X$, *then there is a unique symmetric connection* ∇ *on* T(X) *such that* $\nabla g = 0$.

Proof. First of all, if (∇, g) is any Riemannian bundle structure on T(X) and we calculate

$$g(\nabla_u v, w) + g(\nabla_v w, u) - g(\nabla_w u, v)$$

then we find the identity

$$2g(\nabla_u v, w) = ug(v, w) + vg(w, u) - wg(u, v) - g(u, [v, w]) + g(v, [w, u]) + g(w, [u, v])$$

$$- g(u, T(v, w)) + g(v, T(w, u)) + g(w, T(u, v))$$

for all $u, v, w \in \mathscr{D}(T(X))$. In particular, if ∇ is symmetric the last three terms vanish; because g is non-singular, we see that $\nabla_u v$ is completely determined by g, thereby giving both the uniqueness of ∇ and its existence (with the required properties).

Remark. Observe that this theorem is valid for *pseudo-Riemannian metrics on* X, i.e. for $g \in \mathscr{D}(\odot^2 T^*(X))$ such that each g(x) is non-singular at every $x \in X$.

6. DIFFERENTIAL OPERATORS

The viewpoint (using covariant differentials) in this section is that of Singer [18].

(A) Let X be a Riemannian n-manifold with Levi-Civitå connection D as in Theorem 5F. Let $\xi: E \to X$ be a vector bundle with connection ∇. We take the tensor product connection on $(\otimes^k T^*(X)) \otimes E$ as in Section 4A, and still denote it by ∇. The covariant differential $\nabla: \mathscr{D}(E) \to \mathscr{D}(T^*(X) \otimes E)$ can then be iterated:

$$\nabla^k : \mathscr{D}(E) \to \mathscr{D}(\otimes^k T^*(X) \otimes E)$$

We set ∇^0 = identity on $\mathscr{D}(E)$.

If $\eta: F \to X$ is another vector bundle, we define the *coefficient bundle* $\mathrm{Hom}(\odot^k T^*(X), \mathrm{Hom}(E, F)) = \mathrm{Hom}(\odot^k T^*(X) \otimes E, F)$. We shall sometimes treat its elements as belonging to $\mathrm{Hom}(\otimes^k T(X) \otimes E, F)$ using the canonical projection $\otimes^k T^*(X) \to \odot^k T^*(X)$.

As in Proposition 2D we have the identification:

$$\mathscr{D}(\mathrm{Hom}_R(\otimes^k T^*(X) \otimes E, F) = \mathrm{Hom}_{\mathscr{D}(R)}(\mathscr{D}(\otimes^k T^*(X) \otimes E), \mathscr{D}(F))$$

Definition. *A differential operator from E to F homogeneous of order* k is a homomorphism $A:\mathscr{D}(E)\to\mathscr{D}(F)$ which factor through ∇^k for some coefficient section $\underline{a}$:

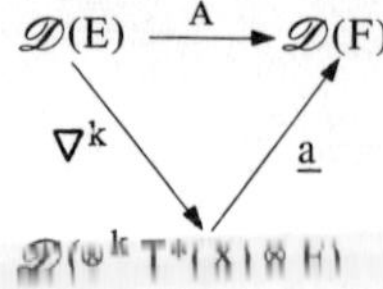

If we replace $\otimes^k T^*(X)$ by

$$\sum_{p=0}^{k} \otimes^p T^*(X)$$

and ∇^k by

$$\sum_{p=0}^{k} \nabla^p$$

we obtain the notion of *differential operator from* E *to* F *of order* k. The totality of differential operators of all orders forms a vector subspace Diff(E, F) of $\mathrm{Hom}_R(\mathscr{D}(E),\mathscr{D}(F))$ and a $\mathscr{D}(R)$-submodule, filtered by $\mathrm{Diff}^k(E,F)$, the space of operators of orders $\leqslant k$.

It is important to note that differential operators are local, i.e. $\mathrm{spt}(A\varphi)\subset\mathrm{spt}(\varphi)$ for all $\varphi\in\mathscr{D}(E)$; in fact that property goes a long way toward characterizing differential operators in the space $\mathrm{Hom}_R(\mathscr{D}(E),\mathscr{D}(F))$.

(B) **Definition.** Suppose that $A\in\mathrm{Diff}^k(E,F)$ has the form

$$A=\sum_{p=0}^{k}\underline{a}_p\nabla^p$$

with leading coefficient $\underline{a}_k\neq 0$. We define the *symbol* of A as the map $\sigma_A:T^*(X)\to\mathrm{Hom}(E,F)$ covering the projection $\pi:T^*(X)\to X$ given by $\sigma_A(w)=(-1)^{k/2}\underline{a}_k(\odot^k w)$ for all $w\in T^*(X)$:

$$\begin{array}{ccc} & & \mathrm{Hom}(E,F)\\ & \overset{\sigma_A}{\nearrow} & \downarrow\\ T^*(X) & \longrightarrow & X\end{array}$$

We make no comment that the factor $(-1)^{k/2}$ may take us momentarily out of the real domain.

Property 1. The composition $B\circ A$ of differential operators is a differential operator, and $\sigma_{B\circ A}=\sigma_B\circ\sigma_A$, where the right member is induced by the bundle pairing $\mathrm{Hom}(F_1,F_2)\times\mathrm{Hom}(E,F_1)\to\mathrm{Hom}(E,F_2)$.

Property 2. If A, B, A + B $\in \mathrm{Diff}^k(E, F)$ but are not in $\mathrm{Diff}^{k-1}(E, F)$, then $\sigma_{A+B} = \sigma_A + \sigma_B$.

Property 3. Each $u \in \mathscr{D}(TX)$ defines an element in $\mathrm{Diff}^1(E, E)$ by ∇_u. The evaluation $\sigma(w)$ on $w \in T^*(X)$ of its symbol is scalar multiplication in $E_{\pi(w)}$ by $\sqrt{-1}\, w \cdot u$.

Definition. We say that A *is elliptic* if fibre dim E = fibre dim F, and each $\sigma_A(w): E_{\pi(w)} \to F_{\pi(w)}$ is an isomorphism.

(C) **Proposition.** For each $A \in \mathrm{Diff}^k(E, F)$ *there is a unique* $A^* \in \mathrm{Diff}^k(F^*, E^*)$ *such that for each* $\varphi \in \mathscr{D}(E)$, $\psi \in \mathscr{D}(F^*)$ *there is a twisted form* $\tau \in \widetilde{\mathscr{D}}(\Lambda^{n-1}T^*(X))$ *satisfying*

$$(A\varphi) \cdot \psi * 1 - \varphi \cdot (A^*\psi) * 1 = d\tau$$

Furthermore, $\mathrm{spt}(\tau) \subset \mathrm{spt}(\varphi)$, so that if $\varphi \in \mathscr{D}_0(E)$ has compact support we have by Theorem 5E

$$\int_X (A\varphi) \cdot \psi * 1 = \int_X \varphi(A^*\psi) * 1$$

Again, if X is oriented, then $\tau \in \mathscr{D}(\Lambda^{n-1}T^*(X))$.

A^* is called *the formal adjoint* of A; compare Lemma 3A of Part II. Note that its uniqueness is an immediate consequence of the integral formula.

Proof. If $A \in \mathrm{Diff}^0(E, F) = \mathscr{D}(\mathrm{Hom}(E, F))$, then we define A^* at each $x \in X$ as the algebraic adjoint $A_x^*: F_x^* \to E_x^*$ of A_x. Then $(A\varphi) \cdot \psi = \varphi \cdot (A^*\psi)$ at every point.

Consider next the special case that E = F and $A = \nabla_u$ for some $u \in \mathscr{D}(T(X))$, i.e. $A = \underline{a}\nabla$, where $\underline{a}: \mathscr{D}(T^*(X) \otimes E) \to \mathscr{D}(E)$ is given by $\underline{a}(w \otimes \varphi) = (w \cdot u)\varphi$. Then there is a unique $f_u \in \mathscr{D}(R)$ such that $d(*P^{-1}u) = f_u * 1$. We define $A^* = -\nabla_u^* - f_u$; then

$$(A\varphi) \cdot \psi - \varphi \cdot (A^*\psi) = u(\varphi \cdot \psi) + f_u(\varphi \cdot \psi)$$

whence

$$(A\varphi) \cdot \psi * 1 - \varphi \cdot (A^*\psi) * 1 = d(\varphi \cdot \psi) \wedge * P^{-1}u + (\varphi \cdot \psi) \wedge d(* P^{-1})u$$

$$= d[(\varphi \cdot \psi) \wedge * P^{-1}u]$$

The last equality follows from the derivation formulas of Part III, Section 4D.

For the next step we observe that if A, B are differential operators satisfying our first identity (for some τ_A, τ_B) then so is $B \circ A$, with $(B \circ A)^* = A^* \circ B^*$ (and $\tau_A + \tau_B$). Thus all linear combinations of compositions of 0th-order operators and ∇^p satisfy the identity. But these include all differential operators, whence the identity is verified in general.

Corollary. *If* E *and* F *are Riemannian bundles, then*

$$\int_X (A\varphi) * \psi = \int_X \varphi * (A^*\psi)$$

for any $\varphi \in \mathscr{D}(E)$, $\psi \in \mathscr{D}(F)$, *at least one of which is compact.* Note that we are interpreting $A^* \in \mathrm{Diff}(F, E)$, as we shall do consistently in case of Riemannian bundles.

Corollary. *If* $A \in \mathrm{Diff}^k(E, F)$, *then* $\sigma_{A^*} = (-1)^k \sigma_A^*$, *where the right member assigns to each* $w \in T^*(X)$ *the adjoint homomorphism of* $\sigma_A(w)$. It is sufficient to prove this for operators of order 0 and 1, where it is trivial.

Definition. If $\xi: E \to X$ is a Riemannian bundle, then $A \in \mathrm{Diff}^{2s}(E, E)$ is *strongly elliptic* if σ_A is positive definite at every point, i.e. for every non-zero $w \in T^*(X)$ and non-zero $v \in E_{\pi(w)}$ we have $\langle \sigma_A(w)\, v, v \rangle_{\pi(w)} > 0$. Note that the order of A is necessarily even if A is strongly elliptic; see Section 6A of Part II.

Proposition. *If* E *and* F *are Riemannian bundles with the same fibre dimension, the* $A \in \mathrm{Diff}^s(E, F)$ *is elliptic if and only if* $(-1)^s A^* A \in \mathrm{Diff}^{2s}(E, E)$ *is strongly elliptic.*

This is immediate, because $\sigma_{A^*A} = \sigma_{A^*}\sigma_A = (-1)^s \sigma_A{}^* \, \sigma_A$.

(D) Let (θ, U) be a chart on X and $\rho: \zeta^{-1}(U) \to U \times R^p$ a trivialization of $\zeta: E \to X$ over U. For each $\varphi \in \mathscr{D}(E)'$ let $\bar{\varphi}: \theta(U) \to R^p$ be defined by $\bar{\varphi}(x) = \pi \rho \varphi \theta^{-1}(x)$, where $\pi: U \times R^p \to R^p$ is projection onto the second factor. We shall use the same notation for local representations for $\psi \subset \mathscr{D}(F)$, relative to a trivialization $\sigma: \eta^{-1}(U) \to U \times R^q$.

Proposition. *An* R*-homomorphism* $A: \mathscr{D}(E) \to \mathscr{D}(F)$ *is a* k^{th}*-order differential operator if and only if each point of* X *is contained in a chart* (θ, U) *over which there are trivializations of* E *and* F, *such* A *has the representation:*

$$\overline{A\varphi}(x) = \sum_{|\alpha| \leqslant k} a_\alpha(x)\, D^\alpha \bar{\varphi}(x)$$

for suitable smooth maps $a_\alpha: \theta(U) \to M(p, q)$, *the vector space of* $p \times q$ *matrices. Furthermore,*

$$\sigma_A(w) = (-1)^{k/2} \sum a_{i_1 \ldots i_k} w_{i_1} \ldots w_{i_k}$$

where $(w_1, \ldots, w_n)$ *represents* w *in* (θ, U).

Proof. First of all, if ∇ and $\tilde{\nabla}$ are any connections on E then we can find $\underline{a}_j \in \mathscr{D}(\odot^j T^*(X) \otimes E, F)$ for $0 \leqslant j < k$ such that

$$\underline{a}\nabla^k = \underline{a}\tilde{\nabla}^k + \sum_{j=0}^{k-1} \underline{a}_j \tilde{\nabla}^j$$

Namely, for $k = 1$, $\nabla - \tilde{\nabla} = w$ assigns to each $\varphi \in \mathscr{D}(E)$ an element $w(\varphi) \in \mathscr{D}(\mathrm{Hom}(T(X), E))$, and furthermore w is $\mathscr{D}(R)$-linear in φ, so that $\underline{a}_0 = \underline{a}w$; the general case follows by induction. In particular, if a homomorphism A can be expressed as a polynomial in ∇ (some connection E), then it can be expressed in terms of any other $\tilde{\nabla}$. The co-ordinate expression for the symbol will follow below from the fact that the coefficients of ∇^k and $\tilde{\nabla}^k$ are equal.

To prove the necessity we take any chart (θ, U) and trivializations ρ, σ of the bundle restrictions $E|U$, $F|U$ respectively. Define the connection:

$$\tilde{\nabla}: \mathscr{D}(T(X)|U) \times \mathscr{D}(E|U) \to \mathscr{D}(E|U)$$

by

$$\tilde{\nabla}_u \varphi = \sum_{\alpha=1}^{n} u^\alpha \frac{\partial \varphi}{\partial x^\alpha}$$

Thus ∇ and $\tilde{\nabla}$ are two connections on $E|U$ each expressible in terms of the other. The restrictions of the coefficients $\underline{a}_j$ of A to U have (using the trivializations ρ, σ) the co-ordinate representatives $(\underline{a}_j(x))_{i_1 \ldots i_j} \in M(p, q)$. Composing these gives the desired representation of $\overline{A\varphi}$ in U.

To prove the sufficiency, let $\mathscr{U} = (\theta_i, U_i)$ be a locally finite covering of X by charts over which A has the stated co-ordinate representation. Let (γ_i) be a partition of unity subordinate to $\mathscr{U}$, and $\tilde{\nabla}_i$ the above connection in each (θ_i, U_i). Then

$$\nabla = \sum_i \gamma_i \tilde{\nabla}_i$$

is a connection on E in terms of which we can express A.

Remark. The proposition shows that the concepts of differential operators, their symbols, and the notion of ellipticity are independent of choice of Riemann structure on X, and of Riemannian metric and connection on the bundle E. Of course the adjoint A^* of A will depend on metrics. An alternative formalism (not involving Riemann structures) which is co-ordinate free is given in Ref.[20], characterizing the kth-order differential operators as those homomorphisms which factor through the kth jet extension $j^k: \mathscr{D}(E) \to \mathscr{D}(J^k(E))$, where $J^k(E) \to X$ is the vector bundle of k-jets of sections of E.

IV. THE EXISTENCE THEOREM AND APPLICATIONS

1. THE EXISTENCE THEOREM

(A) Let X be a Riemannian n-manifold, and $\xi: E \to X$ a Riemannian vector bundle over X. We take the tensor product Riemannian structure in each bundle $\otimes^k T^*(X) \otimes E$, and let $\langle\,\rangle_x$ denote indifferently the inner product in X(x), X*(x), E_x and in their tensor products.

If $\varphi, \psi \in \mathscr{D}_0(E)$ are compact sections we define the number

$$\langle \varphi, \psi \rangle_0 = \int_X \langle \varphi, \psi \rangle * 1$$

where $\langle\varphi,\psi\rangle$ denotes the function $x \to \langle\varphi(x),\psi(x)\rangle_x$ on X. Then for each integer $r \geqslant 0$ let

$$\langle\varphi,\psi\rangle_r = \sum_{k=0}^{r} \langle\nabla^k\varphi, \nabla^k\psi\rangle_0$$

These are clearly inner products on $\mathscr{D}_0(E)$, and we let $H_0^r(E)$ denote the indicated completion.

Suppose henceforth that X is compact, and write $H^r(E) = H_0^r(E)$. Then we have the following properties:

Rellich's theorem. *If* $s < t$, *then the injection* $H^t(E) \to H^s(E)$ *is compact and dense.* That the map is dense is clear, since $\mathscr{D}(E)$ is dense in both. To see that it is compact, let (φ_i) be any bounded sequence in $H^t(E)$, and take $\gamma \in \mathscr{D}(R)$ with $spt(\gamma)$ contained in some chart (θ, U). Then $(\gamma\varphi_i)$ can be viewed as a sequence in $H_0^t(U)$, so that we can appeal to Section 5B of Part II to extrac a subsequence converging in $H^s(E)$. If (γ_p) is a finite squared partition of unity

$$\sum_p \gamma_p^2 = 1$$

then we construct a subsequence of (φ_i) such that for each p the sequence $(\gamma_p\varphi_i)$ converges in $H^s(E$

Let $C^s(E)$ denote the Banach space of sections of E of class C^s, with obvious norm.

Sobolev's theorem. *If* $t > n/2 + s$, *then we have a continuous injection* $H^t(E) \to C^s(E)$. In particul $H^\infty(E) = \cap\{H^t(E) : t \geqslant 0\} = \mathscr{D}(E)$. This follows at once from Section 5B of Part II.

Remark. It is an elementary matter to check that the Hilbertian structure of $H^r(E)$ is independen of the Riemannian structures of X and E. In fact, let $\mathscr{U} = (\theta_1, U_i, \rho_i, \gamma_i)$ be a finite system consisting of a covering of X by charts (θ_i, U_i), trivializations $\rho_i : \xi^{-1}(U_i) \to U_i \times \mathbb{R}^m$, and a squared partition of unity (γ_i) with each $spt(\gamma_i) \subset U_i$. Then we define an inner product on $\mathscr{D}(E)$ by

$$\langle\langle\varphi,\psi\rangle\rangle_r = \sum_i \langle\overline{\gamma_i\varphi}, \overline{\gamma_i\psi}\rangle_{H_0^r(\mathbb{R}^n)}$$

where $\overline{\varphi}$ is defined as in Section 6D (Part III). Any such inner product is easily seen to be equivalent that given using the iterated (through r) covariant differentials defined by Riemannian structures.

(B) *If* $A \in Diff^r(E,F)$, *then* A *has a unique extension to a continuous linear map* $A : H^{s+r}(E) \to H$ *for all* $s \geqslant 0$. This is clear because the coefficients in the representation $A = \sum \underline{a}_k \nabla^k$ are smooth, whence there is a constant such that

$$|A\varphi|_s \leqslant \text{const}|\varphi|_{s+r} \quad \text{for all} \quad \varphi \in \mathscr{D}(E)$$

Suppose now that $F = E$ and $r = 2s$; then A can be represented (not uniquely, in general)

$$A = \sum_k C_k \circ B_k$$

where order $B_k \leqslant s$, order $C_k \leqslant s$. Then for any $\varphi, \psi \in \mathscr{D}(E)$ we have by Corollary 6C (Part III)

$$\langle A\varphi, \psi\rangle_0 = \sum_k \langle B_k\varphi, C_k^*\psi\rangle_0$$

Define

$$\alpha: H^s(E) \to \mathbb{R} \quad \text{by} \quad \alpha(\varphi) = \sum_k \langle B_k\varphi, C_k^*\varphi\rangle_0$$

On the dense subspace $\mathscr{D}(E) \subset H^s(E)$ we have the equality $\alpha(\varphi) = \langle A\varphi, \varphi\rangle_0$, so that α is defined independently of the representation in terms of operators of orders $\leqslant s$.

Theorem (Gårding's inequality). *If* $A \in \mathrm{Diff}^{2s}(E, E)$ *is strongly elliptic, then* α *is coercive on* $H^s(E)$, *i.e. there are numbers* $\lambda_0 > 0$, $c > 0$ *such that* $\alpha(\varphi) + \lambda_0 |\varphi|_0^2 \geqslant c|\varphi|_s^2$ *for all* $\varphi \in H^s(E)$.

Proof. First of all, since X is compact there exists $\lambda_0 > 0$ such that $\langle \sigma_A(x, \xi)v, v\rangle_x \geqslant \lambda_0 |\xi|^{2s}|v|^s$ for all $\xi \in X^*(x)$, $v \in E_x$, and $x \in X$. Secondly, it suffices to verify the inequality for smooth sections. Thirdly, Theorem 5D of Part II shows that the inequality is valid (for suitable λ_0, c) for all φ having support in a sufficiently small chart. Finally, to establish the inequality in general we proceed as in Case 4 of Section 3C of Part II. Let (γ_i) be a squared partition of unity with sufficiently small supports, and write

$$\langle A\varphi, \varphi\rangle_0 = \sum \langle A(\gamma_i\varphi), \gamma_i\varphi\rangle_0 + R$$

Now Gårding's inequality is valid for the sections $\gamma_i\varphi$. Also, for any $\epsilon > 0$ we have $|R| \leqslant \epsilon\, \mathrm{const}|\varphi|_s^2 + \epsilon^{-2s+1}\mathrm{const}|\varphi|_0^2$; by choosing ϵ sufficiently small we can again absorb R to obtain the desired inequality for all $\varphi \in \mathscr{D}(E)$.

(C) We are now in position to apply Theorem 4D of Chapter I. Furthermore, given $\psi \in H^t(E)$, if $\varphi \in H^s(E)$ satisfies $\alpha(\varphi, \zeta) = \langle \phi, \zeta\rangle_0$ for all ζ, then $\varphi \in H^{t+r}(E)$ by Theorem 5E of Chapter II, because differentiability is a local matter.

In summary, we have the fundamental

Theorem. *Let* $A \in \mathrm{Diff}^{2s}(E, E)$ *be strongly elliptic. Then*

(1) A *maps* $\mathscr{D}(E)$ *onto* $K^\perp(A^*)$ *in* $\mathscr{D}(E)$ *with kernel* $K(A)$;
(2) $\dim K(A) = \dim K(A^*) < \infty$;
(3) *There is a* $\lambda_0 > 0$ *such that* $A_\lambda = A + \lambda: \mathscr{D}(E) \to \mathscr{D}(E)$ *is a bijection for all* $\lambda \geqslant \lambda_0$;
(4) $\dim K(A_\lambda) > 0$ *for at most countably many* λ *with no finite accumulation point.*

An immediate consequence is the theorem of the Introduction.

Corollary. *Let* $A \in \mathrm{Diff}^s(E, F)$ *be elliptic. Then for any* $\psi \in \mathscr{D}(F)$ *there is a* $\varphi \in \mathscr{D}(E)$ *such that* $A\varphi = \psi$ *if and only if* $\psi \in K^\perp(A^*)$. *Furthermore,* $\dim K(A) < \infty$, $\dim K(A^*) < \infty$.

Proof. The last statement follows because $A^*A = B$ is strongly elliptic, and $K(A) = K(B)$ since $\langle B\varphi, \varphi\rangle_0 = \langle A\varphi, A\varphi\rangle_0$. The necessity in the first statement is also clear. To prove the sufficiency, take $\psi \in K^\perp(A^*)$; then $A^*\psi$ is orthogonal to $K(B^*) = K(A)$. The fundamental theorem asserts the existence of $\varphi \in \mathscr{D}(E)$ such that $A^*A\varphi = A^*\psi$; i.e. $A\varphi - \psi \in K(A^*)$. On the other hand, both ψ and $A\varphi$ belong to $K^\perp(A^*)$, so that $A\varphi - \psi = 0$.

Remark. In general the index (A) = dim K(A) − dim K(A*) of an elliptic operator is not zero; see Section 2 below. Its evaluation in terms of differential invariants (in particular, the characteristi classes of X, E, F) was a problem formulated by Gelfand and solved in complete generality by Atiyah-Singer [20].

Remark. In the preceding Corollary three integers appear: dim X, fibre dim E = fibre dim F, and order A = s. The existence of an elliptic operator in such a setting with prescribed order and/or with prescribed analytic properties implies drastic topological restrictions on X, E, F. Very little is known about this basic question (but see Ref. [21]).

Extensive exercise. Throughout the last three Parts we have restricted attention to elliptic operators on closed manifolds (or to 0 boundary values in case of a bounded domain in Euclidean space). It would be a most instructive project now to take a compact Riemannian n-manifold with boundary (a smooth Riemannian (n − 1)-manifold) and re-examine our steps with an eye towards: (1) formulating the concept of elliptic boundary value problems; (2) developing the required version of Gårding's inequality so that an analogue of Theorem 1C and its Corollary is obtained. Hints and special cases can be obtained from various sources mentioned in the Bibliography.

2. HODGE'S THEOREM

(A) Suppose that X is a compact Riemannian n-manifold, and $\xi: E \to X$ a Riemannian bundle with connection ∇. Then as in Sections 4C and D of Part III, using the natural pairing $E \times E^* \to \mathbb{R}$ we have the exterior differential $d: \mathscr{A}^p(E) \to \mathscr{A}^{p+1}(E)$ defined on E-valued p-forms on X, and

$$d(\varphi \wedge * \psi) = d\varphi \wedge * \psi + (-1)^p \varphi \wedge d * \psi$$

for all $\varphi \in \mathscr{A}^p(E)$, $\psi \in \mathscr{A}^q(E)$; similarly for twisted E-valued forms. In particular, replacing (p, q) by (p − 1, p) we have

$$\int_X d(\varphi * \psi) = 0$$

so that

$$\int_X d\varphi \wedge * \psi = (-1)^p \int_X \varphi \wedge (d * \psi) = (-1)^{np+n+1} \int_X \varphi * (* d * \psi)$$

i.e.

$$\langle d\varphi, \psi \rangle_0 = \langle \varphi, (-1)^{np+n+1} * d * \psi \rangle_0$$

Thus we obtain the

Lemma. *The formal adjoint* $d^*: \mathscr{A}^p(E) \to \mathscr{A}^{p-1}(E)$ *of* d *relative to* $\langle\,\rangle_0$ is given by $d^*\varphi = (-1)^{np+n+1} * d * \varphi$. Furthermore, $* d^* \varphi = (-1)^p d * \varphi$, $* d \varphi = (-1)^{p+1} d^* * \varphi$ *for* $\varphi \in \mathscr{A}^p(E)$, *where in both right members* d *and* d* *are constructed in* E*.

Definition. The *Dirichlet integral* in $\mathscr{A}(E)$ is the positive quadratic function $D:\mathscr{A}(E) \to \mathbb{R}$ given by

$$D(\varphi) = \tfrac{1}{2}(|d\varphi|_0^2 + |d^*\varphi|_0^2)$$

its Laplace operator $\Delta:\mathscr{A}^p(E) \to \mathscr{A}^p(E)$ is the second-order linear differential operator $\Delta = dd^* + d^*d$. A form $\varphi \in \mathscr{A}^p(E)$ is *harmonic* if $\Delta\varphi = 0$. Let $\mathscr{H}^p(X,E) = \mathrm{Ker}\{\Delta:\mathscr{A}^p(E) \to \mathscr{A}^p(E)\}$; similarly for the space $\tilde{\mathscr{H}}^p(X,E)$ of twisted harmonic E-valued p-forms on X.

We have $\langle\Delta\varphi,\psi\rangle_0 = \langle\varphi,\Delta\psi\rangle_0$, so that Δ *is symmetric;* and $\langle\Delta\varphi,\varphi\rangle_0 = |d\varphi|_0^2 + |d^*\varphi|_0^2 \geqslant 0$, whence Δ is positive. From that we obtain the

Lemma. $\varphi \in \mathscr{A}(E)$ *is harmonic if and only if both* $d\varphi = 0$ and $d^*\varphi = 0$.

Note that the directional derivative of D at φ in the direction ψ

$$\frac{d}{dh}[D(\varphi+h\psi)]_{h=0} = \langle\Delta\varphi,\psi\rangle_0$$

o that $\Delta\varphi$ serves as the gradient of D at φ; *the critical points of* D *in* $\mathscr{A}(E)$ *are just the harmonic orms,* and these are the absolute minima for D.

From the relations $* \, dd^* = d^*d \, *$ and $* \, d^*d = dd^* \, *$ we find $* \Delta = \Delta *$, where in the right member Δ operates in E^*; that implies the important

uality theorem. *Let* X *be a compact Riemannian n-manifold, and* $\xi: E \to X$ *a Riemannian vector undle. Then for each* p $(0 \leqslant p \leqslant n)$ *we have the involutory isomorphism:*

$$*\,\tilde{\mathscr{H}}^p(X,E) \to \mathscr{H}^{n-p}(X,E^*)$$

emark. It would seem that the sequence of eigenvalues of Δ would be useful in studying the opological and differential geometric properties of $\xi: E \to X$. However, beyond knowledge of eir asymptotic distribution and in spite of many attempts, practically nothing is known about at sequence even for the case $E = X \times \mathbb{R}$ and $A = \Delta$.

) For each point $x \in X$ and $w \in X^*(x)$ we use the interior product (certainly a local operator) f Section 4C of Part III to define the endomorphism:

$$i_w = i_{p^{-1}(w)}: \Lambda^p X^*(x) \otimes E_x \to \Lambda^{p-1} X^*(x) \otimes E_x$$

etting ϵ_w its adjoint:

$$\langle\epsilon_w\varphi,\psi\rangle_x = \langle\varphi, i_w\psi\rangle_x$$

r all $\varphi \in \Lambda^p X(x) \otimes E_x$, $\psi \in \Lambda^{p+1} X^*(x) \otimes E_x$, we find that $\sigma_{d+d^*}(w) = \sqrt{-1}(\epsilon_w + i_w)$, so that $+ d^*$ *is an elliptic operator* on E-valued differential forms on X. Also,

$$\langle\sigma_\Delta(w)\varphi,\varphi\rangle = \langle(\sigma_d\sigma_{d^*}(w) + \sigma_{d^*}\sigma_d(w))\varphi,\varphi\rangle = \langle(\epsilon_w i_w + i_w\epsilon_w)\varphi,\varphi\rangle = \langle 2|w|^2\varphi,\varphi\rangle$$

hence Δ *is strongly elliptic.* We can now apply Theorem 1C and its Corollary to assert the existence harmonic E-valued forms.

emark. As we have emphasized all along, both existence and differentiability in Theorem 1C are sed on Gårding's inequality, which in the case of the present section takes the form

$$D(\varphi) + \lambda_0|\varphi|_0^2 \geqslant c|\varphi|_1^2 \qquad \text{for all } \varphi \in \mathscr{A}(E)$$

If we were interested only in harmonic forms we could produce a special simplified proof (avoidi the Fourier transform, and making use of the special character of the Laplacian) of that estimate. That is the basis of the proof of Hodge's theorem (for the special case of E the trivial line bundle and X oriented, but these are superficial simplifications) given by Morrey-Eells (Section 5 of [13]

(C) Now specialize to $E = X \times \mathbb{R}$ with its connection given as in Example 1 of Section 4E of Part III. Let us write $\mathscr{A}^p = \mathscr{A}^p(E)$ and $\mathscr{H}^p = \mathscr{H}^p(E)$. Then the exterior differential $d: \mathscr{A}^p \to \mathscr{A}^{p+1}$ has square zero, as does its adjoint $d^*: \mathscr{A}^p \to \mathscr{A}^{p-1}$. It follows that $\mathrm{Im}(d) = d\mathscr{A}$ is orthogonal to $\mathrm{Ker}(d^*)$; and that $\mathrm{Im}(d^*) = d^*\mathscr{A}$ is orthogonal to $\mathrm{Ker}(d)$. By Theorem 1C we obtain the

Theorem. $\mathscr{A} = \mathrm{Im}(d) \oplus \mathscr{H} \oplus \mathrm{Im}(d^*)$. *Also,* $H^0 = dH^1 \oplus \mathscr{H} \oplus d^*H^1$.

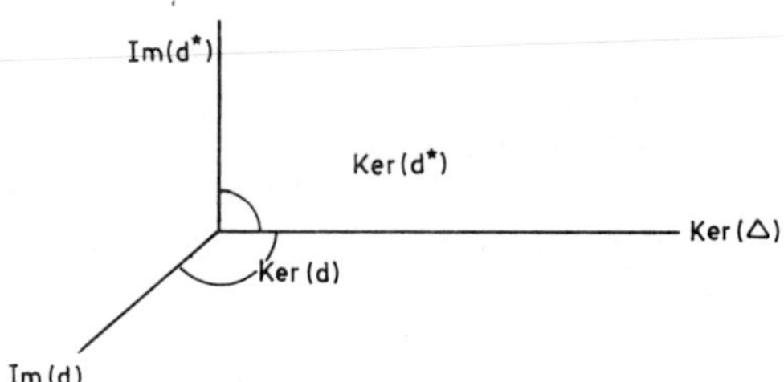

Exercise. Formulate and prove an analogous theorem for a differential operator B with $B^2 = C$ Note that by Proposition 4D of Part III we obtain such differential operators whenever $\xi: E \to X$ is a Riemannian bundle with curvature zero.

Exercise. Study the relations between harmonic E-valued forms and polyharmonic E-valued for (i.e. those which belong to the kernel of some iterated Laplacian Δ^k). Compare these last relatic with those obtained from studying the extremals of the iterated Dirichlet integral

$$D^k(\phi) = \tfrac{1}{2}\|(d+d^*)^k\phi\|_0^2$$

Definition. For any n-manifold X its *de Rham cohomology* is the quotient

$$\mathscr{R} = \mathrm{Ker}(d: \mathscr{A} \to \mathscr{A})/\mathrm{Im}(d: \mathscr{A} \to \mathscr{A})$$

It is clear from the formula $d(\varphi \wedge \psi) = d\varphi \wedge \psi + (-1)^p \varphi \wedge d\psi$ that the denominator is a bilate ideal in the numerator, whence the exterior product induces an associative, commutative (in the graded sense), graded algebra structure

$$\mathscr{R} = \sum_{p=0}^{n} \mathscr{R}^p$$

in the de Rham cohomology.

Corollary (Hodge's theorem). *If X is a compact Riemannian n-manifold then in every de Rhan cohomology class there is precisely one harmonic form.*

In particular, each $\dim \mathscr{R}^p = \dim \mathscr{H}^p$ is finite, and is called the pth *Betti number* $\beta_p(X)$ *of* X. Also, $\mathscr{R}^p = 0$ for $p > n$; and if X is furthermore orientable, then by the duality theorem 2A we have the isomorphism $*: \mathscr{H}^p \to \mathscr{H}^{n-p}$, so that $\beta_p(X) = \beta_{n-p}(X)$ for all $0 \leqslant p \leqslant n$. This relation, when coupled with the homology properties below, is a special case of Poincaré's duality theorem.

Exercise. Suppose X is connected and compact. Then $\beta_n(X) = 1$ or 0 depending on whether X is orientable or not.

(D) We now assume familiarity with the definitions and elementary properties of the singular homology space

$$H(X) = \sum_{p \geqslant 0} H_p(X)$$

and the singular cohomology algebra (with cup product)

$$H^*(X) = \sum_{p \geqslant 0} H^p(X)$$

with real coefficients; also the fact that on X the cohomology based on smooth singular chains and that based on continuous singular chains are canonically isomorphic as algebras (the isomorphism is induced from the inclusion map of the chain spaces).

Given any $\varphi \in \mathscr{A}^p$ and any smooth p-chain c, we have the integral $\int_c \varphi$ defined. If $\varphi', \varphi \in \mathrm{Ker}(d : \mathscr{A}^p \to \mathscr{A}^{p+1})$ and $\varphi' - \varphi = d\psi$, and similarly if c', c are smooth singular p-chains such that $\partial c' = 0 = \partial c$ and $c' - c = \partial b$ (here ∂ denotes the singular chain boundary operator), then by Stokes' theorem

$$\int_{c'} \varphi' = \int_c \varphi$$

Thus the integral $\int_c \varphi$ depends only on the homology class of c and the de Rham cohomology class of φ. That number is called the *period of* φ on c. Now $H^p(X) = \mathrm{Hom}_{\mathbb{R}}(H_p(X), \mathbb{R})$, and the integral permits us to view the elements of $\mathscr{R}^p$ as linear forms on $H_p(X)$, i.e. the integral induces a linear map

$$I : \mathscr{R} \to H^*(X)$$

de Rham's theorem asserts that I *is an isomorphism of algebras.* For a sheaf theoretic proof (which is in essence a general uniqueness theorem for cohomologies on X) see Hirzebruch (Ref.[8], section 2). Since $H^*(X)$ is a homology (in particular, a topological) invariant of X, it follows that $\beta_p(X)$ *is independent of both the Riemannian and the differential structures of* X.

We can put Hodge's theorem in a more classical form as follows. Let $c_1, \ldots, c_\beta$ ($\beta = \beta_p(X) < \infty$) be a set of chains on X with all $\partial c_i = 0$ whose homology classes form a base for $H_p(X)$.

Given any set $\pi_1, \ldots, \pi_\beta$ of β real numbers, *Hodge's theorem asserts that there is one and only one harmonic p-form* φ *having these periods:*

$$\int_{c_i} \varphi = \pi_i \qquad (1 \leqslant i \leqslant \beta)$$

Remarks. If ϕ and ψ are harmonic forms on X, they represent de Rham cohomology classes, as does their exterior product $\phi \wedge \psi$. However, in general $\phi \wedge \psi$ is not itself harmonic. Give an example.

If X and Y are both compact Riemannian manifolds, and $f: X \to Y$ a smooth map, then f induces a degree preserving homomorphism of algebras:

$$\mathscr{R}(f): \mathscr{R}(Y) \to \mathscr{R}(X)$$

In general, however, f does not map harmonic forms into harmonic forms. Even fibre maps (respecting Riemannian structures) do not behave well. But study the special case of projection of a Riemannian product onto one of its factors.

(E) *Example 1.* Let X be oriented, and define

$$E = \sum_{p \geqslant 0} \Lambda^{2p}\, T^*(X), \quad F = \sum_{p \geqslant 0} \Lambda^{2p+1}\, T^*(X)$$

These have the same fibre dimension, because

$$\sum_{p=0}^{n} (-1)^p \binom{n}{p} = 0$$

The elliptic operator $A = d + d^*$ maps forms of even degree into forms of odd degree, and A^* does the opposite. Then

$$\text{index } (d + d^*) = \sum_{p \geqslant 0} (\beta_{2p}(X) - \beta_{2p+1}(X)) = \text{Euler characteristic of X}$$

Suppose for simplicity that X is oriented.

Example 2. The elements of $\mathscr{A}^0$ are the real-valued functions φ on X, and $d\varphi = 0$ if and only i is a constant. Since $d^*|\mathscr{A}^0 = 0$, we find that *the harmonic 0-forms are just the constant functi* (This fact reflects the maximum principle for the second-order elliptic operator Δ on a manifold without boundary.) By duality, the harmonic n-forms are just the constant multiples of the vol form $*1$.

Example 3. Let E be an n-dimensional vector space, and g, g' two *conformally related* inner pro i.e. there is a number γ such that $g' = \exp(2\gamma)g$; this is merely a convenient way of writing a strictly positive multiplicative factor. If $\alpha_0, \alpha_0' \in \Lambda^n E$ define the same orientation of E and are of norm one relative to g, g' respectively, then $\alpha_0' = \exp(-n\gamma)\alpha_0$; thus if $\mathscr{D}$ and $\mathscr{D}'$ are the du isomorphisms, we have $\mathscr{D}' = \exp(-n\gamma)\mathscr{D}$. Similarly, if $P, P': E \to E^*$ are the isomorphism def by g, g' respectively, then $P'(x)y = g'(x, y) = \exp(2\gamma)P(x)y$. It follows that for $\Lambda^{n-p}E \to \Lambda^{n-p}E$ we have $P' = \exp(2(n-p)\gamma)P)$, whence for $\Lambda^p E^* \to \Lambda^{n-p}E^*$ we have the star isomorphism

$$*' = \exp((n-2p)\gamma)*$$

Thus if $n = 2p$, *the automorphism* $*$ *of* $\Lambda^p E^*$ *is conformal invariant.*

Suppose that X has dimension n = 2p. Then $* : \mathscr{H}^p \to \mathscr{H}^p$ is an automorphism depending only on the conformal equivalence class of the Riemannian metric; it is symmetric if p is even, and skew symmetric if p is odd. We conclude:

(1) If $\dim X = 4r$, then the signature of $*$ (= number of positive eigenvalues minus the number of negative eigenvalues) is a significant invariant of X. (It admits an expression in terms of the characteristic classes of X (see Hirzebruch [8])).

(2) If $\dim X = 4r+2$, then $\beta_{2r+1} \equiv 0(2)$.

These properties are especially interesting for complex manifolds of complex dimension one, for it is known in that case that the complex structure defines a conformal equivalence class of Riemannian metrics, whence *the entire space* $\mathscr{H}^1$ *depends only on the complex structure of* X.

Example 4. Let $\pi : \hat{X} \to X$ be a finite regular covering of X. Then a Riemannian metric on X induces a Riemannian metric on $\hat{X}$, and $\pi^* : \mathscr{H}^p(X) \to \mathscr{H}^p(\hat{X})$ is injective. Thus $\beta_p(\hat{X}) \geqslant \beta_p(X)$ for all $0 \leqslant p \leqslant n$. Betti number relations between $\hat{X}$ and X can be given by properties of harmonic forms, and the action of the subgroup of the fundamental group of X corresponding to the covering space $\hat{X}$. We suggest that as an exercise, starting with a reinterpretation of the pleasant paper of Eckmann [3].

For the two-leaved covering $\tilde{X} \to X$ of any compact manifold, we have powerful cohomology relations between $\tilde{X}$ and X (for any system of local coefficients) given by its Gysin sequence (the fibre is a 0-sphere); see Thom (Ref.[22], Chapter 1,III).

REFERENCES

[1] ARONSZAJN, N., "On coercive integro-differential quadratic forms", Conf. on Partial Differential Equations, University of Kansas, 1954, pp.94–106.

[2] BERS, L., JOHN, F., SCHECHTER, M., "Elliptic equations", Lectures in Applied Mathematics, Interscience (1964).

[3] ECKMANN, B., Coverings and Betti numbers, Bull. Am. Math. Soc. **55** (1949) 95–101.

[4] GAFFNEY, M., The heat equation of Milgram and Rosenbloom for open Riemann manifolds, Ann. Math. **60** (1954) 458–66.

[5] GODEMENT, R., Théorie des faisceaux, Hermann (1958).

[6] HESTENES, M., Quadratic variational theory and linear elliptic partial differential equations, Trans. Am. Math. Soc. **101** (1961) 306–50.

[7] HILDEBRANDT, S., Einige konstruktive Methoden ..., J. reine u. angewandte Mathematik **213** (1963) 66–88.

[8] HIRZEBRUCH, F., Neue topologische Methoden in der algebraischen Geometrie, Springer (1956).

[9] KOSZUL, J., Lectures on Fibre Bundles and Differential Geometry, Tata Institute, Bombay (1960).

[10] LIONS, J., Equations différentielles opérationnelles, Springer (1961).

[11] NIRENBERG, L., Remarks on strongly elliptic partial differential equations, Commun. Pure Appl. Math. **8** (1955) 648–74.

[12] MILGRAM, A., ROSENBLOOM, P., Harmonic forms and heat conduction I, Proc. Natl Acad. Sci. USA **37** (1951) 180–84.

[13] MORREY, C., EELLS, J., A variational method in the theory of harmonic integrals I, Ann. Math. **63** (1956) 91–128.

[14] DE RHAM, G., Variétés différentiables, Hermann (1960).

[15] SCHECHTER, M., Various types of boundary conditions for elliptic equations, Commun. Pure Appl. Math. **13** (1960) 407–25.

[16] SCHWARTZ, L., Théorie des distributions I, Hermann (1950).

[17] SEELEY, R., Integro-differential operators on vector bundles, Trans. Am. Math. Soc. **117** (1965) 167–204.

[18] SINGER, I., Topology Seminar, Harvard Univ. (1962).

[19] SPENCER, D., "De Rham theorems and Neumann decompositions associated with linear partial differential equations", Colloque CNRS, 1964.

[20] Seminar on the Index Theorem, by Borel, Palais, Solovay, ..., Institute for Advanced Study, Princeton, 1964 (Ann. Math. Study).

[21] SOLOMJAK, M., Linear elliptic systems of the first order, Dokl. Akad. Nauk SSSR (transl. Sov. Math. **4** (1963) 604–07).

[22] THOM, R., Espaces fibrés en sphères et carrés de Steenrod, Ann. ENS **69** (1952) 109–82.

[23] WARNER, F.W., Foundations of Differentiable Manifolds and Lie Groups, Scott, Foresman (1971).

IAEA-SMR-18/16

SEVERAL COMPLEX VARIABLES

M.J. FIELD*
Mathematics Institute,
University of Warwick,
Coventry, Warwickshire,
United Kingdom

Abstract

SEVERAL COMPLEX VARIABLES.

Topics discussed include the elementary theory of holomorphic functions of several complex variables; the Weierstrass preparation theorem; meromorphic functions, holomorphic line bundles and divisors; elliptic operators on compact manifolds; hermitian connections; the Hodge decomposition theorem.

INTRODUCTION

These notes are based on an introductory course of lectures on "Several Complex Variables"; they are not intended to provide a survey of topics in complex analysis and I have made no attempt to give comprehensive historical references. My main aim is to give a useful introduction to other courses and seminars in these Proceedings as well as to some of the standard references in several variable complex analysis. There is some overlap between Section 1and my earlier notes in "Global Analysis and its Applications" [9]. As far as possible, however, I have tried to make the two sets of notes complementary. If these notes have a theme it is that of the interplay between complex analysis and other branches of mathematics. I have especially stressed the role of elliptic partial differential operators in analysis on complex manifolds.

Section 1 develops the elementary theory of holomorphic functions of more than one variable. It is largely based on Hörmander's text [16], and stresses applications of the existence and regularity theory for $\bar{\partial}$. Section 2 centres round the problem of obtaining a satisfactory definition of a meromorphic function of more than one variable. Included is a proof of the Weierstrass preparation theorem. In Section 3 we show how holomorphic line bundles, divisors and analytic sets naturally enter into the study of meromorphic functions on complex manifolds. In Section 4 we prove the Hodge–Kodaira–de Rham decomposition theorem for Kähler manifolds. Apart from the book of Kodaira and Morrow [20], this section is mainly based on the notes of Vesentini [37], the Palais seminar on the Atiyah-Singer index theorem [28] and various notes from M.Sc. courses that I have given at the University of Warwick. It is much more concise than earlier sections and the reader may need to consult the cited references to fill in details that I have omitted.

1. HOLOMORPHIC FUNCTIONS ON $\mathbb{C}^n$

The aim of this section is to develop the basic theory of holomorphic functions defined on open subsets of $\mathbb{C}^n$, $n > 1$. Our approach will particularly stress comparisons with the 1-variable theory and also the role of the $\bar{\partial}$-operator.

---

* Present address: Department of Pure Mathematics, University of Sydney, Sydney, NSW 2006, Australia.

Notational conventions

$\mathbb{R}^n$ will always denote real n-space and $\mathbb{C}^n$ complex n-space. We shall often identify $\mathbb{C}^n$ and $\mathbb{R}^{2n}$ by letting the point $(z_1, ..., z_n) \in \mathbb{C}^n$ correspond to $(x_1, y_1, x_2, ..., y_n) \in \mathbb{R}^{2n}$, where $z_j = x_j + iy_j$, $1 \leqslant j \leqslant n$.

A function will be said to be C^r if it is r-times continuously differentiable. If the function is C^r for all positive integers r, we shall say that it is C^∞.

If D is an open subset of $\mathbb{C}^n$ (or, more generally, a manifold) we let $C^r(D)$ denote the space of all C^r complex-valued functions on D. If E is a real or complex vector space we shall let $C^r(D,E)$ denote the space of all C^r functions from D to E.

Let $\Omega \subset \mathbb{C}$ be open and suppose that $f: \Omega \to \mathbb{C}$ is C^1. We recall that f is said to be *analytic* or *holomorphic* if f satisfies the Cauchy-Riemann equations on Ω or, equivalently, if f satisfies the partial differential equation

$$\partial f/\partial \bar{z} = 0$$

on Ω, where $\partial/\partial \bar{z} = \frac{1}{2}(\partial/\partial x + i\partial/\partial y)$.

The equation $\partial f/\partial \bar{z} = 0$ is an example of an *elliptic* partial differential equation and many of the remarkable properties of analytic functions can be related to the fact that they are solutions of an elliptic equation (conversely, solutions of elliptic equations have many properties in common with analytic functions). Before we start on the theory of holomorphic functions of more than one variable we shall give one illustration of how the theory of partial differential equations can be applied to problems in analytic functions of 1 variable. The proof of the following existence and regularity theorem may be found in Ref. [16].

Theorem 1.1. Let $\Omega \subset \mathbb{C}$ be open and $f \in C^\infty(\Omega)$. Then the partial differential equation

$$\partial u/\partial \bar{z} = f$$

has a solution $u \in C^\infty(\Omega)$.

Let B be an open neighbourhood of $\zeta \in \mathbb{C}$ and suppose that $m: B \setminus \{\zeta\} \to \mathbb{C}$ is analytic. m is said to be *meromorphic* on B if m can be expanded in a Laurent series

$$m(z) = \sum_{j=-p}^{\infty} a_j (z-\zeta)^j, \quad p \geqslant 0$$

on some neighbourhood of ζ. m is then said to have a *pole* of order p at ζ and

$$\sum_{j=-p}^{-1} a_j (z-\zeta)^j$$

is said to define the *principal part* of m at ζ. More generally, m is said to be meromorphic on an open subset Ω of $\mathbb{C}$ if it is meromorphic, in the above sense, on a neighbourhood of every point of Ω. We shall now use Theorem 1.1 to give an easy proof of the existence of meromorphic functions with specified principal parts.

Theorem 1.2 (the Mittag-Leffler theorem). Let $\Omega \subset \mathbb{C}$ be open and $\Gamma = \{\zeta_1, \zeta_2, \ldots\}$ be a countable subset of Ω with no limit points in Ω (i.e. a discrete subset of Ω). Suppose that for $j = 1, 2, \ldots$ we are given a meromorphic function m_j defined on an open neighbourhood Ω_j of ζ_j such that m_j has principal part P_j at ζ_j and is analytic on $\Omega_j \setminus \{\zeta_j\}$. Then there exists a meromorphic function m on Ω satisfying the following properties:

(1) m is analytic on $\Omega \setminus \Gamma$;

(2) m has principal part P_j at ζ_j for every j.

Proof. Since Γ is a discrete subset of Ω, we may find open neighbourhoods U_j of each ζ_j such that

(a) $\zeta_k \in U_j$ if and only if $k = j$;

(b) $\bigcup_j U_j = \Omega$.

Although the meromorphic function m_j need not be defined on the whole of U_j, its principal part P_j is actually analytic on $\mathbb{C} \setminus \{\zeta_j\}$ and therefore defines a meromorphic function on U_j for every j. It is clearly sufficient to prove the result in the special case $m_j = P_j$ and $\Omega_j = U_j$ and we shall assume these conditions hold.

Suppose that we could find analytic functions $h_j : U_j \to \mathbb{C}$, $1 \leqslant j \leqslant \infty$, such that on the overlap $U_j \cap U_k$ we had the identity

$$A \ldots\ldots\ldots\ldots\ldots\ldots \quad h_j - h_k = P_k - P_j$$

This would then imply that for all j, k we had

$$h_j + P_j = h_k + P_k \text{ on } U_j \cap U_k$$

Hence we could define a meromorphic function m satisfying the conditions of the theorem by

$$m|U_j = h_j + P_j$$

Our proof will proceed by showing that we can find C^∞ functions $g_j : U_j \to \mathbb{C}$ satisfying identity A. We shall then construct a function $u \in C^\infty(\Omega)$ such that $g_j + u$ is analytic on U_j for every j. Setting $h_j = g_j + u$, our result will follow.

Let $\{\phi_a\}$ be a C^∞ partition of unity subordinate to the cover $\{U_j\}$. That is, we have:

(1) For $a = 1, 2, \ldots$, $\phi_a : \Omega \to \mathbb{R}$ is a C^∞ positive function which is identically zero outside a compact subset of some $U_{j(a)}$.

(2) All but a finite number of the ϕ_a vanish on any compact subset of Ω.

(3) $\sum \phi_a = 1$ on Ω.

(For the existence of C^∞ partitions of unity we refer to Refs [5, 21 and 35].)

We define the analytic functions $h_{ij} : U_i \cap U_j \to \mathbb{C}$ by

$$h_{ij} = P_i - P_j$$

and set

$$g_k = \sum_a \phi_a h_{i(a)k}$$

The reader may easily verify that $g_k \in C^\infty(U_k)$ for every k. Moreover,

$$g_k - g_j = \sum \phi_a (h_{i(a)k} - h_{i(a)j})$$

$$= \sum \phi_a h_{jk} \text{ – definition of } h_{ij}$$

$$= h_{jk}$$

Since h_{jk} is analytic, it follows that for all j, k we have

$$\partial g_k/\partial \bar{z} = \partial g_j/\partial g_j/\partial \bar{z} \quad \text{on } U_j \cap U_k$$

Therefore, there exists $\psi \in C^\infty(\Omega)$ satisfying

$$\psi = \partial g_k/\partial \bar{z} \quad \text{on } U_k, \ k = 1, 2, \ldots$$

We now choose a solution $u \in C^\infty(\Omega)$ of the equation

$$\partial u/\partial \bar{z} = -\psi$$

and define $h_k = g_k + u \in C^\infty(U_k)$. Now $\partial h_k/\partial \bar{z} = \psi - \psi = 0$ for every k and so h_k is analytic. By construction, $h_k + P_k = h_j + P_j$ on $U_j \cap U_k$ and so, setting $m|U_j = h_j + P_j$, the result follows. Q.E.D.

Another important result about the existence of analytic and meromorphic functions is

Theorem 1.3 (the Weierstrass theorem). Let $\Omega \subset \mathbb{C}$ be open and suppose that we are given a discrete subset $S \subset \Omega$ and a map $p: S \to \mathbb{Z} \setminus \{0\}$. Then we may find a meromorphic function m on Ω such that

(1) The set of zeroes of m equals $S_+ = \{s \in S : p(s) > 0\}$;
(2) The set of poles of m equals $S_- = \{s \in S : p(s) < 0\}$;
(3) If $s \in S_+$ (S_-) then s is a zero (pole) of m of order p(s) (-p(s)).

Proof. We shall not give a proof of this result here but merely remark that the result follows from Theorem 1.1 together with the topological fact that the cohomology group $H^2(U, \mathbb{Z})$ vanishes for all open subsets $U \subset \mathbb{C}$. We refer to Ref. [23] for a proof along these lines or to Ref. [16] for a more analytic proof.

Corollary 1.3.1. Let $\Omega \subset \mathbb{C}$ be open. Then there exists an analytic function $f: \Omega \to \mathbb{C}$ which cannot be extended as an analytic or meromorphic function to any open subset $\Omega' \subset \mathbb{C}$ which strictly contains Ω.

Proof. We refer to Refs [9, 16] for proofs. Later on in this section we shall give precise definitions of what is meant by analytic or meromorphic extension.

Corollary 1.3.2. Let $\Omega \subset \mathbb{C}$ be open. Given a meromorphic function m on Ω, we may find (essentially unique) analytic functions f and g on Ω such that, off the pole set of m,

$$m = f/g$$

Proof. Suppose m has poles ζ_j with orders p_j. By Weierstrass' theorem, we may find an analytic function g on Ω with zeroes ζ_j of orders p_j. It follows easily (using Laurent series) that mg is analytic on Ω. Setting f = mg, we obtain m = f/g off the pole set of m. Q.E.D.

The Mittag-Leffler and Weierstrass theorems allow us to construct many analytic functions on an arbitrary open subset of $\mathbb{C}$. When we come to study holomorphic functions of several variables we find that the problem of constructing analytic or meromorphic functions with prescribed poles and zeroes becomes very much harder and, moreover, the geometry of the domain (or rather, its boundary) comes into play.

For j = 1, ..., n we define the partial differential operators:

$$\partial/\partial z_j = \tfrac{1}{2}(\partial/\partial x_j - i\partial/\partial y_j)\ ;\quad \partial/\partial \bar{z_j} = \tfrac{1}{2}(\partial/\partial x_j + i\partial/\partial y_j)$$

Definition 1.4. Let $\Omega \subset \mathbb{C}^n$ be open and suppose that $f:\Omega \to \mathbb{C}$ is C^1. We say that f is *analytic* or *holomorphic* on Ω if f is analytic in each variable separately, i.e. if f satisfies the following system of first-order partial differential equations on Ω:

$$\partial f/\partial \bar{z_j} = 0,\ \ j = 1, \ldots, n$$

Remarks

(1) Our conditions on f are unnecessarily strong. We may in fact assume that f is analytic in each variable separately without requiring even the continuity of f. For a proof of this important fact (Hartog's theorem) we refer to Ref. [16].

(2) It follows easily from Definition 1.4 that the derivative of f at $z \in \Omega$ is a *complex* linear form $Df_z \in L(\mathbb{C}^n, \mathbb{C})$ with matrix

$$[\partial f/\partial z_1, \ldots, \partial f/\partial z_n]$$

Conversely, it may be shown that a C^1 map $f:\Omega \to \mathbb{C}$ is analytic if and only if its derivative $Df_z: \mathbb{C}^n \to \mathbb{C}$ is a *complex* linear map for all $z \in \Omega$.

(3) More generally, if $f = (f_1, \ldots, f_m): \Omega \subset \mathbb{C}^n \to \mathbb{C}^m$ is C^1, we say that f is analytic if each of the component functions of f is analytic, i.e. if

$$\partial f_i/\partial \bar{z_j} = 0,\ 1 \leqslant i \leqslant m,\ 1 \leqslant j \leqslant n$$

It then follows that the derivative of f at $z \in \Omega$ is a *complex* linear map $Df_z \in L(\mathbb{C}^n, \mathbb{C}^m)$ with matrix $[\partial f_i/\partial z_j]$.

As in the 1-variable theory, the main technical tool in developing the elementary theory of holomorphic functions of several variables is an integral representation formula. First, however, some technicalities:

We say $D \subset \mathbb{C}^n$ is a *polydisc* if there exist discs $D_1, \ldots, D_n \subset \mathbb{C}$ such that

$$D = \prod_{j=1}^{n} D_j = \{z \in \mathbb{C}^n : z_j \in D_j,\ 1 \leqslant j \leqslant n\}$$

If the discs $D_1, \ldots, D_n$ are all open (closed) we say that D is an open (closed) polydisc. If D_j has centre a_j, $1 \leqslant j \leqslant n$, we say that D has centre $(a_1, \ldots, a_n)$. Let ∂D_j denote the boundary of D_j. We define

$$\partial_0 D = \prod_{j=1}^{n} \partial D_j \subset \mathbb{C}^n$$

$\partial_0 D$ is called the *distinguished boundary* of D.

Theorem 1.5 (Cauchy's integral formula for polydiscs). Let D be an open polydisc and f a continuous function on $\bar{D}$ which is analytic in D. Then

$$f(z) = (2\pi i)^{-n} \int_{\partial_0 D} f(\zeta_1, \ldots, \zeta_n) \prod_{i=1}^{n} (\zeta_i - z_i)^{-1} \, d\zeta_1 \ldots d\zeta_n$$

for all points $z = (z_1, \ldots, z_n) \in D$.

Proof. Repeated application of the Cauchy formula in one variable gives

$$f(z_1, \ldots, z_n) = (2\pi i)^{-n} \int_{\partial D_1} \frac{d\zeta_1}{(\zeta_1 - z_1)} \cdots \int_{\partial D_n} \frac{d\zeta_n}{(\zeta_n - z_n)} f(\zeta_1, \ldots, \zeta_n)$$

For fixed $z = (z_1, \ldots, z_n)$, the integrand is continuous on a compact domain of integration and so Fubini's theorem gives the result. Q.E.D.

Remark. There are other integral representation formulae for holomorphic functions of more than one variable which do not involve such restrictive assumptions on the boundary of the domain. We refer to Refs [11] and [30] for more details.

As in the case n = 1, the Cauchy integral formula easily gives many useful results about analytic functions. First, however, a definition.

Definition 1.6. A series

$$\sum_{m \in N^n} a_m(z)$$

of complex-valued functions defined on a compact subset K of $\mathbb{C}^n$ is said to converge *normally* if

$$\sum \|a_m\|_K < \infty \quad (\|a_m\|_K = \sup_K |a_m(z)|)$$

If $a_m(z)$ are functions defined on an open subset Ω of $\mathbb{C}^n$, we shall say that the above series converges normally on Ω if it converges normally on every compact subset of Ω.

Remark. In the statement of Definition 1.6 we are using "multi-index" notation: N denotes the positive integers, including 0, and $m \in N^n$ is to be thought of as an n-tuple of positive integers $(m_1, \ldots, m_n)$. $a_m(z)$ is then shorthand for $a_{m_1 \ldots m_n}(z)$. Finally the summation is over all n-tuples of positive integers. Later on, we shall use the notation z^m for $z_1^{m_1} \ldots z_n^{m_n}$; $|m|$ for $m_1 + \ldots + m_n$; $m!$ for $m_1! \ldots m_n!$; ∂^m for $\partial^{m_1}/\partial z_1^{m_1} \ldots \partial^{m_n}/\partial z_n^{m_n}$ and r^{-m} for $r_1^{-m_1} \ldots r_n^{-m_n}$.

Lemma 1.7. Suppose that we are given $a_m \in \mathbb{C}$, $m \in N^n$, such that the series $\sum a_m z^m$ converges normally in an open polydisc D centered at 0. Then

$$f(z) = \sum a_m z^m$$

defines an analytic function on D. f is necessarily C^∞.

A similar result holds for a series $\sum a_m (z-a)^m$ and polydisc D centered at a.

Proof. A straightforward generalization of the proof for n = 1.

Let $\Omega \subset \mathbb{C}^n$ be open and $f: \Omega \to \mathbb{C}$. If we can find a (normally convergent) series $\sum a_m (z-a)^m$ such that

$$f(z) = \sum a_m (z-a)^m$$

for all z belonging to an open polydisc $D \subset \Omega$ and centered at a, we say that $\sum a_m (z-a)^m$ is a *power series* representation for f on D. From Lemma 1.7 it follows that if f has a power series representation on D it must be analytic on D. When n = 1, it is well known that every analytic function has a power series representation on a neighbourhood of each point where it is defined. We now prove that this is also true for $n > 1$.

Theorem 1.8. Let $\Omega \subset \mathbb{C}^n$ be open and let $f: \Omega \to \mathbb{C}$ be analytic. Then we may find an open polydisc neighbourhood $D \subset \Omega$ of each point $a \in \Omega$ and a unique power series representation

$$f(z) = \sum a_m (z-a)^m$$

f f on D. The coefficients a_m are given by

$$m! a_m = \partial^m f(a)$$

oof. Without loss of generality, suppose $0 \in \Omega$ and let D be an open polydisc, centre 0, ith $\bar{D} \subset \Omega$. By Theorem 1.5 we have

$$f(z) = (2\pi i)^{-n} \int_{\partial_0 D} f(\zeta_1, \ldots, \zeta_n) \prod_{i=1}^{n} (\zeta_i - z_i) d\zeta_1 \ldots d\zeta_n$$

for all $z \in D$. Now

$$\prod_{i=1}^{n} (\zeta_i - z_i)^{-1} = \sum \frac{z_1^{m_1} \dots z_n^{m_n}}{\zeta_1^{m_1+1} \dots \zeta_n^{m_n+1}}$$

and this sum converges normally on D. Substituting and interchanging orders of summation and integration, we find that

$$f(z) = \sum a_m z^m, \; z \in D$$

with

$$a_{m_1 \dots m_n} = (2\pi i)^{-n} \int_{\partial_0 D} \frac{f(\zeta_1, \dots, \zeta_n)}{\zeta_1^{m_1+1} \dots \zeta_n^{m_n+1}} d\zeta_1 \dots d\zeta_n = (1/m!)\, \partial^m f(0)$$

The uniqueness is obvious since we have explicit formulae for the coefficients. Q.E.D.

Notation. If $\Omega \subset \mathbb{C}^n$ is open, we shall let $A(\Omega)$ denote the set of all analytic functions on Ω.

Corollary 1.8.1. If $\Omega \subset \mathbb{C}^n$ is open and $g \in A(\Omega)$, then all the derivatives $\partial^m g$ belong to $A(\Omega)$.

Proof. Immediate from the power series representation for g at each point of Ω. Q.E.D.

Corollary 1.8.2. If g is analytic and $|g| \leqslant M$ in the polydisc $\{z : |z_j| < r_j,\ 1 \leqslant j \leqslant n\}$ then

$$|\partial^m g(0)| \leqslant M m!\ r^{-m}$$

Proof. As for $n = 1$. We omit details.

As our last application of the Cauchy integral formula we have the most important

Theorem 1.9. Let $\{u_k\}$ be a sequence of holomorphic functions on Ω and suppose that u_k converges uniformly on every compact subset of Ω to a function u. Then $u \in A(\Omega)$.

Proof. $u_k \to u$ uniformly on a neighbourhood of any $z \in \Omega$. Hence u is continuous. If D is a polydisc, centre z, which is contained in Ω, then each u_k verifies Cauchy's formula on D and, since the u_k are uniformly convergent in D, so therefore does u. Since u is continuous, it follows by differentiating under the integral sign that u is C^1 and $\partial u/\partial \bar{z}_j = 0$, $1 \leqslant j \leqslant n$. Hence $u \in A(\Omega)$. Q.E.D.

Examples of analytic functions

(1) Let $P(z) = \sum a_m z^m$, where all but a finite number of the coefficients a_m are zero. P is called a *polynomial* and clearly defines an analytic function on the whole of $\mathbb{C}^n$. If there exists $r \geqslant 0$ such that $a_m = 0$, $|m| \neq r$, P is called a *homogeneous polynomial* (of degree r). Let Z denote the set of zeroes of a polynomial Q:

$$Z = \{z \in \mathbb{C}^n : Q(z) = 0\}$$

Then P/Q is defined as an analytic function on $\mathbb{C}^n \setminus Z$. P/Q is an example of a *rational function.*

(2) If $f \in A(\mathbb{C}^n)$, f is called an *entire* analytic function. Every polynomial is entire; so also are the Laplace transforms of functions (or distributions) with compact support (see, e.g., Refs [17, 38]).

We recall, without proof, the maximal principle

Theorem 1.10. If u is analytic in the connected open subset Ω of $\mathbb{C}^n$ and there exists $z_0 \in \Omega$ such that $|u(z)| \leqslant |u(z_0)|$ for all $z \in \Omega$, then u is constant in Ω.

For the next few paragraphs we wish to consider the problem of the extension of analytic functions. The next lemma is fundamental in what follows.

Lemma 1.11. Let $\Omega \subset \mathbb{C}^n$ be open and connected and let $f \in A(\Omega)$. Suppose that there exists a point $z_0 \in \Omega$ such that all the coefficients in the power series representation of f at z_0 vanish. Then f is identically zero on Ω.

Proof. Let

$$V = \{z : \text{All coefficients of the power series of f at z vanish}\}$$

V is clearly a closed subset of Ω. But, taking the power series expansion for f at $y \in V$, we see that V is also open. Since Ω is connected and $V \neq \emptyset$, it follows that $V = \Omega$ and that $f \equiv 0$. Q.E.D.

Suppose that U and V are open subsets of $\mathbb{C}^n$ with $U \subseteq V$ and that $f \in A(U)$. We say that $h \in A(V)$ is an *analytic continuation* (or *analytic extension*) of f to V if $h|U = f$.

Theorem 1.12 (Uniqueness of analytic continuation). Let $U, V \subset \mathbb{C}^n$ be open and suppose that $U \subseteq V$ and that every connected component of V meets U. If $h \in A(V)$ is an analytic continuation of $f \in A(U)$ then h is unique.

Proof. Without loss of generality we may assume that V is connected. Suppose that $h' \in A(V)$ is another analytic extension of f. Then $h - h'$ vanishes identically in U. Since V is assumed connected it follows by Lemma 1.11 that $h - h'$ vanishes identically in V and so $h = h'$. Q.E.D.

Suppose that $\Omega \subset \mathbb{C}^n$ is open and connected and that $f \in A(\Omega)$. It is natural to try and find the "largest" connected open subset $\Omega' \subset \mathbb{C}^n$ to which f can be extended as an analytic function. As indicated earlier, given any $\Omega \subset \mathbb{C}$, we can always find at least one $f \in A(\Omega)$ which cannot be analytically continued to any open subset $\Omega' \subset \mathbb{C}$ which strictly contains Ω. The situation for $n > 1$ is very different as the following example shows.

Example. Let Ω be a connected open subset of $\mathbb{C}^n$ containing the origin, set $\Omega' = \Omega \setminus \{0\}$ and suppose $n > 1$. Then any $f \in A(\Omega')$ has an analytic continuation to Ω. To prove the existence of a continuation of f to Ω, we take a polydisc $D = \{z : |z_j| < r,\ 1 \leqslant j \leqslant n\}$, such that $\bar{D} \subset \Omega$, and define

$$h(z_1, \ldots, z_n) = (2\pi i)^{-1} \int_{|\zeta_1| = r} \frac{f(\zeta_1, z_2, \ldots, z_n)}{(\zeta_1 - z_1)}\, d\zeta_1$$

Since $\partial h/\partial \bar{z}_j = 0$, $1 \leqslant j \leqslant n$, it follows that $h \in A(D)$. But if $(z_1, ..., z_n) \in D$ and $(z_2, ..., z_n) \neq 0$, the 1-variable Cauchy integral formula implies that

$$f(z_1, ..., z_n) = h(z_1, ..., z_n)$$

Hence $h = f$ on a non-empty open subset of D and hence, by the uniqueness of analytic continuation, $f = h$ on $D \setminus \{0\}$. It follows immediately that the map $F \in A(\Omega)$ defined by $F|D = h$ and $F|\Omega \setminus \{0\} = f$ is an analytic continuation of f to Ω.

Remark. It follows from the above example that an analytic function defined on an open subset of $\mathbb{C}^n$, $n > 1$, can never have an isolated zero. Indeed, suppose the contrary and let $f \in A(\Omega)$ have an isolated zero at z_0. Choose a polydisc neighbourhood D of z_0 such that the only zero of $f|D$ is z_0. Set $D' = D \setminus \{z_0\}$. Now $1/f \in A(D')$. The previous example shows that $1/f$ has an analytic continuation F to the whole of D. Hence $1/f$ must be bounded on some open neighbourhood of z_0. But this is absurd if f has a zero at z_0.

Our immediate aim is now to develop our techniques further so as to prove a rather stronger version of the extension theorem given in the above example. The extension theorem that we shall prove will also give us more information about the zero set of a holomorphic function. Our proof will use an existence and regularity theorem for a family of first-order linear partial differential equations (cf. Theorem 1.1). First we review some complex linear algebra which we shall also need in Section 4.

Let $\{e_1, ..., e_n\}$ denote the standard basis of $\mathbb{C}^n$ (i.e. $e_j \in \mathbb{C}^n$ has all co-ordinates zero, save the jth which equals 1). We let $\mathbb{C}^{n*}$ denote the (complex) dual space of $\mathbb{C}^n$. Thus $\mathbb{C}^{n*}$ is the complex vector space consisting of all complex linear maps $A: \mathbb{C}^n \to \mathbb{C}$.

We shall say that a real linear map $A: \mathbb{C}^n \to \mathbb{C}$ is *conjugate complex linear* if for all $c \in \mathbb{C}$ and $z \in \mathbb{C}^n$ we have

$$A(cz) = \bar{c}A(z)$$

We let $\bar{\mathbb{C}}^{n*}$ denote the set of all conjugate complex linear forms on $\mathbb{C}^n$. $\bar{\mathbb{C}}^{n*}$ is usually referred to as the *conjugate dual* space of $\mathbb{C}^n$. $\bar{\mathbb{C}}^{n*}$ is easily shown to have the natural structure of a complex vector space.

Scalar multiplication by i defines a complex linear endomorphism J of $\mathbb{C}^n$,

$$J(z) = iz, z \in \mathbb{C}^n$$

We refer to J as a *complex structure* on $\mathbb{C}^n$. We remark that $J^2 = -I$. (Conversely it is easily shown that if J is an endomorphism of the real vector space E that satisfies the property $J^2 = -I$, then E may be given the structure of a complex vector space by defining

$$(a + ib)e = ae + bJ(e)$$

for all $a, b \in \mathbb{R}$ and $e \in E$.)

Let $L_{\mathbb{R}}(\mathbb{C}^n, \mathbb{C})$ denote the set of all real linear maps from $\mathbb{C}^n$ to $\mathbb{C}$. $L_{\mathbb{R}}(\mathbb{C}^n, \mathbb{C})$ has the natural structure of a complex vector space (scalar multiplication induced from $\mathbb{C}$) and

$$L_{\mathbb{R}}(\mathbb{C}^n, \mathbb{C}) \cong \mathbb{C}^{n*} \oplus \bar{\mathbb{C}}^{n*}$$

Indeed, if $A \in L_{\mathbb{R}}(\mathbb{C}^n, \mathbb{C})$ and we define

$$A_1 = \tfrac{1}{2}(A - iA \cdot J)\ ;\ A_2 = \tfrac{1}{2}(A + iA \cdot J)$$

the reader may easily verify that $A_1 \in \mathbb{C}^{n*}$, $A_2 \in \overline{\mathbb{C}}^{n*}$ and that the map $A \mapsto (A_1, A_2)$ defines the required isomorphism between $L_{\mathbb{R}}(\mathbb{C}^n, \mathbb{C})$ and $\mathbb{C}^{n*} \oplus \overline{\mathbb{C}}^{n*}$.

We define $dz_j \in \mathbb{C}^{n*}$ by

$$dz_j \left(\sum_{k=1}^{n} c_k e_k \right) = c_j, \quad 1 \leqslant j \leqslant n$$

The set $\{dz_1, ..., dz_n\}$ is then the basis for $\mathbb{C}^{n*}$ dual to the basis $\{e_1, ..., e_n\}$ of $\mathbb{C}^n$.

We similarly construct a basis $\{dz_{\bar{1}}, ..., dz_{\bar{n}}\}$ for $\overline{\mathbb{C}}^{n*}$ by defining

$$dz_{\bar{j}} \left(\sum_{k=1}^{n} c_k e_k \right) = \bar{c}_j, \quad 1 \leqslant j \leqslant n$$

This basis is usually referred to as the basis of $\overline{\mathbb{C}}^{n*}$ *conjugate dual* to the basis $\{e_1, ..., e_n\}$ of $\mathbb{C}^n$.

We let $\Lambda^r \mathbb{C}^{n*}$ and $\Lambda^s \overline{\mathbb{C}}^{n*}$ denote the r- and s-fold complex exterior powers of $\mathbb{C}^{n*}$ and $\overline{\mathbb{C}}^{n*}$ respectively, $r, s \geqslant 0$. We form the complex tensor product

$$\Lambda^r \mathbb{C}^{n*} \otimes \Lambda^s \overline{\mathbb{C}}^{n*}$$

Elements of this space are called *complex exterior forms of type* (r,s). A basis for $\Lambda^r \mathbb{C}^{n*} \otimes \Lambda^s \overline{\mathbb{C}}^{n*}$ is given by the set

$$\{dz_{\alpha_1} \wedge ... \wedge dz_{\alpha_r} \otimes dz_{\bar{\beta}_1} \wedge ... \wedge dz_{\bar{\beta}_s} : 1 \leqslant \alpha_1 < ... < \alpha_r \leqslant n, \ 1 \leqslant \bar{\beta}_1 < ... < \bar{\beta}_s \leqslant n\}$$

For more details about these spaces we refer to Refs [19, 20, 26, 31]. We recall, however, that

$$\Lambda^0 \mathbb{C}^{n*} = \Lambda^0 \overline{\mathbb{C}}^{n*} = \mathbb{C}$$

$$\Lambda^n \mathbb{C}^{n*} \cong \mathbb{C}; \qquad \Lambda^n \overline{\mathbb{C}}^{n*} \cong \mathbb{C}$$

though the latter isomorphisms are not canonical.

If $\Omega \subset \mathbb{C}^n$ is open we call an element of $C^\infty(\Omega, \Lambda^r \mathbb{C}^{n*} \otimes \Lambda^s \overline{\mathbb{C}}^{n*})$ a *complex differential form of type* (r, s). If $\phi \in C^\infty(\Omega, \Lambda^r \mathbb{C}^{n*} \otimes \Lambda^s \overline{\mathbb{C}}^{n*})$, ϕ may be written uniquely in the form

$$\phi = \sum_{\alpha, \beta} \phi_{\alpha_1 ... \alpha_r \bar{\beta}_1 ... \bar{\beta}_s} dz_{\alpha_1} \wedge ... \wedge dz_{\alpha_r} \otimes dz_{\bar{\beta}_1} \wedge ... \wedge dz_{\bar{\beta}_s}, \quad \text{i.e. } 1 \leqslant \alpha_1 < ... < \alpha_r \leqslant n$$

where the sum is over $1 \leqslant \alpha_1 ... < \alpha_r \leqslant n$ and $1 \leqslant \bar{\beta}_1 < ... < \bar{\beta}_s \leqslant n$ and, for all α and β, $\phi_{\alpha_1 \cdots \alpha_r \bar{\beta}_1 ... \bar{\beta}_s} \in C^\infty(\Omega)$.

Theorem 1.13. Let $\Omega \subset \mathbb{C}^n$ be open. Then for $r = 0, 1, \ldots, n$ there exist linear differential operators:

$$\bar{\partial} : C^\infty(\Omega, \Lambda^r \bar{\mathbb{C}}^{n*}) \to C^\infty(\Omega, \Lambda^{r+1} \bar{\mathbb{C}}^{n*})$$

which are uniquely characterized by the following properties:

(1) $\bar{\partial} f = \sum_{j=1}^{n} \partial f / \partial \bar{z}_j \, d\bar{z}_j$ for all $f \in C^\infty(\Omega)$

(2) $\bar{\partial}^2 = 0$

(3) $\bar{\partial}(f \wedge g) = \bar{\partial} f \wedge g + (-1)^r f \wedge \bar{\partial} g$, for all $f \in C^\infty(\Omega, \Lambda^r \bar{\mathbb{C}}^{n*})$, $g \in C^\infty(\Omega, \Lambda^k \bar{\mathbb{C}}^{n*})$

Proof. Any $f \in C^\infty(\Omega, \Lambda^r \bar{\mathbb{C}}^{n*})$ may be written uniquely in the form

$$f = \sum_{\beta} f_{\bar{\beta}_1 \ldots \bar{\beta}_r} \, d z_{\bar{\beta}_1} \wedge \ldots \wedge d z_{\bar{\beta}_r}$$

where $f_{\bar{\beta}_1 \ldots \bar{\beta}_r} \in C^\infty(\Omega)$. We define

$$\bar{\partial} f = \sum_{j=1}^{n} \sum_{\beta} \partial f_{\bar{\beta}_1 \ldots \bar{\beta}_r} / \partial z_{\bar{j}} \, d z_{\bar{j}} \wedge d z_{\bar{\beta}_1} \wedge \ldots \wedge d z_{\bar{\beta}_r}$$

To prove that property 2 holds we use the symmetry of partial derivatives

$$\partial^2 f / \partial z_{\bar{i}} \partial z_{\bar{j}} = \partial^2 f / \partial z_{\bar{j}} \partial z_{\bar{i}}$$

Property 3 is straightforward computation. The fact that $\bar{\partial}$ is uniquely characterized by Properties 1, 2 and 3 follows easily using property 1 and property 3 and we omit details.

Example. Let us compute $\bar{\partial} g$ for

$$g = \sum_{k=1}^{n} g_{\bar{k}} \, d z_{\bar{k}} \in C^\infty(\Omega, \bar{\mathbb{C}}^{n*})$$

By definition we have

$$\bar{\partial} g = \sum_{j=1}^{n} \sum_{k=1}^{n} \partial g_{\bar{k}} / \partial z_{\bar{j}} \, d z_{\bar{j}} \wedge d z_{\bar{k}}$$

Using the relation $dz_{\bar{j}} \wedge dz_{\bar{k}} = -dz_{\bar{k}} \wedge dz_{\bar{j}}$ and re-arranging this sum, we find that

$$\bar{\partial} g = \sum_{1 \leqslant j < k \leqslant n} (\partial g_{\bar{k}}/\partial z_{\bar{j}} - \partial g_{\bar{j}}/\partial z_{\bar{k}})\, dz_{\bar{j}} \wedge dz_{\bar{k}}$$

The significance of the $\bar{\partial}$-operator is indicated by the following trivial

Proposition 1.14. Let $\Omega \subset \mathbb{C}^n$ be open. Then the kernel of $\bar{\partial} : C^\infty(\Omega) \to C^\infty(\Omega, \bar{\mathbb{C}}^{n*})$ is precisely $A(\Omega)$, i.e. $f \in C^\infty(\Omega)$ is analytic if and only if $\bar{\partial} f = 0$.

For $n = 1$, the existence and regularity theorem 1.1 implies that the sequence

$$0 \longrightarrow A(\Omega) \longrightarrow C^\infty(\Omega) \xrightarrow{\bar{\partial}} C^\infty(\Omega, \bar{\mathbb{C}}^*) \longrightarrow 0$$

is *exact* for all open subsets $\Omega \subset \mathbb{C}$, i.e. since $\partial g/\partial \bar{z} = f$ is always solvable, the map $\bar{\partial}$ in the above sequence is *onto*. For dimensions greater than one the analysis of the $\bar{\partial}$-operator turns out to be highly non-trivial and is closely related to the study of elliptic partial differential equations on open subsets of $\mathbb{C}^n$. We now wish to give an application of a rather elementary existence theorem for the $\bar{\partial}$-operator. First, however, a technical definition.

Definition 1.15. Let $\Omega \subset \mathbb{C}^n$ be open and $f : \Omega \to \mathbb{C}^m$. We define the (closed) support of f, $\overline{\text{supp}}(f)$, by

$$\overline{\text{supp}}(f) = \overline{\{z \in \Omega : f(z) \neq 0\}}$$

We let $C^r_c(\Omega, \mathbb{C}^m)$ denote the subset of $C^r(\Omega, \mathbb{C}^m)$ consisting of functions with *compact* support.

We shall now state an elementary existence and uniqueness theorem for the $\bar{\partial}$-operator.

Theorem 1.16. Let $n > 1$ and suppose that $f \in C^\infty_c(\mathbb{C}^n, \bar{\mathbb{C}}^{n*})$ satisfies $\bar{\partial} f = 0$. Then there exists

$g \in C^\infty_c(\mathbb{C}^n)$ satisfying

$$\bar{\partial} g = f$$

Proof. We defer the proof until after we have given an application of the theorem to a problem on analytic extension. We remark, however, that the theorem is false for $n = 1$, as is easily seen by taking f to be of compact support and satisfying

$$\int_{\mathbb{C}} f \neq 0$$

Theorem 1.17 (Hartog's theorem). Let $\Omega \subset \mathbb{C}^n$ be open, $n > 1$. Suppose that $K \subset \Omega$ is compact and $\Omega \setminus K$ is connected. Then every $f \in A(\Omega \setminus K)$ has a unique analytic continuation to $\bar{f} \in A(\Omega)$.

Proof. Let $\psi \in C_c^\infty(\Omega)$ be equal to 1 on some neighbourhood of K. We define $f_0 \in C^\infty(\Omega)$ by

$$f_0 | K = 0$$

$$f_0 | \Omega \setminus K = (1-\psi) f$$

We shall construct $v \in C^\infty(\mathbb{C}^n)$ such that $\bar{f} = f_0 + v$ is our required extension. We wish to choose v so that $\bar{f}$ is analytic, i.e.

$$0 = \partial \bar{f} = \bar{\partial} f_0 + \bar{\partial} v$$

Hence v must satisfy the partial differential equation:

$$\bar{\partial} v = -\bar{\partial} f_0$$

$$= f \bar{\partial} \psi$$

Since $f \bar{\partial} \psi \in C_c^\infty(\Omega)$ we may define $q \in C_c^\infty(\mathbb{C}^n)$ by

$$q | \Omega = f \bar{\partial} \psi$$

$$q | \mathbb{C}^n | \Omega = 0$$

Since $\bar{\partial} q = 0$, we may apply Theorem 1.16 to deduce that there exists a solution $v \in C_c^\infty(\mathbb{C}^n)$ of the equation

$$\bar{\partial} v = q$$

Now since v is of compact support, v vanishes on the unbounded component of the support of q by the uniqueness of analytic continuation. Since $\overline{\text{supp}(q)} \subset \Omega \setminus K$, there exists an open set in $\Omega \setminus K$ where $v = 0$ and $f = f_0$. Hence $f = \bar{f}$ on some open subset of $\Omega \setminus K$. Since $\Omega \setminus K$ is connected, $f = \bar{f}$ on $\Omega \setminus K$ and so $\bar{f}$ is an analytic continuation of f to Ω. Q.E.D.

The following trivial corollary of Hartog's theorem shows that the zero set of an analytic function always propagates to the boundary of the domain on which the function is defined.

Corollary 1.17.1. Let $\Omega \subset \mathbb{C}^n$ be connected and let $f \in A(\Omega)$ be non-constant, $n > 1$. Set $Z = \{z \in \Omega : f(z) = 0\}$. Then Z is never a compact subset of Ω.

We now turn to the proof of Theorem 1.16. First we need to prove a stronger version of the Cauchy integral formula

$$f(z) = (2\pi i)^{-1} \int_{\partial\Omega} \frac{f(\zeta)}{(\zeta - z)} d\zeta$$

Theorem 1.18. Let $\Omega \subset \mathbb{C}$ be bounded, with $\partial\Omega$ a finite union of C^1 Jordan curves. Then, if $f \in C^1(\bar{\Omega})$, we have at all points $\zeta \in \Omega$

$$f(\zeta) = (2\pi i)^{-1}\left\{\int_{\partial\Omega} \frac{f(z)}{(z-\zeta)}\,dz + \int_{\Omega} \frac{\partial f/\partial \bar{z}}{(z-\zeta)}\,dz\,d\bar{z}\right\}$$

Proof. $C^1(\bar{\Omega})$ denotes the set of functions which are continuous on $\bar{\Omega}$ and C^1 on Ω. $dz\,d\bar{z}$ denotes the "complex" measure $-2i dx\,dy$.

We recall the complex form of Stokes' theorem in the plane

$$\int_{\partial\Omega} g dz = \int_{\Omega} \partial g/\partial \bar{z}\; d\bar{z}\,dz, \quad g \in C^1(\bar{\Omega})$$

which follows easily, by taking real and imaginary parts, from the real version of Stokes' or Green's theorem.

Let $0 < \epsilon < d(\zeta, \mathbb{C}\backslash\Omega)$. Set $\Omega_\epsilon = \{z \in \Omega : |z-\zeta| > \epsilon\}$. Taking $g(z) = f(z)/(z-\zeta)$ and applying Stokes' theorem to g on Ω_ϵ, we find that

$$\int_{\Omega_\epsilon} \partial f/\partial \bar{z}\,(z-\zeta)^{-1}\, d\bar{z}\, dz = \int_{\partial\Omega} f(z)\,(z-\zeta)^{-1}\, dz - \int_0^{2\pi} f(\zeta - \epsilon e^{i\theta})\, i d\theta$$

Letting $\epsilon \to 0$, the result follows (the limiting procedure is valid since $(z-\zeta)^{-1}$ is integrable on Ω and f and $\partial f/\partial \bar{z}$ are continuous at ζ). Q.E.D.

Corollary 1.18.1. Let $g \in C_c^\infty(\mathbb{C})$. Define

$$u(\zeta) = (2\pi i)^{-1} \int_{\mathbb{C}} (z-\zeta)^{-1}\, g(z)\, dz\, d\bar{z}$$

Then

(1) $u \in C^\infty(\mathbb{C})$;

(2) $\partial u/\partial \bar{\zeta} = g$, and u is analytic outside of the support of g.

Proof. Changing variables, we have

$$u(\zeta) = -(2\pi i)^{-1} \int g(\zeta - z)\; z^{-1}\, dz\, d\bar{z}$$

Since z^{-1} is integrable on compact sets, it is legitimate to differentiate under the sign of integration indefinitely often, and it follows that u is C^∞. Also

$$\partial u/\partial\bar\zeta = -(2\pi i)^{-1} \int \partial g(\zeta - z)/\partial\bar\zeta \, z^{-1} \, dz \, d\bar z$$

$$= (2\pi i)^{-1} \int (z-\zeta)^{-1} \, \partial g/\partial\bar z \, dz \, d\bar z$$

We now apply Theorem 1.10 with f = g and Ω a disc containing the support of g to obtain $\partial u/\partial\bar\zeta = g$. Q.E.D.

Proof of Theorem 1.16. Notice first that since $\bar\partial^2 = 0$, a necessary condition for the solvability of the equation $\bar\partial g = f$ is $\bar\partial f = 0$ (of course, this condition is trivially satisfied if n = 1). Suppose that

$$f = \sum_{j=1}^{n} f_{\bar j} \, dz_{\bar j}$$

where $f_{\bar j} \in C_c^\infty(\mathbb{C}^n)$, $1 \leqslant j \leqslant n$. We have to solve the system of first-order partial differential equations:

$$\partial g/\partial z_{\bar j} = f_{\bar j}, \; 1 \leqslant j \leqslant n$$

subject to the condition $\bar\partial f = 0$ or, equivalently, the conditions

$$\partial f_{\bar j}/\partial \bar z_{k} = \partial f_{\bar k}/\partial z_{\bar j}, \quad 1 \leqslant j, k \leqslant n$$

(see the example following Theorem 1.13). We define

$$g(z) = (2\pi i)^{-1} \int_{\mathbb{C}} (\zeta - z_1)^{-1} f_{\bar 1}(\zeta, z_2, \ldots, z_n) \, d\zeta \, d\bar\zeta$$

$$= -(2\pi i)^{-1} \int_{\mathbb{C}} \zeta^{-1} f_{\bar 1}(z_1 - \zeta, z_2, \ldots, z_n) \, d\zeta \, d\bar\zeta$$

The second formula for g implies that $g \in C^\infty(\mathbb{C}^n)$. Clearly g(z) = 0 if $|z_2| + \ldots + |z_n|$ is sufficiently large. Corollary 1.18.1 implies that $\partial g/\partial z_{\bar 1} = f_{\bar 1}$. Differentiating under the integral sign with respect to $z_{\bar j}$ and using the equality $\partial f_{\bar j}/\partial z_{\bar 1} = \partial f_{\bar 1}/\partial z_{\bar j}$ we obtain

$$\partial g/\partial z_{\bar j} = (2\pi i)^{-1} \int_{\mathbb{C}} (\zeta - z_1)^{-1} \, \partial f_{\bar j}/\partial\bar\zeta \, (\zeta, z_2, \ldots, z_n) \, d\zeta \, d\bar\zeta$$

$$= f_{\bar j}(z) \text{ (Theorem 1.18)}$$

Hence $\bar\partial g = f$. We must show that g has compact support. But we know that $\bar\partial g = 0$ outside of $\overline{\text{supp}(f)}$ and hence g is analytic on $\mathbb{C}^n \setminus \overline{\text{supp}(f)}$. Since $g = 0$ for $|z_2| + \ldots + |z_n|$ sufficiently large, it follows by the uniqueness of analytic continuation that g vanishes identically on the unbounded component of $\mathbb{C}^n \setminus \overline{\text{supp}(f)}$. Q.E.D.

Let Ω be an open subset of $\mathbb{C}$. As indicated earlier, we may find $f \in A(\Omega)$, which does not extend across any point of the boundary of Ω. The corresponding type of domain in $\mathbb{C}^n$, $n > 1$, is of special interest.

Definition 1.19. Let $\Omega \subset \mathbb{C}^n$ be open. We shall say that Ω is a *domain of holomorphy* if the following condition is satisfied:

> Let $z \in \Omega$ and U be a connected open neighbourhood of z (in $\mathbb{C}^n$). Suppose that $f|U \cap \Omega$ continues analytically to U for all $f \in A(\Omega)$. Then $U \subset \Omega$.

There are many different characterizations of domains of holomorphy. We refer the reader to Refs [9, 11, 12, 16, 26, 32] for more details. For a proof of the following deep result about domains of holomorphy we refer to Ref. [16].

Theorem 1.20. $\Omega \subset \mathbb{C}^n$ is a domain of holomorphy if and only if the $\bar\partial$-sequence is *exact*. That is, if $f \in C^\infty(\Omega, \Lambda^r \bar{\mathbb{C}}^{n*})$ and $\bar\partial f = 0$, then there exists $g \in C^\infty(\Omega, \Lambda^{r-1} \bar{\mathbb{C}}^{n*})$ such that $\bar\partial g = f$, $r > 0$.

Remark. The non-exactness of the $\bar\partial$-sequence gives a measure of how far Ω is from being a domain of holomorphy. We refer the reader to Refs [2] and [14], where this type of question is pursued further.

2. LOCAL PROPERTIES OF ANALYTIC FUNCTIONS

In this section we investigate the local theory of analytic functions of more than one variable and, in particular, power series rings. This theory turns out to be algebraically interesting and, apart from showing the interplay between algebra and complex analysis, is essential for the proper development of the theory of meromorphic functions of more than one variable. Much of this section may be read as being complementary to the section on sheaf theory in the 1972 ICTP lecture notes "Holomorphic function theory and complex manifolds" [9].

The set $\mathbb{C}\{z_1 - a_1, \ldots, z_n - a_n\}$ of convergent power series at $(a_1, \ldots, a_n) \in \mathbb{C}^n$ has the structure of a commutative ring with identity under the usual operations of addition and multiplication. In the sequel we shall use the more abbreviated notation O_a for the ring $\mathbb{C}\{z_1 - a_1, \ldots, z_n - a_n\}$, where $a = (a_1, \ldots, a_n) \in \mathbb{C}^n$.

Let $\mathscr{U}_a$ denote the set of all open neighbourhoods of the point $a \in \mathbb{C}^n$. For any $U \in \mathscr{U}_a$ we have a natural map:

$$p_U^a : A(U) \to O_a$$

defined as follows.

Let $f \in A(U)$ have the power series representation

$$f(z) = \sum a_m (z-a)^m, \quad z \in D$$

on some open polydisc D, centre a. Then we define

$$p_U^a(f) = \sum a_m(z-a)^m$$

A(U) has the structure of a commutative ring with identity under the usual operations of addition and multiplication of functions and we trivially have

Proposition 2.1. For all $U \in \mathcal{U}_a$, the map $p_U^a : A(U) \to O_a$ is a ring homomorphism.

Suppose that $W \in \mathcal{U}_a$ is *connected.* By the uniqueness of analytic continuation the map $p_U^a : A(W) \to O_a$ is *injective.* It follows that if $p_U^a(f) = p_V^a(g)$ for $f \in A(U)$, $g \in A(V)$ and $U, V \in \mathcal{U}_a$ then $f = g$ on some open neighbourhood of a.

Since $O_a = \bigcup_{U \in \mathcal{U}_a} p_U^a(A(U))$, any point $\psi \in O_a$ defines an essentially unique analytic function on some neighbourhood of a. Indeed, if $p_U^a(f) = p_V^a(g) = \psi$, the above arguments show that $f = g$ on some neighbourhood of a. It follows that *locally* we may think of points in O_a as analytic functions (defined on some neighbourhood of a). For example, given any $\psi \in O_a$, the *value* of ψ at a, $\psi(a)$, may be defined to be f(a), where $f \in A(U)$ satisfies $p_U^a(f) = \psi$. In the sequel we shall often confuse points in O_a with analytic functions.

If $f \in A(U)$, $U \in \mathcal{U}_a$, then $p_U^a(f)$ is usually referred to as the *germ* of f at a. O_a is often called the ring of germs of analytic functions at a.

For all $a, b \in \mathbb{C}^n$, $O_a \cong O_b$. Indeed, we may define a natural isomorphism between the two rings by mapping $\sum a_m(z-a)^m$ to $\sum a_m(z-b)^m$. In the sequel we shall often take $a = 0$. Because $O_a \cong O_0$ for all $a \in \mathbb{C}^n$, this will not result in any loss of generality.

Our first algebraic result about the rings O_a is the elementary

Proposition 2.2. O_a is an *integral domain.* That is, if $f, g \in O_a$ and $fg = 0$, then either $f = 0$ or $g = 0$.

Proof. Without loss of generality take $a = 0$. We may suppose that $f = p_D^0(\tilde{f})$, $g = p_D^0(\tilde{g})$, where $\tilde{f}, \tilde{g} \in A(D)$ and D is an open polydisc neighbourhood of zero. Now $fg = 0$ implies that $\tilde{f}\tilde{g} \equiv 0$ on D. Suppose $\tilde{f} \not\equiv 0$. Then $\tilde{f} \neq 0$ on some non-empty open subset U of D. Hence $\tilde{g} \equiv 0$ on U. By uniqueness of analytic continuation, $\tilde{g} \equiv 0$ on D. Hence $g = 0$. Q.E.D.

Next we wish to recall some algebraic definitions.

Definition 2.3. Let R be a commutative integral domain with identity.

(1) $u \in R$ is said to be a *unit* if there exists $u^{-1} \in R$ such that

$$uu^{-1} = 1 = u^{-1}u$$

u^{-1} is called the *inverse* of u.

(2) $f \in R$ is said to be *irreducible* or *prime* if any relation

$$f = gh, \quad g, h \in R$$

implies that either g or h is a unit.

Example. We may easily characterize the units in O_a:

$u \in O_a$ is a unit if and only if $u(a) \neq 0$. Indeed suppose that u is a unit. Then we may find a polydisc neighbourhood D of a and $f, g \in A(D)$ such that $p_D^a(f) = u$, $p_D^a(g) = u^{-1}$. The condition $uu^{-1} = 1$ implies that $fg = 1$ on D. Hence $u(a) = f(a) \neq 0$. The converse is similarly proved.

To proceed further with our algebraic study of O_a we need the following basic and extremely important result.

Theorem 2.4 **(the Weierstrass preparation theorem).** Let f be analytic on some neighbourhood Ω of $0 \in \mathbb{C}^n$ and suppose that $f(0, 0, ..., 0, z_n)$ is not identically zero and has a zero of multiplicity $p \geqslant 0$ at $z_n = 0$. Then we may find a polydisc neighbourhood $D \subset \Omega$ of 0 such that if g is analytic and bounded on D, there exist unique $q, r \in A(D)$, with r a polynomial in z_n of degree $<p$ with coefficients depending analytically on $(z_1, ..., z_{n-1})$, such that

$$g = qf + r$$

Moreover, there exists $M \geqslant 0$, depending only on D and f, such that

$$\|q\|_D, \|r\|_D \leqslant M\|g\|_D$$

Proof. The proof we give is that of Grauert and Remmert (see Ref.[25]). We may write

$$f(z_1, ..., z_n) = \sum_{j=0}^{\infty} f_j(z_1, ..., z_{n-1}) z_n^j$$

where $f_p(0, ..., 0) \neq 0$. We may find polydisc neighbourhoods

$$D = \{z \in \mathbb{C}^n : |z_j| < r_j, \ 1 \leqslant j \leqslant n\}$$

$$D' = \{z \in \mathbb{C}^{n-1} : |z_j| < r_j, \ 1 \leqslant j \leqslant n-1\}$$

such that $f_p \neq 0$ on D', $1/f_p$ is bounded on D' and f is bounded on D. Multiplying f by $1/f_p$ we may, without loss of generality, assume that $f_p \equiv 1$ on D'. Let

$$E = \left\{\sum a_m z^m \in O_0 : \sum |a_m| r^m < \infty\right\}$$

E is the set of all bounded analytic functions on D. If we define

$$\left\|\sum a_m z^m\right\| = \sum |a_m| r^m$$

it is easily verified that $\| \ \|$ is a norm on E and that, with respect to $\| \ \|$, E is complete.

Since our series are absolutely convergent we have

$$\|ab\| \leqslant \|a\| \, \|b\| \qquad \text{for all } a, b \in E$$

Let us set $d = \max\{r_1, ..., r_{n-1}\}$ and $s = r_n$. Since $f_j(0, ..., 0) = 0$, $0 \leqslant j \leqslant p-1$, there exists $M_1 \geqslant 0$, independent of d and s, such that

$$\|f_j(z_n \ldots z_{n-1})z_n^j\| \leqslant M_1 ds^j/p, \quad 0 \leqslant j \leqslant p-1$$

On the other hand, there exists $M_2 > 0$ such that

$$\|\sum_{j>p} f_j(z_1, ..., z_{n-1})z_n^j\| \leqslant M_2 s^{p+1}$$

Hence

$$\|f - z_n^p\| = \|f - f_p(z_1, ..., z_{n-1})z_n^p\| \quad \leqslant M_2 s^{p+1} + M_1 d(1 + s^p)$$

Given $0 < \epsilon < 1$, we may choose d and s sufficiently small so that

$$M_1 d(1 + s^p) + M_2 s^{p+1} \leqslant \epsilon s^p$$

and so

$$\|f - z_n^p\| \leqslant \epsilon s^p$$

For the rest of the proof, d and s (and the corresponding polydiscs D and D′) will remain fixed.

Now let $g \in A(D)$ be bounded on D and set

$$g = q(g)z_n^p + h(g)$$

where $q(g), h(g) \in E$ and $h(g)$ is a polynomial in z_n of degree less than or equal to $p-1$. Notice that q and h are uniquely determined by g. We observe that

$$\|g\| = \|q(g)\|s^p + \|h(g)\|$$

and so $\|g\| \geqslant \|q(g)\|s^p$

We define a (continuous) linear map $A: E \rightarrow E$ by

$$Ag = q(g)f + h(g)$$

Now

$$\begin{aligned} \|(A-I)g\| &= \|q(g)f + h(g) - g\| \\ &\leqslant \|q(g)(f - z_n^p)\| \\ &\leqslant \epsilon s^p \|q(g)\| \\ &\leqslant \epsilon\|g\| \end{aligned}$$

It follows that A is a linear isomorphism (for a proof of this standard result from Banach space theory see Ref. [7]). Hence, for any bounded $g \in A(D)$, there exists $\psi \in E$ such that

$$g = A\psi = q(\psi)f + h(\psi)$$

This proves the first part of the theorem.

Now

$$\|q(\psi)\| s^p \leqslant \|\psi\|$$

$$= \|A^{-1} A(\psi)\|$$

$$\leqslant \|A^{-1}\| \, \|g\|$$

and so

$$\|q(\psi)\| \leqslant s^{-p} \|A^{-1}\| \, \|g\|$$

We also have

$$\|h(\psi)\| \leqslant \|\psi\| + \|q(\psi)\| s^p$$

$$\leqslant \|A^{-1} A(\psi)\| + s^{-p} \|A^{-1}\| \, \|g\|$$

$$\leqslant \|A^{-1}\| \, (1 + s^{-p}) \, \|g\|$$

Since $\|f\|_D \leqslant \|f\|$, for all $f \in E$, the last statement of the theorem follows with $M = \|A^{-1}\| \, (1 + s^{-p})$.

The uniqueness of q and r follows easily from the injectivity of A and we leave details to the reader. Q.E.D.

Remarks

(1) Notice that if f is any analytic function defined on a neighbourhood of $0 \in \mathbb{C}^n$, we may always find a linear change of co-ordinates of $\mathbb{C}^n$ such that the conditions of Theorem 2.4 hold for f.

(2) The Weierstrass preparation theorem gives us a procedure for *dividing* (germs of) analytic functions. For this reason it is sometimes referred to as a division theorem. In recent years an important differentiable analogue of the Weierstrass preparation theorem has been proved by B. Malgrange. This result (the Malgrange preparation theorem) has been developed and applied by Malgrange, Mather and many others to prove important results about differentiable functions. We refer to Refs [22, 24, 36] for more details.

Our first application of the Weierstrass preparation theorem will be to put the germ of an analytic function in some sort of canonical form.

Definition 2.5. $P \in O_0$ is said to be a *Weierstrass polynomial* (of degree p) if

$$P(z_1, \ldots, z_n) = z_n^p + \sum_{j=0}^{p-1} a_j(z_1, \ldots, z_{n-1}) z_n^j$$

where the a_j are analytic functions defined on some neighbourhood of $0 \in \mathbb{C}^{n-1}$ and vanishing at 0 for every j.

For notational economy we shall now denote the variable $(z_1, \ldots, z_{n-1}) \in \mathbb{C}^{n-1}$ in abbreviated form by z' and let O_0' denote the ring of germs of analytic functions at $0 \in \mathbb{C}^{n-1}$.

Corollary 2.4.1. Let $f \in O_0$ and suppose that $f(0, ..., 0, z_n)$ does not vanish identically and has a zero of order p at $z_n = 0$. Then there exist a unique unit $h \in O_0$ and Weierstrass polynomial W of degree p such that $f = hW$, i.e. on some neighbourhood of $0 \in \mathbb{C}^n$ we have

$$f(z) = h(z)\left(z_n^p + \sum_{j=0}^{p-1} a_j(z')z_n^j\right)$$

where $h(0) \neq 0$ and $a_j(0) = 0$, $0 \leqslant j \leqslant p-1$.

Proof. Take $g(z) = z_n^p$ in Theorem 2.4. Q.E.D.

We shall now use Theorem 2.4, together with Corollary 2.4.2, to prove some important results about the structure of O_0. First we recall

Definition 2.6. Let R be a commutative integral domain with identity.

(1) R is said to be a *unique factorization domain* if every $f \in R$ can be written as a finite product of irreducible factors and if this decomposition of f is unique up to the order of the factors and units.

(2) R is said to be *Noetherian* if every ideal $I \lhd R$ is finitely generated, i.e. there exist $g_1, ..., g_k \in I$ such that

$$I = \left\{\sum \alpha_j g_j : \text{all } \alpha_j \in R\right\}$$

(3) If $M \subset R^m$ (R^m denotes the m-fold product of R), M is said to be a *submodule* of R^m if M is closed under (co-ordinatewise) addition and scalar multiplication by elements of R.

We recall the algebraic

Lemma 2.7. Let R be Noetherian and M be a submodule of R^m. Then M is finitely generated.

Proof. By induction. Trivial for $m = 1$, since M is then an ideal of R. Suppose true for $m-1$. Let $\pi : R^m \to R$ denote the projection on the first factor and set $M_1 = \pi M \subset R$, $M_2 = \{(r_2, ..., r_{m-1}) \in R^m : (0, r_2, ..., r_m) \in M\}$. M_1 is an ideal and so we may find $f_1, ..., f_k \in M$ such that the set $\pi(f_1), ..., \pi(f_k)$ generates M_1. We let $\widetilde{M}_1$ be the submodule of M generated by $f_1, ..., f_k$. We leave it to the reader to verify that $M = \widetilde{M}_1 + M_2$. Applying the inductiv hypothesis to M_2 gives the result. Q.E.D.

Theorem 2.8. O_0 is Noetherian and a unique factorization domain.

Proof. We shall prove that O_0 is Noetherian. The proof that O_0 is a unique factorization domain is similar and we refer to Refs [12, 16] for full details.

For $n = 1$, the result is trivially true since every ideal in O_0 is generated by a power of z_1. Suppose the result is true for $n-1$ and let I be a non-zero ideal of O_0. Changing co-ordinates if necessary, we may find $W \in I$ such that W is a Weierstrass polynomial:

$$W(z) = z_n^p + \sum_{j=0}^{p-1} a_j(z')z_n^j$$

For any $f \in I$ we then have, by Theorem 2.4,

$$f = q(f)W + r(f)$$

where $r(f)$ is a polynomial in z_n of degree $<p$. Let $M = \{r(f) : f \in I\}$. If we write

$$r(f) = a_1(z')z_n^{p-1} + \ldots + a_p(z')$$

where $a_j \in O'_0$, we see that the set of all coefficients $(a_1, \ldots, a_p)$ corresponding to polynomials $r(f)$, $f \in I$, defines a submodule of O'^p_0. By Lemma 2.7 and our inductive hypothesis we may find a set $(h_{11}, \ldots, h_{1p}), \ldots, (h_{k1}, \ldots, h_{kp})$ of generators for this submodule. In other words, the set of polynomials

$$p_j(z', z_n) = h_{j1}(z')z_n^{p-1} + \ldots + h_{jp}(z'), \quad 1 \leqslant j \leqslant k$$

generates M over O'_0. Therefore $\{W, p_1, \ldots, p_k\}$ gives a finite set of generators of I. Q.E.D.

Remark. The technique of the proof of Theorem 2.8 is fairly standard: we use the Weierstrass preparation theorem to reduce a problem about analytic functions in n variables to one about polynomials whose coefficients are analytic functions in n–1 variables and then use induction on n.

Example. Let us see what the above algebraic results tell us about the local structure of the zero set of an analytic function of more than one variable. Let $f \in O_0$. By uniqueness of factorization, we may decompose f into a product of primes. Thus let us set

$$f = p_1^{s_1} \ldots p_k^{s_k}$$

where $p_1, \ldots, p_k \in O_0$ are prime. Let us suppose that $p_1, \ldots, p_k$, f are defined as analytic functions on the polydisc neighbourhood $D = \{z : |z_1| < r_1, \ldots, |z_n| < r_n\}$. Let

$$X = \{z \in D : f(z) = 0\}$$

Since $f = p_1^{s_1} \ldots p_k^{s_k}$, it follows immediately that

$$X = \bigcup_{j=1}^{k} X_j$$

where $X_j = \{z \in D : p_j(z) = 0\}$, $1 \leqslant j \leqslant k$.

It follows that the local study of the zero set of $f \in O_0$ may be reduced to that of the zero sets of *primes* in O_0. Let us therefore suppose that $p \in A(D)$ and that p defines a prime of O_0. Let $D' = \{z \in \mathbb{C}^{n-1} : |z_1| < r_1, \ldots, |z_{n-1}| < r_{n-1}\}$. Changing co-ordinates if necessary and multiplying by a unit, we may suppose that p is a Weierstrass polynomial (on a possibly smaller D):

$$p(z, z_n) = z_n^p + \sum_{j=0}^{p-1} a_j(z')z_n^j$$

where $a_j \in A(D')$ vanish at zero. Let $\delta \in A(D')$ denote the *discriminant* of p (δ is a polynomial in the coefficients $a_0, \ldots, a_{p-1}$ of p and $\delta(z') = 0$ if and only if the polynomial $p(z', z_n)$ has a multiple root at z'). The irreducibility of p implies that $\delta \not\equiv 0$. Let $\Sigma = \{z \in D' : \delta(z') = 0\}$.

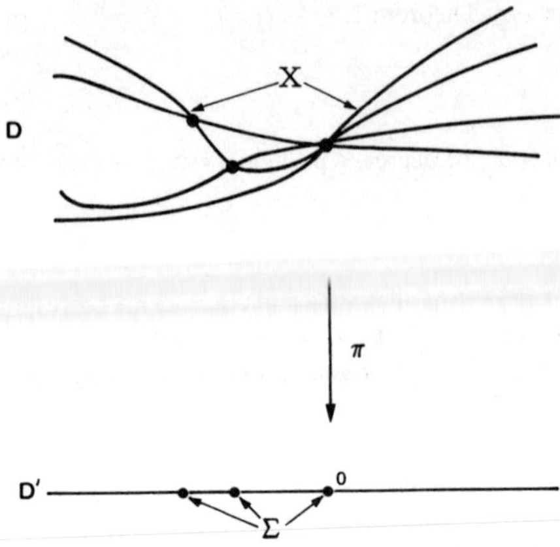

FIG.1. Local description of zero set.

We let $\pi : D \to D'$ denote the projection onto the first n−1 co-ordinates. Figure 1 describes $X = p^{-1}(0)$.

At each point $z' \in D' \setminus \Sigma$

$$z_n^p + \sum_{j=0}^{p-1} a_j(z') z_n^j$$

has precisely p distinct roots (in Fig. 1, p = 4). π clearly restricts to a p-fold covering map of X over $D' \setminus \Sigma$ and $X \setminus \pi^{-1}(\Sigma)$ is a complex manifold of dimension n−1. "Singularities" in X occur over Σ. Notice though that since Σ is the zero set of $\delta \in A(D')$, Σ is of dimension $\leqslant n-2$ and so the singularities of $X\ (\subset \pi^{-1}(\Sigma))$ are of dimension $\leqslant n-2$ (we deliberately do not define "dimension" here – we refer to the references for more details).

How exactly does the *uniqueness* part of unique factorization enter into the geometry of zero sets? Recall that above we obtained a decomposition

$$X = \bigcup_{j=1}^{k} X_j$$

of the zero set of an arbitrary $f \in O_0$ in terms of zero sets of primes. We claim that this decomposition is *unique*. That is, suppose that we also have

$$X = \bigcup_{j=1}^{k'} X_j'$$

where X_j' is the zero set of some prime $p_j' \in O_0$, we claim $k = k'$ and, up to units, $\{p_1, ..., p_k\} = \{p_1',$

To prove this uniqueness it is sufficient to verify the following special case:

Let $P, Q \in O_0$ be primes and suppose that $P^{-1}(0) \subset Q^{-1}(0)$.
Then $P = uQ$ for some unit $u \in O_0$.

For the proof of the special case, we may suppose that P is an irreducible Weierstrass polynomial of degree p. By the division theorem

$$Q = hP + r$$

where r is a polynomial of degree $\leqslant p-1$. If δ denotes the discriminant of P and $\delta(z') \neq 0$, then P has p distinct roots at z'. So, therefore, does r. Since r is of degree $\leqslant p-1$, it follows that $r = 0$ where $\delta \neq 0$. By uniqueness of analytic continuation $r \equiv 0$. For the general case, we may show that each p_i', $1 \leqslant i \leqslant k'$, must vanish identically on at least one X_j and then apply the special case. For full details and also details on the applications of the Noetherian properties of O_0, we refer to Refs [12, 25, 32].

Our final applications of the preparation theorem will enable us to define meromorphic functions of more than one variable. First, we have some technicalities. We shall let $O_0'[z_n]$ denote the ring of polynomials in z_n with coefficients in O_0'.

Lemma 2.9. Suppose $P = QR$, $P, Q, R \in O_0$, and that P is a polynomial in z_n and R a Weierstrass polynomial (in z_n). Then $Q \in O_0'[z_n]$.

Proof. Since the highest-order coefficient of R is a unit in O_0' (actually 1, since R is a Weierstrass polynomial), we may apply the usual algebraic division algorithm (for the ring $O_0'[z_n]$) to obtain

$$P = \tilde{Q}R + \tilde{S}$$

where $\tilde{Q}, \tilde{S} \in O_0'[z_n]$.

The uniqueness part of the Weierstrass preparation theorem implies that $Q = \tilde{Q}$ and $\tilde{S} = 0$. Q.E.D.

Lemma 2.10. Let $P, Q, R \in O_0'[z_n]$ satisfy $P = QR$. If P is a Weierstrass polynomial then so are Q and R.

Proof. Let p = degree(P), q = degree(Q), r = degree(R). Then $p = q+r$ and

$$z_n^p = P(0,z_n) = Q(0,z_n)\, R(0,z_n)$$

Hence $Q(0,z_n)/z_n^q$ and $R(0,z_n)/z_n^r$ are non-zero constants. This clearly implies that, up to units in O_0', Q and R are Weierstrass polynomials. Q.E.D.

Let $U \subset \mathbb{C}^n$ be open and $f \in A(U)$. We use the abbreviated notation f_a to denote the germ $p_U^a(f) \in O_a$ of f at a.

Theorem 2.11. Let f and g be analytic on some neighbourhood of $0 \in \mathbb{C}^n$. Suppose that $f_0, g_0 \in O_0$ are relatively prime, i.e. have no common irreducible factor. Then we may find an open neighbourhood D of $0 \in \mathbb{C}^n$ such that

$f_a, g_a \in O_a$ are relatively prime for all $a \in D$

Proof. Without loss of generality we may assume that f and g are Weierstrass polynomials. By Lemma 2.10, f and g are relatively prime in $O'_0[z_n]$. It follows from Gauss' lemma that f and g are relatively prime in $\mathcal{M}'_0[z_n]$, where $\mathcal{M}'_0$ denotes the quotient field of O'_0. Hence there exist $a, b \in O'_0[z_n]$ and $h \in O'_0$ such that on some neighbourhood of 0 we have

$$h(z') = a(z)f(z) + b(z)g(z)$$

Since f and g are Weierstrass polynomials we may find an open polydisc $D' = \{z' \in \mathbb{C}^{n-1}: |z_1| < r_1, \dots, |z_{n-1}| < r_{n-1}\}$ such that for all $z' \in D'$ the polynomials

$$f(z',z_n), g(z',z_n) \in \mathbb{C}[z_n]$$

do not vanish identically. Choose $r_n > 0$ such that $f, g, a, b, h \in A(D)$, where $D = \{z \in \mathbb{C}^n: |z_j| < r_j\ 1 \leqslant j \leqslant n\}$. Let $\zeta = (\zeta', \zeta_n) \in D$. Since $f(\zeta', (z_n - \zeta_n)), g(\zeta', (z_n - \zeta_n))$ do not vanish identically, we may assume that any common factor of f_ζ, g_ζ is the germ of a Weierstrass polynomial in $O_{\zeta'}[z_n - \zeta_n]$. But this polynomial must divide h_ζ and so, by Lemma 2.9, it must be of degree zero. Hence it is a unit. Q.E.D.

Remark. Theorem 2.11 is very characteristic of local results in complex analysis: an algebraic result that holds at a point tends to hold in a neighbourhood of the point. In other words we have a transition from algebra (results holding at a point) to topology (results holding on an open set). The full development of this relation between algebra and topology requires the machinery of sheaf cohomology theory (see Refs [12, 16, 25, 32]).

O_a is an integral domain and so we may form the field of fractions (quotient field) $\mathcal{M}_a$ of O_a. Thus

$$\mathcal{M}_a = \{f_a/g_a : f_a, g_a \in O_a, g_a \neq 0\}$$

Since O_a is a unique factorization domain, each $m \in \mathcal{M}_a$ may be written uniquely (up to multiplication by units) in the form f_a/g_a where f_a and g_a are relatively prime.

Let us set

$$\mathcal{M} = \bigcup_{a \in \mathbb{C}^n} \mathcal{M}_a$$

$\mathcal{M}$ is called the *sheaf of germs of meromorphic functions* on $\mathbb{C}^n$.

Definition 2.12. Let $U \subset \mathbb{C}^n$ be open. A map $m: U \to \mathcal{M}$ is said to be a *meromorphic function* on U if

(1) $m(z) \in \mathcal{M}_z$ for all $z \in U$.

(2) For every $z \in U$, there exists an open neighbourhood $V \in \mathcal{U}_z$ and $f, g \in A(V)$ such that $m(a) = f_a/g_a$ for all $a \in V$.

We shall denote the set of meromorphic functions on U by $\mathcal{M}(U)$.

Example. Let $U \subset \mathbb{C}^n$ be open and let $f, g \in A(U)$, where g is not identically zero on any connected component of U. We may define $m \in \mathcal{M}(U)$ by

$$m(z) = f_z/g_z, \quad z \in U \quad (\text{“}m = f/g\text{”})$$

However, unless U satisfies additional conditions, not every meromorphic function on U may be written in this form.

Proposition 2.13. Let $U \subset \mathbb{C}^n$ be open and let $m \in \mathscr{M}(U)$. We may find an open cover $\{U_j\}$ of U and $m_j \in \mathscr{M}(U_j)$ such that

(1) $m_j = m|U_j$.

(2) For every j, there exist $f_j, g_j \in A(U_j)$ such that $m_j = f_j/g_j$ and $f_{j,a}$ and $g_{j,a}$ are relatively prime for all $a \in U_j$ ($f_{j,a}$ denotes the germ of f_j at a).

Proof. Immediate from Theorem 2.11.

Remark. Proposition 2.13 is relatively trivial if $n = 1$: every meromorphic function m on $U \subset \mathbb{C}$ may be written in the form $u(z)(z-a)^p$ on some neighbourhood of each point $a \in U$ ($u(a) \neq 0$ and $p \in Z$). Moreover, it follows easily from the Weierstrass theorem that we may find $f, g \in A(U)$ such that $m = f/g$ and f_a and g_a are relatively prime for all $a \in U$. This representation of m is then unique up to units.

Notation. If $f_a, g_a \in O_a$ are relatively prime we shall write $(f_a, g_a) = 1$. More generally, if $f, g \in A(U)$ and $(f_a, g_a) = 1$ for all $a \in U$, we shall write $(f, g) = 1$.

The local description of meromorphic functions given by Proposition 2.13 is essentially unique as the next lemma shows.

Lemma 2.14. Let $U \subset \mathbb{C}^n$ be open and $m \in \mathscr{M}(U)$. Suppose that for some open subset $V \subset U$ we may find $f, f', g, g' \in A(V)$ such that

(A) $m|V = f/g = f'/g'$.

(B) $(f, g) = 1$, $(f', g') = 1$.

Then there exists $u \in A(V)$ such that

(1) $f = uf'$, $g = ug'$.

(2) u does not vanish anywhere on u, i.e. u_a is a unit for all $a \in V$.

Proof. At each point $a \in V$ we have $f_a/g_a = f'_a/g'_a$. Hence $f_a g'_a = g_a f'_a$. Since $(f'_a, g'_a) = 1$, f'_a must divide f_a and so there exists a unique $u_a \in O_a$ such that

$$f_a = u_a f'_a$$

Substituting, we see at once that $g_a = u_a g'_a$. Since $(f_a, g_a) = 1$, there exists $v_a \in O_a$ such that

$$f'_a = v_a f_a ; \; g'_a = v_a g_a$$

and so $u_a v_a = 1$ and u_a is a unit. So far the argument is purely algebraic. Now we observe that since u_a is defined uniquely for all $a \in V$ it is the germ of some $u \in A(V)$. u clearly satisfies conditions 1 and 2 of the lemma. Q.E.D.

Next we wish to show how the definition of pole and zero sets generalizes to meromorphic functions of more than one variable. First we recall

Definition 2.15. Let $U \subset \mathbb{C}^n$ be open and $X \subset U$. X is said to be an *analytic set* if for each point $z \in U$ we may find an open neighbourhood W_z of z and an analytic map $h^z : W_z \to \mathbb{C}^p$ such that $(h^z)^{-1}(0) = X \cap W_z$. The integer p will, in general, depend on z. (For more details about analytic sets we refer the reader to Refs [12, 25, 32].)

Let $U \subset \mathbb{C}^n$ be open and $m \in \mathscr{M}(U)$. By Proposition 2.13 we may find an open cover $\{U_j\}$ of U and $m_j \in \mathscr{M}(U_j)$ such that $m_j = m|U_j$ and for every j there exist $f_j, g_j \in A(U_j)$ such that $m_j = f_j/g_j$ and $(f_j, g_j) = 1$. We define the subsets $Z(m), P(m), T(m) \subset U$ by

$$Z(m) \cap U_j = \{z \in U_j : f_j(z) = 0\}$$

$$P(m) \cap U_j = \{z \in U_j : g_j(z) = 0\}$$

$$T(m) \cap U_j = \{z \in U_j : f_j(z) = g_j(z) = 0\}$$

We claim that Z(m), P(m), T(m) are well defined analytic subsets of U which depend only on m (not on the local representation of m as f_j/g_j). Let us choose another open cover $\{U'_k\}$ of U together with relatively prime $f'_k, g'_k \in A(U'_k)$ such that for every k, $m|U'_k = f'_k/g'_k$. On the overlap $U_j \cap U'_k$ we have

$$f_j/g_j = f'_k/g'_k$$

and so, by Lemma 2.14, there exists a nowhere zero $u \in A(U_j \cap U'_k)$ such that on $U_j \cap U'_k$

$$f_j = uf'_k;\ g_j = ug'_k$$

But this implies immediately that $f_j(z) = 0$ if and only if $f'_k(z) = 0$ and $g_j(z) = 0$ if and only if $g'_k(z) = 0$. Hence Z(m) and P(m) are well defined analytic sets. Since $T(m) = Z(m) \cap P(m)$, T(m) is also a well defined analytic set.

Definition 2.16. Let $U \subset \mathbb{C}^n$ be open and $m \in \mathcal{M}(U)$. The sets Z(m), P(m), T(m) constructed above are respectively called the *zero set of m,* the *pole set of m* and the *indeterminancy set of m.*

Remarks
(1) Notice the key role played by Theorem 2.11 and the Weierstrass preparation theorem in the development of the elementary theory of meromorphic functions of more than one variable.
(2) If m is a meromorphic function on $U \subset \mathbb{C}$, $T = \emptyset$. However, as we shall soon see, T need not vanish for meromorphic functions of more than one variable.

Example. Let $f(z_1, z_2) = z_1^2 - z_2^2$ and $g(z_1, z_2) = z_1 z_2$ and define $m = f/g$. Then $m \in \mathcal{M}(\mathbb{C}^2)$ and is an example of an *entire meromorphic function* (in fact a *rational function,* since f and g are polynomials). The reader may easily verify that

$$Z(m) = \{(z_1, z_2) : z_1 = \pm z_2\}$$

$$P(m) = \{(z_1, z_2) : z_1 = 0 \text{ or } z_2 = 0\}$$

$$T(m) = \{(0,0)\}$$

We shall point out the significance of the indeterminancy set of a meromorphic function after the next proposition.

Proposition 2.17. Let $U \subset \mathbb{C}^n$ be open and $m \in \mathcal{M}(U)$. Then m defines an analytic function on $U \backslash P(m)$.

Proof. Let $\{U_j\}$ be an open cover of U and $f_j, g_j \in A(U_j)$ be relatively prime and satisfy $m|U_j = f_j$
We may define

$$m : U \backslash P(m) \to \mathbb{C}$$

by

$$m(z) = f_j(z)/g_j(z), \; z \in U_j$$

We leave it to the reader to check that m gives a well defined analytic function on $U \setminus P(m)$. Q.E.D.

Remarks. If m is a meromorphic function on an open subset U of $\mathbb{C}$, the pole set of m is a discrete subset of U and the analytic map $m : U \setminus P(m) \to \mathbb{C}$ naturally extends to an analytic map $m : U \to P^1(\mathbb{C})$ by defining $m(z) = \infty$ for $z \in P(m)$ ($P^1(\mathbb{C})$ denotes one-dimensional complex projective space, i.e. the Riemann sphere). Therefore, the study of meromorphic functions of one variable is equivalent to the study of holomorphic curves in $P^1(\mathbb{C})$. There is no such way of "compactifying" a meromorphic function of more than one variable to give a holomorphic function into some complex manifold. Indeed, it may be shown that if $z \in T(m)$, then m takes all values, including ∞, infinitely often in any neighbourhood of z and so there is no way of assigning a value to m at z. We refer the reader to Refs [12, 16] for proofs of this fact.

To conclude this section, we shall return to the problem of constructing meromorphic functions with specified principal parts or, alternatively, with specified pole and zero sets.

Definition 2.18. Let $U \subset \mathbb{C}^n$ be open and suppose that we are given an open cover $\{U_j\}$ of U and $m_j \in \mathscr{M}(U_j)$ such that $m_j - m_k \in A(U_j \cap U_k)$ for all j, k. The "Cousin I problem" is to find $m \in \mathscr{M}(U)$ such that for every j

$$m - m_j \in A(U_j)$$

It is known that the Cousin I problem is always solvable if U is a domain of holomorphy (more generally, a Stein manifold – see Refs [12, 16, 32]). In fact it is easy to solve if we know that the following portion of the $\bar{\partial}$-sequence is exact:

$$0 \to A(U) \to C^\infty(U) \xrightarrow{\bar{\partial}} C^\infty(U, \Lambda^1 \bar{\mathbb{C}}^{n*}) \xrightarrow{\bar{\partial}} C^\infty(U, \Lambda^2 \bar{\mathbb{C}}^{n*})$$

As remarked at the end of Section 1, the $\bar{\partial}$-sequence is exact for domains of holomorphy (more generally, Stein manifolds). The proof that the Cousin I problem is always solvable if the above portion of the $\bar{\partial}$-sequence is exact is the same as the proof we gave for the Mittag-Leffler theorem and we shall not repeat it.

Next we turn to the generalization of the Weierstrass theorem.

Notation. Let $U \subset \mathbb{C}^n$ be open. We let $A^*(U)$ denote the set of analytic functions on U which do not vanish anywhere on U. We let $\mathscr{M}^*(U)$ denote the set of meromorphic functions on U which are not identically zero on any connected component of U. Notice that if $f \in A^*(U)$ then $1/f \in A^*(U)$, and $m \in \mathscr{M}^*(U)$ implies $1/m \in \mathscr{M}^*(U)$. In other words, both $A^*(U)$ and $\mathscr{M}^*(U)$ are (multiplicative) groups and $A^*(U)$ is a subgroup of $\mathscr{M}^*(U)$.

Definition 2.19. Let $U \subset \mathbb{C}^n$ be open and suppose that we are given an open cover $\{U_j\}$ of U and $m_j \in \mathscr{M}^*(U_j)$ such that $m_j m_k^{-1} \in A^*(U_j \cap U_k)$ for all j, k. The Cousin II problem is to find $m \in \mathscr{M}(U)$ such that for every j

$$m m_j^{-1} \in A^*(U_j)$$

It is known that the Cousin II problem is always solvable if U is a domain of holomorphy (more generally, a Stein manifold) and $H^2(U,\mathbb{Z}) = 0$. In fact the proof follows easily from the exactness of the $\bar{\partial}$-sequence for domains of holomorphy. We shall not give details here but refer to Refs [12, 16, 32]. See also Sections 3 and 4.

As in the case of meromorphic functions of one complex variable, the solution of the Cousin II problem gives much information about meromorphic functions. For example, if U is a domain of holomorphy, every $m \in \mathcal{M}(U)$ may be written in the form f/g, for $f,g \in A(U)$. Moreover, if $H^2(U,\mathbb{Z}) = 0$, we may require that $(f,g) = 1$.

3. MEROMORPHIC FUNCTIONS, DIVISORS AND LINE BUNDLES

In this section we turn our attention to (compact) complex manifolds and, in particular, to generalizations of the Weierstrass theorem. We recall the definition of a complex manifold (see also Ref. [31], these Proceedings).

Definition 3.1. A paracompact Hausdorff topological space M is said to be a *complex manifold* of dimension n if we can find an atlas of charts $\mathcal{A} = \{(U_\alpha, \phi_\alpha)\}_{\alpha\in\Lambda}$ for M such that for all $\alpha, \beta \in$

$$\phi_\alpha \dot{\phi}_\beta^{-1} : \phi_\beta(U_\alpha \wedge U_\beta) \subset \mathbb{C}^n \to \mathbb{C}^n$$

is analytic.

Let M be a complex manifold of dimension n and fix $z \in M$. We let $\mathcal{U}_z$ denote the set of all open neighbourhoods of z in M. We may define an equivalence relation $\sim_z$ on the set of all analytic functions defined on some neighbourhood of z by:

Let $f \in A(U)$, $g \in A(V)$, $U, V \in \mathcal{U}_z$. Then $f \sim_z g$ if there exists $W \in \mathcal{U}_z$ such that $f|W$

That is, $f \sim_z g$ if and only if f = g on some neighbourhood of z. Obviously $\sim_z$ is an equivalence relation and we let $O_{M,z}$ denote the set of equivalence classes of $\sim_z$. If $f \in A(U)$, $U \in \mathcal{U}_z$, we let f_z denote the $\sim_z$ equivalence class of f. f_z is called the *germ* of f at z.

$O_{M,z}$ naturally has the structure of a commutative ring induced from the commutative ring $\{A(U) : U \in \mathcal{U}_z\}$. Indeed, if $f \in A(U)$, $g \in A(V)$, we define

$$f_z + g_z = (f+g)_z \quad ; \quad f_z g_z = (fg)_z$$

We refer to Refs [9, 12] for more details.

Definition 3.2. $O_{M,z}$ is called the *ring of germs of analytic functions* at z.

The relationship between $O_{M,z}$ and the rings O_a, $a \in \mathbb{C}^n$, considered in Section 2 is given b

Proposition 3.3. Let M be a complex manifold of dimension n. Then $O_{M,z}$ is isomorphic (as a ring) to O_0, $0 \in \mathbb{C}^n$, for all $z \in M$.

Proof. Let us take a fixed chart (U_α, ϕ_α) of M with $U_\alpha \in \mathcal{U}_z$ and $\phi_\alpha(z) = 0 \in \mathbb{C}^n$. We define a map

$$\gamma_U : A(U) \to O_0$$

by

$$\gamma_U(f) = \text{power series of } f\phi_\alpha^{-1} \text{ at } 0$$

We leave it to the reader to verify that the set $\{\gamma_U : U \in \mathscr{U}_z\}$ induces the required isomorphism between $O_{M,z}$ and O_0. Notice, though, that the isomorphism depends on the choice of chart (U_α, ϕ_α).

The above proposition shows that the algebraic structure of $O_{M,z}$ is that of the ring O_0 studied in Section 2. In particular, $O_{M,z}$ is a Noetherian unique factorization domain. We set

$$O_M = \bigcup_{z \in M} O_{M,z}$$

O_M is called the *sheaf of germs of analytic functions* on M or the *Oka sheaf* of M.

Suppose $s: M \to O_M$ satisfies:
(1) $s(z) \in O_{M,z}$ for all $z \in M$ (s is a "section" of O_M).
(2) For each $z \in M$, $\exists U \in \mathscr{U}_z$, $f \in A(U)$ such that $s(y) = f_y$ for all $y \in U$ (s is continuous, relative to the "sheaf topology" on O_M – see Refs [9 , 16]).

We claim that s defines a unique analytic function $\tilde{s}$ on M. Indeed, we may define $\tilde{s} \in A(M)$ by $\tilde{s}_z = s(z)$, $z \in M$, or, with the notation of (2) above, by

$$\tilde{s}|U = f$$

The reader may easily verify that $\tilde{s}$ is a well defined analytic function on M. Conversely, if $\tilde{s} \in A(M)$ we may define a unique map $s: M \to O_M$ satisfying (1) and (2) above by setting $s(z) = \tilde{s}_z$ for all $z \in M$. Hence we may think of an analytic function on M as a continuous section of the sheaf O_M.

Although we shall not give definitions or make any systematic study of sheaves in these notes, most of what we do in the rest of this section can be written most elegantly in the formalism of sheaves (we refer to Refs [9, 12, 16, 32] for basic sheaf theory relevant to complex analysis).

For the rest of this section we shall drop the subscript M from O_M and use the notation O for the Oka sheaf of M.

Since O_z is an integral domain for all $z \in M$, we may form the quotient field $\mathscr{M}_z$ of O_z. We set

$$\mathscr{M} = \bigcup_{z \in M} \mathscr{M}_z$$

$\mathscr{M}$ is called the *sheaf of germs of meromorphic functions* on M.

Definition 3.4. A map $m: M \to \mathscr{M}$ is said to define a *meromorphic function* on M if:
(1) $m(z) \in \mathscr{M}_z$, for all $z \in M$.
(2) For each $z \in M$, there exist $U \in \mathscr{U}_z$, $f, g \in A(U)$ such that

$$m(y) = f_y/g_y, \text{ for all } y \in U$$

We denote the set of meromorphic functions on M by $\mathscr{M}(M)$.

Remark. It follows from Theorem 2.11 that if $m \in \mathscr{M}(M)$, we may find an open cover $\{U_j\}$ of M and corresponding $f_j, g_j \in A(U_j)$ such that $m|U_j = f_j/g_j$ and $(f_j, g_j) = 1$ for every j. Using this fact, we may define the pole, zero and indeterminancy sets of $m \in \mathscr{M}(M)$ just as we did in Section 2. We let P(m), Z(m), T(m) denote the pole, zero and indeterminancy sets of $m \in \mathscr{M}(M)$ respectively. As in Section 2, P(m), Z(m) and T(m) are analytic subsets of M.

Example 1. Let M denote a Riemann surface (i.e. a complex manifold of dimension 1) and let $m \in \mathscr{M}(M)$. We claim that Z(m) and P(m) are discrete subsets of M and that $T(m) = \emptyset$. This follows easily from the known result when M is an open subset of $\mathbb{C}$. Indeed, for any $z \in M$, we may find a chart (U, ϕ) for M, which contains z, and $f, g \in A(U)$ such that $m|U = f/g$. We have

$$Z(m) \cap U = \phi^{-1}(Z(f\phi^{-1}/g\phi^{-1})) \;;\; P(m) \cap U = \phi^{-1}(P(f\phi^{-1}/g\phi^{-1}))$$

Moreover, we may define the multiplicity of poles and zeroes of m. Let $z \in Z(m)$. Then we may choose a chart (U,ϕ) containing z and $f, g \in A(U)$ such that

$$m|U = f/g \text{ and } (f_z, g_z) = 1$$

We define the multiplicity of the zero z of m to be the multiplicity of the zero $\phi(z)$ of $f\phi^{-1} : \phi(U) \subset \mathbb{C} \to \mathbb{C}$. Similarly for poles. The reader may verify that these definitions do not depend on our choice of chart.

Example 2

Let us construct some meromorphic functions on $P^n(\mathbb{C})$. ($P^n(\mathbb{C})$ denotes complex projectiv space of dimension n, i.e. the set of all compex lines through the origin of $\mathbb{C}^{n+1}$ – see Ref. [31], these Proceedings, for construction of the complex structure on $P^n(\mathbb{C})$). Let $P, Q \in A(\mathbb{C}^{n+1})$ be homogeneous, relatively prime, polynomials of the same degree r. That is, P and Q are polynomials with no common factor and for all $z \in \mathbb{C}^{n+1}$ and $\lambda \in \mathbb{C}$ we have

$$P(\lambda z) = \lambda^r P(z) \;;\; Q(\lambda z) = \lambda^r Q(z)$$

Let Z(Q) denote the zero set of Q. Then Z(Q) is a union of complex lines through the origin of $\mathbb{C}^{n+1}$ since $z \in Z(Q)$ if and only if $\lambda z \in Z(Q)$, $\lambda \in \mathbb{C}^*$. Hence Z(Q) defines a subset of $P^n(\mathbb{C})$ which we shall continue to denote by Z(Q). (Z(Q) is an example of a projective algebraic variety see Ref. [9] for a more complete definition) Now P/Q is homogeneous of degree *zero* on $\mathbb{C}^{n+1} \setminus$ i.e.

$$(P/Q)(\lambda z) = (P/Q)(z), \text{ for all } z \in \mathbb{C}^{n+1} \setminus Z(Q), \lambda \in \mathbb{C}$$

But this implies that P/Q is constant on lines in $\mathbb{C}^{n+1}$ and so P/Q drops down to a map

$$P/Q : P^n(\mathbb{C}) \setminus Z(Q) \to \mathbb{C}$$

We shall now show that P/Q gives a well-defined meromorphic function on $P^n(\mathbb{C})$ with pole set Z(Q) and zero set Z(P). To do this we recall that $P^n(\mathbb{C})$ has atlas $\{(U_i, \phi_i) : i = 1, ..., n+1\}$ where

$$U_i = \{(z_1, ..., z_{n+1}) : z_i \neq 0\}$$

$$\phi_i : U_i \to \mathbb{C}^n \;;\; (z_1, ..., z_{n+1}) \mapsto (z_1/z_i, ..., \widehat{z_i/z_i}, ..., z_{n+1}/z_i)$$

"$\widehat{\ }$" denotes omission.

We define $f_i, g_i \in A(U_i)$ by

$$f_i(z_1, ..., z_{n+1}) = P(z_1, ..., z_{n+1}) z_i^{-r}$$

$$g_i(z_1, ..., z_{n+1}) = Q(z_1, ..., z_{n+1}) z_i^{-r}$$

Notice that f_i and g_i are well defined analytic functions on U_i since they are homogeneous of degree zero.

We claim that we may define $m \in \mathscr{M}(P^n(\mathbb{C}))$ by

$$m|U_i = f_i/g_i, \ 1 \leqslant i \leqslant n+1$$

All we must check is that $f_i/g_i|U_j \cap U_i = f_j/g_j|U_j \cap U_i$. But this is obvious since $f_i/g_i = P/Q$.

In other words, P/Q induces a meromorphic function on $P^n(\mathbb{C})$. In the sequel we shall denote this meromorphic function on $P^n(\mathbb{C})$ by P/Q. Meromorphic functions on $P^n(\mathbb{C})$ which can be written as a quotient of polynomials are called *rational* functions.

Finally we show that $Z(P/Q) = Z(P)$; $P(P/Q) = Z(Q)$. We certainly have $Z(P/Q) \subset Z(P)$; $Z(P/Q) \subset Z(Q)$. The result follows once we have shown that $(P,Q) = 1$ (in the polynomial ring) implies $(P_z, Q_z) = 1$ for all $z \in \mathbb{C}^{n+1}$. We leave the verification of this fact as an easy exercise.

For each open subset U of M we shall define:

(1) $A^*(U) = \{f \in A(U)\colon$ f is non-zero on U$\}$.

(2) $\mathscr{M}^*(U) = \{m \in \mathscr{M}(U)\colon$ m is not identically zero on any component of U$\}$.

It follows, as in Section 2, that $A^*(U)$, $\mathscr{M}^*(U)$ are multiplicative groups and that $A^*(U)$ is a subgroup of $\mathscr{M}^*(U)$.

Definition 3.5. A *divisor*, $d = \{m_j \in \mathscr{M}^*(U_j)\}$, on the complex manifold M consists of an open cover $\{U_j\}$ of M together with a corresponding set of meromorphic functions $\{m_j \in \mathscr{M}^*(U_j)\}$ such that for all j, k we have

$$m_j m_k^{-1} \in A^*(U_j \cap U_k)$$

Remarks

(1) The key point of Definition 3.5 is that m_j and m_k have the same poles and zeroes on $U_j \cap U_k$ (with the same "multiplicities").

(2) We shall regard the divisors $d = \{m_j \in \mathscr{M}^*(U_j)\}$ and $d' = \{m_j' \in \mathscr{M}^*(U_j')\}$ as being *equal* if $m_j(m_k')^{-1} \in A^*(U_j \cap U_k')$ for all j, k.

(3) For a sheaf-theoretic definition of divisor, see Refs [10, 23, 32].

Example. The data of the Weierstrass theorem in Section 1 and of the Cousin II problem in Section 2 define divisors on the appropriate domains in $\mathbb{C}^n$.

Notation. We shall denote the set of all divisors on M by $\mathscr{D}(M)$.

Proposition 3.6. $\mathscr{D}(M)$ has the natural structure of an Abelian group.

Proof. Let $d = \{d_j \in \mathscr{M}^*(U_j)\}$, $d' = \{d_k' \in \mathscr{M}^*(U_k')\}$ denote divisors on M. We define

$$d + d' = \{d_j d_k' \in \mathscr{M}^*(U_j \cap U_k')\}$$

$$-d = \{d_j^{-1} \in \mathscr{M}^*(U_j)\}$$

The reader can verify that these definitions of + and – give $\mathscr{D}(M)$ the structure of an Abelian group with zero element defined by

$$0 = \{1 \in A(M)\}$$

where 1 denotes the function identically equal to 1 on M. Q.E.D.

Example

Let M denote a compact Riemann surface. We let $\widetilde{\mathscr{D}}(M)$ denote the free Abelian group generated by the points of M. That is, every element of $\widetilde{\mathscr{D}}(M)$ may be thought of as a formal sum

$$\sum_{z \in M} p(z) \cdot z$$

where $p(z) \in Z$ for all $z \in M$ and $p(z) = 0$ for all but a *finite* number of z. $\widetilde{\mathscr{D}}(M)$ clearly has the structure of an Abelian group. Indeed, if $\tilde{d} = \sum \tilde{p}(z) \cdot z$ and $\tilde{d}' = \sum \tilde{p}'(z) \cdot z$, we may define

$$\tilde{d} + \tilde{d}' = \sum (\tilde{p}(z) + \tilde{p}'(z)) \cdot z$$

$$-\tilde{d} = \sum -\tilde{p}(z) \cdot z \quad ; \quad 0 = \sum 0 \cdot z$$

(Those familiar with homology theory will recognize that $\widetilde{\mathscr{D}}(M)$ is none other than $C_0(M, Z)$ – the set of 0-cycles on M.) We shall now show that, as groups, $\mathscr{D}(M)$ and $\widetilde{\mathscr{D}}(M)$ are naturally isomorphic.

Let $d = \{d_j \in \mathscr{M}^*(U_j)\} \in \mathscr{D}(M)$. As described in the example following Definition 3.4, $Z(d_j)$, $P(d_j)$ are well defined discrete subsets of U_j. For each $z \in U_j$ we define:

$$p_j(z) = 0 \text{ if } z \notin Z(d_j) \cup P(d_j)$$

$$p_j(z) = \text{order of zero of } d_j \text{ at } z \text{ if } z \in Z(d_j)$$

$$p_j(z) = -\text{order of pole of } d_j \text{ at } z \text{ if } z \in P(d_j)$$

Now notice that since $d_j d_i^{-1} \in A^*(U_i \cap U_j)$, we must have

$$p_j | U_i \cap U_j = p_i | U_i \cap U_j$$

So the above procedure gives us a well defined map $p : M \to Z$ which is non-zero on a discrete subset of M. Since M is assumed compact, p is non-zero on a finite subset of M and we may define $\tilde{d} \in \widetilde{\mathscr{D}}(M)$ by

$$\tilde{d} = \sum p(z) \cdot z$$

This clearly defines us a group homomorphism $\mathscr{D}(M) \to \widetilde{\mathscr{D}}(M)$. We shall now construct the inverse and so show that $\mathscr{D}(M)$ and $\widetilde{\mathscr{D}}(M)$ are naturally isomorphic. Suppose that

$\tilde{d} = \sum q(z) \cdot z \in \tilde{\mathscr{D}}(M)$ and $q(z) \neq 0$ on the set $\{z_1, ..., z_r\} \subset M$. Choose open co-ordinate neighbourhoods $U_1, ..., U_r$ of $z_1, ..., z_r$ respectively such that

$$z_j \notin U_k, \ j \neq k$$

For each j we pick $d_j \in \mathscr{M}^*(U_j)$ such that d_j has a zero (pole) at z_j of order $q(z_j)$ $(-q(z_j))$ and no other poles or zeroes on U_j. We then choose an open subset U_{r+1} of M, which does not contain any of the points $z_1, ..., z_r$, so that $\{U_1, ..., U_{r+1}\}$ gives an open cover of M. We define $d_{r+1} \in \mathscr{M}^*(U_{r+1})$ to be identically 1. Set

$$d = \{d_j \in \mathscr{M}^*(U_j), \ 1 \leqslant j \leqslant r+1\} \in \mathscr{D}(M)$$

We leave it to the reader to verify that this construction defines a group homomorphism $\tilde{\mathscr{D}}(M) \to \mathscr{D}(M)$ which is the inverse to the homomorphism $\mathscr{D}(M) \to \tilde{\mathscr{D}}(M)$ constructed above. In the sequel we set $\tilde{\mathscr{D}}(M) = \mathscr{D}(M)$ and use whichever characterization of divisors is most convenient. (A similar description of divisors may be given for non-compact Riemann surfaces, except that we define $\tilde{\mathscr{D}}(M) = \{\sum p(z) \cdot z : p$ does not vanish on a discrete subset of $M\}$.) Let $m \in \mathscr{M}^*(M)$. Then m defines a divisor $\operatorname{div}(m) \in \mathscr{D}(M)$ in the obvious way. That is, we set

$$\operatorname{div}(m) = \sum p(z) \cdot z$$

where p(z) is non-zero on $Z(m) \cap P(m)$ and takes integral values corresponding to the order of the poles and zeroes of m. Clearly

$$\operatorname{div} : \mathscr{M}^*(M) \to \mathscr{D}(M)$$

is a group homomorphism. The Weierstrass theorem implies that if $U \subset \mathbb{C}$ is open and $d \in \mathscr{D}(U)$, we can find $m \in \mathscr{M}^*(U)$ such that $\operatorname{div}(m) = d$, i.e. div is a surjective map. It is natural to ask whether the Weierstrass theorem holds for all Riemann surfaces in this form, i.e. is div always surjective? In fact, it does hold for open, i.e. non-compact, Riemann surfaces (for then $H^2(M, Z) = 0$ and the $\bar{\partial}$-sequence is exact – cf. the proof of the Weierstrass theorem in Section 1). However, div is never surjective if M is compact and there are topological and analytic obstructions to finding a meromorphic function with given divisor. More precisely, suppose M is compact and let $d = \sum p(z) \cdot z \in \mathscr{D}(M)$. We define the *degree* of d, deg(d), by

$$\deg(d) = \sum_{z \in M} p(z) \in Z$$

$\deg : \mathscr{D}(M) \to Z$ is a group homomorphism. A necessary condition for the existence of $m \in \mathscr{M}^*(M)$ satisfying $\operatorname{div}(m) = d$ is that $\deg(d) = 0$. However, unless $M = P^1(\mathbb{C})$, the vanishing of the degree of d will not be a sufficient condition for d to be the divisor of a meromorphic function. We shall return to this question later in this section and also give references.

Assuming the fact that $\deg(\operatorname{div}(m)) = 0$ for all $m \in \mathscr{M}^*(M)$, let us show that every $m \in \mathscr{M}^*(P^1(\mathbb{C}))$ is *rational*. Write $\operatorname{div}(m) = d^+ - d^-$, where

$$d^+ = \sum_{i=1}^{r} p(y_i) \cdot y_i, \quad \text{all } p(y_i) > 0 \quad ; \quad d^- = \sum_{j=1}^{s} q(y_j) \cdot y_j, \quad \text{all } q(y_j) > 0$$

Since $\deg(\mathrm{div}(m)) = 0$,

$$\sum_{i=1}^{r} p(y_i) = \sum_{j=1}^{s} q(y_j) = t, \text{ say}$$

We define homogeneous polynomials $P, Q \in A(\mathbb{C}^2)$ of degree t by

$$P(z_1,z_2) = \prod_{i=1}^{r} (a_i z_1 - b_i z_2)^{p(y_i)} \quad ; \quad Q(z_1,z_2) = \prod_{j=1}^{s} (c_j z_1 - d_j z_2)^{q(y_j)}$$

where the lines $a_i z_1 - b_i z_2$, $c_j z_1 - d_j z_2$ correspond to the points $y_i, y_j \in P^1(\mathbb{C})$, $1 \leqslant i \leqslant r$, $1 \leqslant j \leqslant s$. We claim that $m = c(P/Q)$ for some constant $c \neq 0$. Indeed, our construction ensures that $m(Q/P) \in A^*(P^1(\mathbb{C}))$. Now it follows easily from Theorem 1.10 that every analytic function on a compact Riemann surface must be constant. Hence $m(Q/P)$ is non-zero and constant on $P^1(\mathbb{C})$. In fact, it can be shown that every meromorphic function on $P^n(\mathbb{C})$, $n > 1$, is rational – see the remarks at the end of the section.

We next want to show how the theory of divisors on a complex manifold is related to the study of holomorphic line bundles. We recall

Definition 3.7. Let M be a complex manifold and suppose that we are given an open cover $\{U_i\}$ of M and analytic maps $\phi_{ij}: U_i \cap U_j \to \mathbb{C}^*$ such that for all i,j,k we have:
(1) $\phi_{ii} = 1$.
(2) $\phi_{ij}\phi_{jk} = \phi_{ik}$ on $U_i \cap U_j \cap U_k$.
Then $\{\phi_{ij}\}$ is said to define the *transition functions* of a *holomorphic line bundle* on M.

A rather more geometric description of a holomorphic line bundle may be given as follows. Suppose that $\{\phi_{ij}\}$ are the transition functions of a holomorphic line bundle on M. Then we may find a complex manifold E and a surjective holomorphic map $\pi: E \to M$ such that:
(1) $\pi: E \to M$ is *locally trivial:* for every i, there exists a holomorphic diffeomorphism $\phi_i: \pi^{-1}(U_i) \subset E \to U_i \times \mathbb{C}$ which satisfies $\pi_1\phi_i = \pi$ ($\pi_1: U_i \times \mathbb{C} \to U_i$ denotes the projection on the first factor).
(2) For all i, j the transition functions are given in terms of the maps $\{\phi_i\}$ by

$$\phi_{ij}(x) = \pi_2\phi_i\phi_j^{-1}(x,1), \quad x \in U_i \cap U_j$$

For more details about the general theory of complex and holomorphic vector bundles and a proof of the above statement we refer to Ref. [29]. We remark that our transition functions take values in $\mathbb{C}^*$ (as opposed to $GL(1;\mathbb{C})$). Our definitions are equivalent, however, since $\mathbb{C}^* \approx GL(1;\mathbb{C})$.

We shall usually use the notation $\pi: E \to M$ or just E for holomorphic line bundles. However, most of our work with holomorphic line bundles will be done in terms of transition functions.

Let E be a holomorphic line bundle on M. For each $z \in M$, we set

$$E_z = \pi^{-1}(z)$$

E_z has the structure of a complex vector space of dimension 1 (E is a "complex line bundle").

Suppose that $p: E \to M$, $q: F \to M$ are holomorphic line bundles on M with transition functions $\{\phi_{ij}\}$, $\{\psi_{ij}\}$ respectively (we may assume, without loss of generality, that the cover $\{U_j\}$ of M is the same for $\{\phi_{ij}\}$ and $\{\psi_{ij}\}$). We say that a holomorphic map $D: E \to F$ is a *holomorphic line bundle isomorphism* if

(1) $$\begin{array}{ccc} E & \xrightarrow{\;D\;} & F \\ & {\scriptstyle p}\searrow \quad \swarrow{\scriptstyle q} & \\ & M & \end{array}$$ commutes, i.e. $D(E_z) \subset F_z$ for all $z \in M$.

(2) D is a holomorphic diffeomorphism.

(3) $D|E_z : E_z \to F_z$ is a complex linear isomorphism for all $z \in M$.

In terms of the transition functions of E and F, D is given by a family $\{d_i \in A^*(U_i)\}$ which satisfies the compatibility conditions

A .... $\phi_{ij} d_j = d_i \psi_{ij}$

The functions $\{d_i\}$ are given explicitly in terms of D by Fig.2.

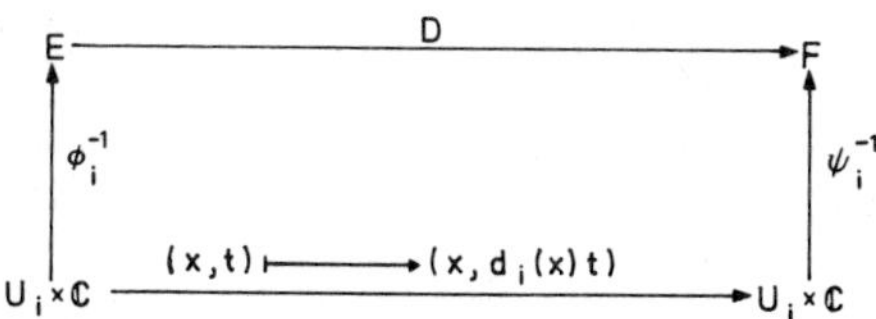

FIG.2. Local form for vector bundle map.

Remark. In the sequel we shall only be considering *isomorphism classes* of holomorphic line bundles and we therefore regard isomorphic line bundles as being equal. For future reference, we remark that the transition functions $\{\phi_{ij}\}$, $\{\psi_{ij}\}$ define the same line bundle if and only if there exists $\{d_i \in A^*(U_i)\}$ such that condition A above holds.

Definition 3.8. Let $\pi: E \to M$ be a holomorphic line bundle. A holomorphic map $s: M \to E$ is said to be a holomorphic *section* of E if πs = identity map on M. That is, if $s(z) \in E_z$ for all $z \in M$. We denote the set of sections of E by O(E).

Remarks

(1) O(E) has the structure of a complex vector space induced from the complex structure on the fibres of E.

(2) In terms of transition functions $\{\phi_{ij}\}$ for E, a section of E is given by a family $\{s_i \in A(U_i)\}$ which satisfies the compatibility conditions

$$\phi_{ij} s_j = s_i$$

on the overlaps $U_i \cap U_j$.

We may also define meromorphic sections of a holomorphic line bundle. For technical reasons we prefer to give a definition in terms of transition functions.

Definition 3.9. Let $\{\phi_{ij}\}$ be transition functions for the holomorphic line bundle E. We say that $m = \{m_j \in \mathscr{M}(U_j)\}$ defines a *meromorphic section* of E if

$$\phi_{ij} m_j = m_i$$

on the overlaps $U_i \cap U_j$. We denote the set of meromorphic sections of E by $\mathscr{M}(E)$.

Remark. We leave it to the reader to verify that if $m \in \mathscr{M}(E)$, then we have well defined zero, pole and indeterminacy sets of m which are analytic subsets of M. We continue to denote these sets by Z(m), P(m), T(m) respectively. If $s \in O(E)$, Z(s) is usually referred to as the *zero set* of s.

We shall denote the set of holomorphic line bundles on the complex manifold M by HLB(M).

We shall now briefly describe some constructions that can be made in HLB(M).

1. Tensor product

Let E, F $\in$ HLB(M) have transition functions $\{\phi_{ij}\}, \{\psi_{ij}\}$ respectively. The *tensor product* of E and F, $E \otimes F$, is the holomorphic line bundle with transition functions $\{\phi_{ij} \cdot \psi_{ij}\}$. The fibre of $E \otimes F$ at z is naturally isomorphic to $E_z \otimes F_z$ (which is of complex dimension 1, since E_z and F_z are of complex dimension 1).

2. Dual bundle

Let E $\in$ HLB(M) have transition functions $\{\phi_{ij}\}$. Then E* is the holomorphic line bundle with transition functions $\{\phi_{ij}^{-1} = \phi_{ji}\}$. $(E^*)_z$ is naturally isomorphic to $(E_z)^* = L(E_z, \mathbb{C})$.

3. Trivial line bundle

The trivial holomorphic line bundle $\underline{\mathbb{C}}$ is the product bundle $M \times \mathbb{C} \xrightarrow{\pi_1} M$.

We remark that if E $\in$ HLB(M), then $E \otimes E^* = \underline{\mathbb{C}}$. Indeed, this may be easily proved by either looking at transition functions or by observing that $E_z \otimes E_z^* \approx \mathbb{C}$ by the map $(e_z, e_z^*) \mapsto e_z^*(e_z)$ (this map is usually referred to as "trace"). From this remark follows

Proposition 3.10 HLB(M) has the natural structure of an Abelian group. Group composition is defined to be tensor product, the identity to be $\underline{\mathbb{C}}$ and the inverse of E to be E*.

Example. We shall define the transition functions of a holomorphic line bundle on $P^n(\mathbb{C})$. Taking the usual atlas $\{(U_i, \phi_i), 1 \leqslant i \leqslant n+1\}$ for $P^n(\mathbb{C})$, we define

$$\phi_{ij}(z_1, \ldots, z_{n+1}) = z_j/z_i$$

The reader may verify that $\{\phi_{ij}\}$ define the transition functions of a holomorphic line bundle on $P^n(\mathbb{C})$ which we shall in future denote by H. If we let L = H*, we see that L has the transition functions

$$\psi_{ij} = z_i/z_j$$

For $k \in Z$, we set

$$\begin{aligned} H^k &= \otimes^k H, \ k > 0 \\ &= \otimes^k L, \ k < 0 \\ &= \underline{\mathbb{C}}, \ k = 0 \end{aligned}$$

H^k has transition functions $\{z_j^k/z_i^k\}$. It may be shown that

$$HLB(P^n(\mathbb{C})) = \{H^k : k \in Z\}$$

That is, H generates the group $HLB(P^n(\mathbb{C}))$ and so $HLB(P^n(\mathbb{C}))$ is isomorphic to the additive group of integers, Z (for a proof of this statement (and much more!) we refer to Ref. [33]).

Let $s = \{s_i \in A(U_i)\}$ be defined by

$$s_i(z_1, \ldots, z_{n+1}) = z_1/z_i, \quad 1 \leqslant i \leqslant n+1$$

Then $(z_j/z_i)s_j = s_i$ on $U_i \cap U_j$ and so s is a holomorphic section of H. Observe that $Z(s) = \{(z_1, \ldots, z_{n+1}) : z_1 = 0\}$. Hence Z(s) is the complex submanifold of $P^n(\mathbb{C})$ which is the image of the hyperplane $z_1 = 0$ in $\mathbb{C}^{n+1}$. Clearly Z(s) is analytically diffeomorphic to $P^{n-1}(\mathbb{C})$. Any submanifold of $P^n(\mathbb{C})$ that is the image of an n-dimensional hyperplane in $\mathbb{C}^{n+1}$ is called a *hyperplane* in $P^n(\mathbb{C})$. Every such hyperplane in $P^n(\mathbb{C})$ is analytically diffeomorphic to $P^{n-1}(\mathbb{C})$. We shall show next that every section of H has zero set a hyperplane and, for this reason, H is usually called the *hyperplane section bundle* of $P^n(\mathbb{C})$.

We shall now describe $O(H^k)$, $k \in Z$. We distinguish three cases.

1. $k = 0$

$H^0 = \mathbb{C}$ and so $O(H^0) = A(P^n(\mathbb{C})) = \mathbb{C}$ (constant functions).

2. $k > 0$

Let $s = \{s_i \in A(U_i) : 1 \leqslant i \leqslant n+1\}$ be a section of H^k. Then

$$(z_j/z_i)^k s_j(z_1, \ldots, z_{n+1}) = s_i(z_1, \ldots, z_{n+1})$$

and so

$$z_j^k s_j(z_1, \ldots, z_{n+1}) = z_i^k s_i(z_1, \ldots, z_{n+1})$$

Hence we may define $P \in A(\mathbb{C}^{n+1} \backslash \{0\})$ by

$$P|\{z_j \neq 0\} = z_j^k s_j$$

By Hartog's theorem (Theorem 1.17), P extends uniquely to an analytic function on $\mathbb{C}^{n+1}$. Now s_j is homogeneous in $(z_1, \ldots, z_{n+1})$ of degree zero and so P is homogeneous of degree k. Taking the power series of P at $0 \in \mathbb{C}^{n+1}$, we see immediately that P is a homogeneous polynomial of degree k. Conversely, any homogeneous polynomial P of degree k on $\mathbb{C}^{n+1}$ defines a unique section $s = \{s_i \in A(U_i)\}$ of H^k by taking

$$s_i = P/z_i^k$$

Hence

$O(H^k) \cong$ space of homogeneous polynomials of degree k on $\mathbb{C}^{n+1}$.

It follows that if $s \in O(H^k)$, then $Z(s) \subset P^n(\mathbb{C})$ is a projective algebraic variety and the zero locus of a homogeneous polynomial of degree k.

3. $k < 0$

We leave it as an exercise to show that $O(H^k) = \{0\}$, $k < 0$.

The reader will already have noticed a certain similarity in the definitions of holomorphic line bundles and divisors and we now wish to describe the relationships between HLB(M) and $\mathscr{D}(M)$ more precisely.

We first define a group homomorphism $\mathscr{D}(M) \to HLB(M)$. Let $d = \{d_i \in \mathscr{M}^*(U_i)\} \in \mathscr{D}(M)$. We define $[d] \in HLB(M)$ by requiring $[d]$ to have the transition functions

$$\phi_{ij} = d_i/d_j$$

Since $d_i/d_j \in A^*(U_i \cap U_j)$ and $(d_i/d_j)(d_j/d_k) = d_i/d_k$, the conditions of Definition 3.7 are satisfied and so $\{d_i/d_j\}$ is a set of transition functions for a holomorphic line bundle on M. The reader may verify that $d \mapsto [d]$ is a group homomorphism.

Proposition 3.11. The kernel of the homomorphism $\mathscr{D}(M) \to HLB(M)$ is precisely the set of divisors of meromorphic functions on $M: \{\text{div}(m) : m \in \mathscr{M}^*(M)\}$.

Proof. Suppose $[d] = \underline{\mathbb{C}}$. The transition functions for $[d]$ are equal to $\{d_i/d_j \in A^*(U_i \cap U_j)\}$ and those of $\underline{\mathbb{C}}$ are given by $\{1 \in A^*(U_i \cap U_j)\}$. To say that $[d] = \underline{\mathbb{C}}$ means that the bundles $[d]$ and $\underline{\mathbb{C}}$ are isomorphic line bundles. Hence there exist $g_i \in A^*(U_i)$ such that

$$(d_i/d_j)g_j = g_i \cdot 1$$

That is, $d_i/g_i = d_j/g_j$ on $U_i \cap U_j$ and so we may define $m \in \mathscr{M}^*(M)$ by

$$m|U_i = d_i/g_i$$

Clearly $\text{div}(m) = d$. We leave as an easy exercise the verification that $[\text{div}(m)] = \underline{\mathbb{C}}$ for all $m \in \mathscr{M}^*(M)$.

Definition 3.12. We say that $d, d' \in \mathscr{D}(M)$ are *linearly equivalent* if $\exists m \in \mathscr{M}^*(M)$ such that

$$d - d' = \text{div}(m)$$

We let $\mathscr{L}(M)$ denote a group of linear equivalence classes of divisors on M.

Since $[d] = [d']$ if and only if d is linearly equivalent to d′ (Proposition 3.11), it follows that the map $d \mapsto [d]$ induces a group homomorphism:

$$\Phi : \mathscr{L}(M) \to HLB(M)$$

We continue to write $\Phi(d) = [d]$, $d \in \mathscr{L}(M)$. By Proposition 3.11, Φ is injective. It is clearly important to ascertain when Φ is an *isomorphism.* Suppose that $E \in HLB(M)$ has a non-zero meromorphic section. We claim that we have a well defined group homomorphism:

$$\text{div} : \mathscr{M}^*(E) \to \mathscr{D}(M)$$

Let E have transition functions $\{\phi_{ij}\}$ and suppose that $s = \{s_i \in \mathscr{M}^*(U_i)\} \in \mathscr{M}^*(E)$. Now $\phi_{ij}s_j = s_i$ and so for all i,j we have

$$s_i/s_j = \phi_{ij} \in A^*(U_i \cap U_j)$$

We define $\mathrm{div}(s) \in \mathscr{D}(M)$ by

$$\mathrm{div}(s) = \{s_i/s_j \in A^*(U_i \cap U_j)\}$$

It is easily checked that div is a group homomorphism (div(s) is usually called the *divisor* of the section s). We now show that the linear equivalent class of div(s) depends only on E and not on the particular choice of section s. Suppose $s' = \{s_i' \in \mathscr{M}^*(U_i)\} \in \mathscr{M}^*(E)$. We define $m \in \mathscr{M}^*(M)$ by

$$m|U_i = s_i/s_i'$$

The compatibility conditions $\phi_{ij}s_j = s_i$, $\phi_{ij}s_j' = s_i'$, ensure that $s_i/s_i' = s_j/s_j'$ on the overlap $U_i \cap U_j$ and so m is well defined. We leave it to the reader to verify that

$$\mathrm{div}(s) - \mathrm{div}(s') = \mathrm{div}(m)$$

Proposition 3.13. $\Phi: \mathscr{L}(M) \to \mathrm{HLB}(M)$ is an isomorphism if and only if every holomorphic line bundle on E has a non-zero meromorphic section.

Proof. Let $E \in \mathrm{HLB}(M)$. We define $(E) \in \mathscr{L}(M)$ by

$$(E) = \text{linear equivalence class of div(s)}$$

where s is any non-zero meromorphic section of E. The arguments preceding the statement of the proposition show that (E) is well defined and we clearly have

$$[(E)] = E \text{ for all } E \in \mathrm{HLB}(M)$$

Since we already know that Φ is injective it follows that Φ is an isomorphism with inverse the map $E \mapsto (E)$. Q.E.D.

The justification for our lengthy discussion of holomorphic line bundles comes from the following deep theorem.

Theorem 3.14. Let M be either a projective algebraic manifold, i.e. a compact submanifold of $P^n(\mathbb{C})$, or a Stein manifold – equivalently (see Ref. [16]) a closed submanifold of $\mathbb{C}^n$. Then every holomorphic line bundle on M has a non-trivial meromorphic section.

Proof. We shall not prove this result here but instead refer the reader to Refs [32, 33] for the projective algebraic case, and to Refs [12, 16] for the Stein manifold case.

It follows from Theorem 3.14 and Proposition 3.13 that on Stein or projective algebraic manifolds every divisor is the divisor of a meromorphic section of some line bundle. However, this result is *not* true for all complex manifolds (see Refs [6, 13]).

Lemma 3.15. Let $E \in \mathrm{HLB}(M)$ and suppose that $s \in \mathscr{M}^*(E)$. Then s induces an isomorphism

$$D_s: \mathscr{M}^*(E) \to \mathscr{M}^*(M)$$

Proof. Let $s' \in \mathscr{M}^*(E)$. We define $D_s(s') \in \mathscr{M}^*(M)$ by the condition

$$\mathrm{div}(D_s(s')) = \mathrm{div}(s') - \mathrm{div}(s)$$

That D_s is a well defined isomorphism follows as in the proof of Proposition 3.13 and we omit details.

Holomorphic sections of $E \in HLB(M)$ are of particular interest. Suppose $s, s' \in O(E)$ and s is not identically zero. Then $m = D_s(s')$ is a meromorphic function on M with $Z(m) \subset Z(s')$ and $P(m) \subset Z(s)$.

Example. As an exercise the reader may study the map

$$D_s : O(H^k) \to \mathscr{M}^*(P^n(\mathbb{C}))$$

where $s = \{z_1^k/z_i^k \in A^*(U_i)\}$ and H is the hyperplane section bundle of $P^n(\mathbb{C})$ (cf. the example following Proposition 3.10).

If M is a *compact* complex manifold, $\dim_{\mathbb{C}} O(E) < \infty$ (see the next section for more details and references). Computation of $\dim_{\mathbb{C}} O(E)$ in terms of other invariants of M is an important problem – for example, it gives us information about meromorphic functions on M related to the bundle E. If M is a compact Riemann surface we have the celebrated Riemann-Roch theorem:

$$\dim_{\mathbb{C}} O(E) = \dim_{\mathbb{C}} O(TM^* \otimes E^*) + c_1(E) + 1 - g$$

(TM^* denotes the holomorphic cotangent bundle of M; $c_1(E)$ is the degree of the divisor of any meromorphic section of E; and g is the genus of M.)

The Riemann-Roch theorem gives much information about the existence of meromorphic functions with specified poles and zeroes on a given compact Riemann surface. The Riemann-Roch theorem has been generalized to general compact complex manifolds. We refrain from making any applications of these theorems here but instead refer the reader to Refs [4, 15].

Now is perhaps the best time to make a brief digression to show how the theory of divisors can be developed using sheaf cohomological methods. What we say here is not necessary for the understanding of the rest of this paper and we refer to Refs [6, 23] for more details.

We have the exact sheaf sequences

$$0 \to O^* \to \mathscr{M}^* \to \mathscr{D} \to 0$$

$$0 \to Z \to O \overset{e}{\to} O^* \to 0$$

O^* is the sheaf of germs of invertible analytic functions on M (defined by the presheaf $\{A^*(U), r_{VU}\}$), Z is the constant sheaf of integers and e is the exponential map defined by $e(f) = \exp(2\pi i f)$, $f \in O_x$. We have the following portions of the long exact cohomology sequences of these sequences:

$$0 \to O^*(M) \to \mathscr{M}^*(M) \overset{a}{\to} \mathscr{D}(M) \overset{b}{\to} H^1(M, O^*) \to H^1(M, \mathscr{M}^*) \to \ldots$$

$$\to H^1(M, Z) \to H^1(M, O) \overset{\delta}{\to} H^1(M, O^*) \to H^2(M, Z) \to H^2(M, O) \to \ldots$$

The transition functions of a holomorphic line bundle on M define a cohomology class in $H^1(M,O^*)$ and it is easily shown that $H^1(M,O^*) \cong HLB(M)$. The map $a:\mathcal{M}^*(M) \to \mathcal{D}(M)$ is just the map "div" we defined earlier, whilst the map $b:\mathcal{D}(M) \to H^1(M,O^*) = HLB(M)$ is the map $d \mapsto [d]$. Now $\mathcal{L}(M) = \mathcal{D}(M)/\mathcal{M}^*(M)$ and so $\mathcal{L}(M) \cong HLB(M)$ if and only if $H^1(M,\mathcal{M}^*) = 0$. (The vanishing of $H^1(M,\mathcal{M}^*)$ for a compact Riemann surface is a key part of the proof of the Riemann-Roch theorem referred to above.) Let us examine the second sequence. The map $\delta: H^1(M,O^*) = HLB(M) \to H^2(M,Z)$ is $-$the first Chern class map (or +, depending on conventions). We shall define $c_1(E) = -\delta(E)$, $E \in HLB(M)$. For a compact Riemann surface, $H^2(M,Z) \cong Z$ (integration of differential forms) and c_1 defines a homomorphism from HLB(M) to Z. In this case $c_1(E) = \deg(\mathrm{div}(s))$, for all $s \in \mathcal{M}^*(E)$. For an arbitrary complex manifold M and $E \in HLB(M)$, $c_1(E) = 0$ if and only if E is differentiably trivial (as a complex line bundle). However, a differentiably trivial holomorphic line bundle need not be *holomorphically* trivial and we define

$$\mathrm{Pic}(M) = \{E \in HLB(M) : c_1(E) = 0\}$$

If M is projective algebraic, Pic(M) has the structure of a compact connected complex Lie group, i.e. a complex torus, and Pic(M) gives important invariants of the complex/algebraic structure of M. If M is a Stein manifold, $\mathrm{Pic}(M) = \{\underline{\mathbb{C}}\}$ (on Stein manifolds "a differentiably trivial holomorphic vector bundle is holomorphically trivial [1]). Let us turn to other terms in the second sequence. If M is a Stein manifold, the exactness of the $\bar\partial$-sequence on M implies that $H^1(M,O) = H^2(M,O) = 0$. Hence $c_1: H^1(M,O^*) = HLB(M) \to H^2(M,Z)$ is an isomorphism and so we can solve the Cousin II problem for a divisor d on a Stein manifold if and only if $c_1([d]) = 0$. For projective algebraic manifolds the problem of finding when a divisor is the divisor of a meromorphic function is rather difficult and we shall restrict attention to compact Riemann surfaces. If M is a compact Riemann surface, $H^2(M,O) = 0$ (essentially for dimensional reasons) and so we have the following exact sequence:

$$H^1(M,Z) \to H^1(M,O) \to H^1(M,O^*) \overset{c_1}{\to} H^2(M,Z) \to 0$$

Now, by Serre duality, $H^1(M,O) \cong O(TM^*)$. By the finiteness theorem referred to above, $\dim_{\mathbb{C}} O(TM^*) < \infty$. We set $g = \dim_{\mathbb{C}} O(TM^*)$ (g is actually a topological invariant of M: the genus of M). A holomorphic section of TM^* is called an *Abelian differential.* We are interested in finding those elements of Pic(M) which are divisors of meromorphic functions on M (recall a necessary condition for d to be the divisor of a meromorphic function was $c_1[d] = 0$, i.e. $[d] \in \mathrm{Pic}(M)$. Now, by the exactness of the above sequence,

$$\mathrm{Pic}(M) \cong O(TM^*)/H^1(M,\mathbb{Z})$$

Now topology shows us that $H^1(M,Z)$ is isomorphic to Z^{2g}. Since $O(TM^*) \cong \mathbb{C}^g$, Pic(M) is a complex torus of dimension g. d will be the divisor of a meromorphic function on M if and only if [d] defines a lattice point of this torus. Abel's theorem gives explicit conditions for this to happen in terms of a basis for the Abelian differentials of M. For a full statement, with proofs, we refer to Ref. [23]. We end this digression by remarking that $H^q(P^n(\mathbb{C}),O) = 0$, $q \geq 1$, and so the analysis of divisors given above for Stein manifolds also goes through for the complex projective spaces (see Ref. [33]).

So far in this section we have emphasized the relationships between the theory of divisors and that of meromorphic functions and line bundles. We shall conclude the section by pointing out some geometric aspects of the theory of divisors.

We first wish to develop a little of the *global* theory of analytic sets. We shall give no proofs but refer the reader to Refs [12, 25] for expositions of the theory of analytic sets.

Definition 3.16. An analytic subset X of the complex manifold M is said to be *irreducible* if X cannot be written as the union of two distinct proper non-empty analytic subsets of X. That is, if $X = Y \cup Z$, where Y and Z are analytic subsets of M, then either Y and Z must equal X.

Remark. If X is irreducible then it may be shown that X is connected.

Examples

(1) An isolated point of M is an irreducible analytic set. Two or more distinct points of M define a reducible analytic set, but not irreducible.

(2) Any complex closed submanifold of M which is connected is an irreducible analytic set. In particular, the hyperplane $\{(z_1, \ldots, z_{n+1}) \in P^n(\mathbb{C}) : z_1 = 0\}$ is irreducible.

Definition 3.17. Let X be an analytic subset of M. A point $x \in X$ is said to be a *regular* or *non-singular* point of X if there exists an open neighbourhood U of x in X which is complex-analytically a submanifold of M. x is said to be a *singular* point of X if it is not regular.

We denote the set of singular points of X by Sing(X) and the set of regular points of X by X' Sing(X) is clearly a closed subset of X (actually it is an analytic subset of X) and X' is open and dense in X. We have the following basic decomposition theorem for analytic sets.

Theorem 3.18. Let X be an analytic subset of the compact complex manifold M. Then X has a unique (up to order) finite decomposition

$$X = X_1 \cup \ldots \cup X_p$$

where X_i is an irreducible analytic set, $1 \leqslant i \leqslant p$, and no X_i is contained in

$$\bigcup_{j \neq i} X_j, \quad 1 \leqslant i \leqslant p$$

Moreover X_i' is connected, $1 \leqslant i \leqslant p$ (in fact this characterizes the irreducibility of X_i).

In the case when M is non-compact we have a similar theorem with "locally finite" replacing "finite".

We are particularly interested in the study of analytic sets of co-dimension 1.

Definition 3.19. An analytic subset X of M is said to be of co-dimension 1 if X is locally defined as the zero set of an analytic function, i.e. if for each $x \in M$ we may find an open neighbourhood U of x and $s \in A(U)$ such that $X \cap U = s^{-1}(0)$.

Remark. If X is of co-dimension 1 then X' is a co-dimension 1 complex submanifold of M. The converse is also true if X is irreducible.

Example. Let $d = \{d_i \in \mathscr{M}^*(U_i)\}$ be a divisor on M. We define analytic subsets $|d|^+$, $|d|^-$ and $|d|$ of M by

$$|d|^+ \cap U_i = Z(d_i)$$

$$|d|^- \cap U_i = P(d_i)$$

$$|d| \cap U_i = Z(d_i) \cup P(d_i)$$

Since $d_i d_j^{-1} \in A^*(U_i \cap U_j)$, $|d|^+$, $|d|^-$ and $|d|$ are well-defined analytic subsets of M and we clearly also have $|d|^+ \cup |d|^- = d$.

In an earlier example we showed that the group of divisors of a compact Riemann surface was isomorphic to the free Abelian group generated by the points of M, i.e. to the free Abelian group generated by the set of all irreducible analytic subsets of M of co-dimension 1. We are now going to generalize this result to complex manifolds of dimension greater than 1. First we state, without proof, the following non-trivial "coherence" result.

Proposition 3.20. Let X be an irreducible analytic subset of M of co-dimension 1. Then, for each $x \in M$, we may find an open neighbourhood U of x in M and $s \in A(U)$ such that
(1) $X \cap U = s^{-1}(0)$.
(2) $s_z \in O_z$ is irreducible for all $z \in X \cap U$.

Theorem 3.21. Let M be a compact complex manifold. Then $\mathscr{D}(M)$ is isomorphic to the free Abelian group generated by the set of all irreducible analytic subsets of M of co-dimension 1.

If M is non-compact a similar result holds with the appropriate local finiteness condition.

Proof. Suppose that X is an irreducible analytic subset of M of co-dimension 1. By (2) of Proposition 3.20, we may choose for each point $x \in M$ a neighbourhood U of x and $s \in A(U)$ such that s_z is irreducible for all $z \in X \cap U$. Repeating this construction for all $x \in M$, we may choose a cover $\{U_i\}$ of M and corresponding $\{d_i \in A(U_i)\}$ such that $(d_i)_z \in O_z$ is always irreducible if $z \in X$. Now the unique factorization property of O_z implies that if $z \in X \cap U_i \cap U_j$ then $(d_j)_z$ is equal to $(d_i)_z$ up to units. It follows that for all i,j

$$d_i d_j^{-1} \in A^*(U_i \cap U_j)$$

and so $\{d_i \in A(U_i)\}$ defines a divisor. If we denote this divisor by d, we note that for $n \in Z$ the divisor nd is given by $\{d_i^n \in \mathscr{M}^*(U_i)\}$. We may now define a group homomorphism from the free Abelian group on the irreducible analytic subsets of M of co-dimension 1 to $\mathscr{D}(M)$ by mapping $\sum n_i X_i$ to $\sum n_i d_i$, where the divisor d_i corresponds to the irreducible analytic set X_i. This map is clearly injective.

Suppose now that $d \in \mathscr{D}(M)$. As in the example following Definition 3.19, d defines an analytic subset |d| of M. Suppose that

$$|d| = Z_1 \cup \dots \cup Z_s$$

is the unique decomposition of |d| into irreducible analytic sets of co-dimension 1 given by Theorem 3.18. Applying Proposition 3.20 to Z_j, we may find an open cover $\{U_i^j\}$ of M and $\{s_i^j \in A(U_i^j)\}$ such that $(s_i^j)_z$ is irreducible for $z \in Z_j$. As in the first part of the proof, $\{s_i^j \in A(U_i^j)\}$ defines a divisor $d_j \in \mathscr{D}(M)$ for $j = 1, \dots, s$. Clearly $|d_j| = Z_j$.

Suppose $d = \{d_k \in \mathscr{M}^*(V_k)\}$. If $y \in V_k$ we have a unique factorization

$$(d_k)_y = p_1^{r_1} \dots p_s^{r_s}$$

where $p_1, \dots, p_s \in O_y$ are irreducible and $r_1, \dots, r_s \in Z$. Since $d_k d_j^{-1} \in A^*(U_j \cap U_k)$, it follows that, up to units, this factorization is uniquely determined by d and we set

$$d_y = p_1^{r_1} \dots p_s^{r_s}$$

Let $y \in Z_j$. Then, by Proposition 3.20, one of the factors of d_y, say p_j, defines Z_j on some neighbourhood of y. That is, we may find an open subset U of y such that

$$Z_j \cap U = p_j^{-1}(0)$$

and $(p_j)_z$ is irreducible for $z \in Z_j$. It follows (since Z_j is connected) that r_j is constant on Z_j.

We have shown that associated to each Z_j we have a well defined integer r_j. It is clear from our construction that $d - \sum r_j d_j$ is none other than the zero divisor. Therefore, the map $d \mapsto \sum r_j Z_j$ defines the inverse to the map constructed in the first part of the proof. Q.E.D.

Remark. In this section we have stressed relationships between the theory of divisors and meromorphic functions and holomorphic line bundles rather than the geometric or algebraic aspects of the theory. In fact divisors figure prominently in algebraic geometry and it is in the context of algebraic geometry that divisors ("linear systems") have received most attention. (There is a most extensive literature on this subject but see, e.g., Ref. [40]). We only mention here that it follows from Chow's theorem that every analytic subset of a projective algebraic variety is a projective algebraic subset. As a consequence, every meromorphic function on a projective algebraic manifold is actually rational and the set of divisors of, say $P^n(\mathbb{C})$, is isomorph[ic] to the free Abelian group generated by the set of irreducible homogeneous polynomials on $\mathbb{C}^{n+1}$.

4. ELLIPTIC OPERATORS IN COMPLEX ANALYSIS

In Section 1 we showed how the existence and regularity theory of the $\bar{\partial}$-operator could be used to prove results about analytic and meromorphic functions defined on open subsets of $\mathbb{C}^n$. In this section we want to give one or two applications of the theory of elliptic partial differential operators on compact complex manifolds (for the non-compact case see Refs [3, 16, We find that for compact complex manifolds the topological and differential geometric structure of the manifold enters into the analysis of the $\bar{\partial}$-operator. We shall show that every complex Hermitian manifold M has a natural connection (the hermitian connection of M) which not only embodies the complex differential geometry of M but also the complex structure of M in the form of the operator $\bar{\partial}$. If M is Kähler, then the hermitian connection is "equal" to the Rieman[n] connection and we may find topological conditions that the manifold must satisfy using Hodge theory.

In these notes we assume familiarity with Ref. [31], these Proceedings. Our exposition will be somewhat more concise than in previous sections and we shall generally omit computational details. We refer the reader to Refs [8, 28, 29] for the basic theory of elliptic operators on manifolds, to Refs [19, 20, 41] for full details on complex differential geometry and to Refs [20 for applications of elliptic theory to the topology of Kähler manifolds.

We shall divide the material of this section into five subsections:

(A) Partial differential operators on compact manifolds
(B) Contractions and linear algebra
(C) Complex Laplace-Beltrami operator
(D) Connections
(E) Introduction to the Hodge theory of Kähler manifolds

A. Partial differential operators on compact manifolds

In this sub-section M will denote a differentiable manifold of dimension n with tangent bundle TM. We shall let ${}_cTM$ and ${}_cTM'$ denote the complex tangent and complex contangent bundles of M respectively, i.e.

$${}_cTM = TM \otimes_{\mathbb{R}} \underline{\mathbb{C}}$$

$${}_cTM' = TM' \otimes_{\mathbb{R}} \underline{\mathbb{C}} = L_{\mathbb{R}}(TM, \underline{\mathbb{C}})$$

Both ${}_cTM$ and ${}_cTM'$ have the natural structure of complex vector bundles induced from the complex structure on the trivial bundle $\underline{\mathbb{C}}$.

For r = 0, 1, ... we form the complex exterior powers $\Lambda^r{}_c TM'$. $\Lambda^r{}_c TM'$ is called the bundle of *complex r-forms* on M. We clearly have

$$\Lambda^r{}_c TM' \cong \Lambda^r TM' \otimes_{\mathbb{R}} \mathbb{C}$$

and indeed the operation of complexification always commutes with the operations of tensor and exterior powers.

Definition 4.1. Let E and F be complex vector bundles on M. A *linear partial differential operator* (LPDO for short) is a complex linear map

$$D: C^\infty(E) \to C^\infty(F)$$

which is local:

$$\overline{\text{supp}}(Ds) \subset \overline{\text{supp}}(s), \text{ for all } s \in C^\infty(E)$$

Remark. The fact that an LPDO is local implies that the value of Ds at x is determined by the values of s on arbitrarily small open neighbourhoods of x. Any operator that involves integration will not usually be local. Another consequence of the local property of an LPDO is that we can determine it uniquely by its restrictions to co-ordinate charts.

Example 1. The (complex) exterior differentiation operators

$$d: C^\infty(\Lambda^r{}_c TM') \to C^\infty(\Lambda^{r+1}{}_c TM'), \; r \geqslant 0$$

are complex linear and certainly local. Hence d is an LPDO. (Notice that d is often even *defined* locally and then proved to give a globally defined operator on M).

Example 2. Let $X \in C^\infty({}_cTM)$. Lie differentiation $L_X: C^\infty({}_cTM) \to C^\infty({}_cTM)$ is an LPDO (L_X is here defined as the complexification of the real Lie derivative).

Example 3. Let $A: E \to F$ be a vector bundle map between the complex vector bundles E and F. Then A induces an LPDO (of order zero):

$$A_*: C^\infty(E) \to C^\infty(F)$$

defined by

$$A_*(s)(x) = A(s(x)), \;\; s \in C^\infty(E), \;\; x \in M$$

Definition 4.1 is intended to give a globalization of the usual definition of a partial differential operator defined on an open subset of $\mathbb{R}^m$. We shall now give a local description of an LPDO defined on a manifold to justify this claim. Suppose the fibre dimensions of E and F are p and q respectively and fix $x \in M$. We may find an open neighbourhood U of x over which E and F are trivial complex bundles:

$$E|U \cong U \times \mathbb{C}^p \; ; \;\; F|U \cong U \times \mathbb{C}^q$$

It then follows that $C^\infty(E|U) \cong C^\infty(U, \mathbb{C}^p)$ and $C^\infty(F|U) \cong C^\infty(U, \mathbb{C}^q)$ and so D restricts to a map

$$\tilde{D}: C^\infty(U, \mathbb{C}^p) \to C^\infty(U, \mathbb{C}^q)$$

Assuming that U is a co-ordinate chart, it is no loss of generality to suppose that U is an open subset of $\mathbb{R}^n$, n = dimension(M). Taking U smaller, if necessary, it may then be proved using the theory of distributions that there exists $k \geqslant 0$ such that for all $\phi \in C^\infty(U, \mathbb{C}^p)$,

$$(\tilde{D}\phi)(y) = \sum_{|m| \leqslant k} a_m(y)\, \partial^m \phi(y), \; y \in U$$

where $a_m : U \to L_{\mathbb{C}}(\mathbb{C}^p, \mathbb{C}^q)$ are C^∞ maps and $\partial^m \phi = \partial^m \phi / \partial x_1^{m_1} \ldots \partial x_n^{m_n}$.

Since D is local, it follows that the above formula gives the local form for Ds for any section s of E. However, unless M is compact, k need not be bounded on M. For the operators we shall be interested in, k will be less than or equal to 2.

We now wish to give an "invariant" definition of the order of an LPDO.

Definition 4.2. Let $D: C^\infty(E) \to C^\infty(F)$ be an LPDO. We say that D is of order k if, at every point $x \in M$,

$$D(f^{k+1} s)(x) = 0$$

for all $s \in C^\infty(E)$ and $f \in C^\infty(M)$ with $f(x) = 0$ and if k is the smallest integer for which this is true.

Example 4. Exterior differentiation $d: C^\infty(\Lambda^r{}_c TM') \to C^\infty(\Lambda^{r+1}{}_c TM')$ is of order 1, $r \geqslant 0$. Indeed, let $\phi \in C^\infty(\Lambda^r{}_c TM')$, $f \in C^\infty(M)$ and suppose $f(x) = 0$. Then

$$d(f^2\phi) = d(f^2) \wedge \phi + f^2 d\phi$$

$$= 2f df \wedge \phi + f^2 d\phi$$

$$= 0 \text{ at } x$$

On the other hand,

$$d(f\phi) = df \wedge \phi + f d\phi$$

$$= df(x) \wedge \phi(x) \text{ at } x$$

In general $df(x) \wedge \phi(x) \neq 0$ and so d is not of order zero.

Example 5. The LPDO induced by a vector bundle map (see Example 3) is always of order zero.

Notation. We shall let $\mathrm{Diff}^k(E,F)$ denote the set of all LPDO from E to F of order less than or equal to k.

For our purposes we shall be most interested in the highest-order terms of an LPDO. We shall now define an invariant of an LPDO which gives us the highest-order terms in a particularly convenient form.

Definition 4.3. Let $D \in \mathrm{Diff}^k(E,F)$. The (k-) *symbol* of D is the map

$$\sigma(D): TM' \to L(E,F)$$

covering the identity on M, which is defined by

$$\sigma(D)(\phi_x)(e_x) = 1/k!\, D(f^k s)(x)$$

where $\phi_x \in TM'_x$, $e_x \in E_x$ and $f \in C^\infty(M)$, $s \in C^\infty(E)$ are chosen such that

$$f(x) = 0,\ df(x) = \phi_x \text{ and } s(x) = e_x$$

Remarks
(1) It is not hard to verify that $\sigma(D)$ is well defined and does not depend on our choice of functions f and s.
(2) $\sigma(D)$ is not a vector bundle map (unless k = 1) but it is homogeneous of degree k in cotangent vectors:

$$\sigma(D)(\lambda v) = \lambda^r \sigma(D)(v),\ \lambda \in \mathbb{R},\ v \in TM'$$

Example 6. $d : C^\infty(\Lambda^r{}_c TM') \to C^\infty(\Lambda^{r+1}{}_c TM')$ has symbol "wedge product". In fact it follows from Example 4 that

$$\sigma(d) : TM' \to L(\Lambda^r{}_c TM',\ \Lambda^{r+1}{}_c TM')$$

is the map defined by

$$\sigma(d)(\xi)(\phi) = \xi \wedge \phi,\ \xi \in TM'_x,\ \phi \in \Lambda^r{}_c TM'_x$$

We shall now give, without proof, some elementary properties of symbols and differential operators.
(1) If E, F, G are complex vector bundles on M and $D \in \mathrm{Diff}^k(E, F)$, $D' \in \mathrm{Diff}^{k'}(F, G)$ then $D'D \in \mathrm{Diff}^{k+k'}(E,G)$ and $\sigma(D'D) = \sigma(D')\cdot\sigma(D)$, where the latter composition is induced from the composition $L(F, G) \times L(E, F) \to L(E, G)$.
(2) If $D \in \mathrm{Diff}^k(E, F)$ then $\sigma(D) = 0$ if and only if D is of order k-1 (σ here denotes the k-symbol). Moreover, if we let $\mathrm{Smbl}^k(E, F)$ denote the set of all C^∞ maps $TM' \to L(E, F)$ which cover the identity map on M and restrict to homogeneous polynomials of degree k on the fibres of TM', then the "symbol sequence"

$$0 \to \mathrm{Diff}^{k-1}(E, F) \xrightarrow{i} \mathrm{Diff}^k(E, F) \xrightarrow{\sigma} \mathrm{Smbl}^k(E, F) \to 0$$

is *exact* (i denotes the inclusion map and σ the k-symbol).
(3) If $D \in \mathrm{Diff}^k(E, F)$ has the local form

$$D\phi = \sum_{|m| \leqslant k} a_m \partial^m \phi$$

then $\sigma(D)$ is given locally by

$$\sigma(D)(\xi) = \sum_{|m| = k} a_m(x)\xi^m \in L(\mathbb{C}^p, \mathbb{C}^q)$$

where $\xi \in (\mathbb{R}^n)^*$.

Definition 4.4. Let $D \in \mathrm{Diff}^k(E,F)$. We say that D is *elliptic* if $\sigma(D)(\xi)$ is a linear isomorphism for all non-zero cotangent vectors ξ of M.

Remark. A *necessary* condition for D to be elliptic is that the fibre dimensions of E and F are equal. Finding *sufficient* conditions for the existence of an elliptic LPDO in $\mathrm{Diff}^k(E,F)$ is much harder – see Refs [4, 28].

Definition 4.5. Suppose that $E_0, \ldots, E_n$ are complex vector bundles on M and that we are given LPDOs $D_j \in \mathrm{Diff}^k(E_{j-1}, E_j)$ which satisfy $D_{j+1}D_j = 0$, $1 \leq j < n$. We shall say that the sequence

$$0 \to C^\infty(E_0) \xrightarrow{D_1} C^\infty(E_1) \xrightarrow{D_2} \ldots \xrightarrow{D_n} C^\infty(E_n) \to 0$$

defines an *elliptic complex* on M if the corresponding symbol sequences are exact, i.e. if for all $x \in M$ and $\xi \in TM'_x$, $\xi \neq 0$, the sequence

$$0 \longrightarrow (E_0)_x \xrightarrow{\sigma(D_1)(\xi)} (E_1)_x \xrightarrow{\sigma(D_2)(\xi)} \cdots \xrightarrow{\sigma(D_n)(\xi)} (E_n)_x \longrightarrow 0$$

of vector spaces and linear maps is *exact.*

Remark. Definition 4.4 is a special case of Definition 4.5. Take $E_0 = E$, $E_1 = F$, $n = 1$ and $D_1 = D$.

Before stating the fundamental existence and regularity theorem for elliptic complexes on compact manifolds we shall give some examples.

Example 7 (the de Rham complex). If M is of dimension n we have the de Rham sequence:

$$0 \longrightarrow C^\infty(M) \xrightarrow{d} C^\infty(\Lambda^1{}_c TM') \xrightarrow{d} \ldots \xrightarrow{d} C^\infty(\Lambda^n{}_c TM') \longrightarrow 0$$

Now $d^2 = 0$ and we have already shown that

$$\sigma(d)(\xi)(\phi) = \xi \wedge \phi,\ \xi \in TM'_x,\ \phi \in \Lambda^r{}_c TM'_x,\ x \in M$$

To prove that the de Rham sequence is elliptic it is therefore sufficient to prove the following result.

Let E be a complex vector space of dimension n with dual E^* and let $\xi \in E^*$. Then the sequence

$$0 \longrightarrow \Lambda^0 E^* \xrightarrow{\xi\wedge} \Lambda^1 E^* \xrightarrow{\xi\wedge} \cdots \xrightarrow{\xi\wedge} \Lambda^n E^* \longrightarrow 0$$

is exact. We leave the proof of this fact as an exercise for the reader. It follows that the de Rham complex is elliptic.

Example 8 (the Dolbeault complex). Let M be a complex manifold of complex dimension n and associated almost complex structure $J: TM \to TM$. The complex cotangent bundle of M splits as a sum of complex vector bundles:

$${}_c TM' = TM^* \oplus \overline{TM^*}$$

where TM^* is the *holomorphic cotangent* bundle of M and $\overline{TM}^*$ is the *anti-holomorphic cotangent* bundle of M. TM^* and $\overline{TM}^*$ are characterized by

$$TM^* = L_{\mathbb{C}}(TM, \mathbb{C}) \; ; \; \overline{TM}^* = L_{\mathbb{C}}(\overline{TM}, \mathbb{C})$$

Here $\overline{TM}$ denotes the tangent bundle of M but with complex structure $-J$. In other words, $\overline{TM}^*$ is the bundle of *conjugate* complex linear forms on TM. For $r \geqslant 0$ we have

$$\Lambda^r{}_c TM' = \Lambda^r (TM^* \oplus \overline{TM}^*)$$

$$\approx \bigoplus_{p+q=r} \Lambda^p TM^* \otimes \Lambda^q \overline{TM}^*$$

The latter isomorphism is defined by mapping $f_1 \wedge \ldots \wedge f_p \otimes g_{\bar{1}} \wedge \ldots \wedge g_{\bar{q}} \in \Lambda^p TM^* \otimes \Lambda^q \overline{TM}^*$ to $f_1 \wedge \ldots \wedge f_p \wedge g_{\bar{1}} \wedge \ldots \wedge g_{\bar{q}} \in \Lambda^r{}_c TM'$. We shall denote the image of $\Lambda^p TM^* \otimes \Lambda^q \overline{TM}^*$ in $\Lambda^r{}_c TM'$ by $\Lambda^{p,q}(M)'$. $\Lambda^{p,q}(M)'$ is a complex vector bundle on M (the bundle of "complex (p,q)-forms"). $\Lambda^{p,q}(M)'$ is a holomorphic vector bundle if (and only if) $q = 0$.

C^∞ sections of $\Lambda^r{}_c TM'$ are called complex differential r forms and we shall denote the space of such sections by $C^r(M)$. C^∞ sections of $\Lambda^{p,q}(M)'$ are called complex differential forms of type (p,q). We denote the space of such sections by $C^{p,q}(M)$. Since

$$\Lambda^r{}_c TM' = \bigoplus_{p+q=r} \Lambda^{p,q}(M)'$$

we have

$$C^r(M) = \bigoplus_{p+q=r} C^{p,q}(M)$$

We denote the space of *holomorphic* sections of $\Lambda^{p,0}(M)' \cong \Lambda^p TM^*$ by $\Omega^p(M)$.

For $p = 0, \ldots, n = \dim_{\mathbb{C}} M$, we have the *Dolbeault sequences*

$$0 \longrightarrow C^{p,0}(M) \xrightarrow{\bar{\partial}} C^{p,1}(M) \xrightarrow{\bar{\partial}} \ldots \xrightarrow{\bar{\partial}} C^{p,n}(M) \longrightarrow 0$$

and $\bar{\partial}^2 = 0$. We shall not construct these sequences here but instead refer the reader to Refs [19,31]. In case $p = 0$, the local form for $\bar{\partial}$ is given in Section 1. We claim that the Dolbeault sequences are elliptic complexes. The proof is similar to that given for the de Rham complex and follows easily once we have shown that

$$\sigma(\bar{\partial})(\xi) = \tfrac{1}{2}(\xi + iJ\xi)\Lambda, \; \xi \in TM'$$

We leave the details of the calculation of $\sigma(\bar{\partial})$ to the reader. We remark that we have similar elliptic complexes for ∂.

Example 9. Let E be a holomorphic vector bundle on the complex manifold M. We shall let $\Lambda^{p,q}(M,E)'$ denote the complex vector bundle $\Lambda^{p,q}(M)' \otimes E$. $\Lambda^{p,q}(M,E)'$ is called the bundle of E-valued (p,q)-forms on M. We let $C^{p,q}(M,E)$ denote the space of C^∞ sections of $\Lambda^{p,q}(M,E)'$ and $\Omega^p(M,E)$ denote the space of holomorphic sections of the holomorphic vector bundle $\Lambda^{p,0}(M,E)' \approx \Lambda^p TM^* \otimes E$.

We claim that $\bar{\partial}$ extends to E-valued forms and that for p = 0, ..., n we have elliptic complexes

$$0 \longrightarrow C^{p,0}(M,E) \xrightarrow{\bar{\partial}} C^{p,1}(M,E) \xrightarrow{\bar{\partial}} \dots \xrightarrow{\bar{\partial}} C^{p,n}(M,E) \longrightarrow 0$$

In particular, the complex

$$0 \longrightarrow C^{\infty}(E) \xrightarrow{\bar{\partial}} C^{0,1}(M,E) \xrightarrow{\bar{\partial}} \dots \xrightarrow{\bar{\partial}} C^{0,n}(M,E) \longrightarrow 0$$

is elliptic.

Let us start by defining $\bar{\partial}: C^{\infty}(E) \longrightarrow C^{\infty}(\overline{TM}^* \otimes E)$. Let $s \in C^{\infty}(E)$. Relative to a trivialization. (U,ϕ) of E, let s have the local representation $s^{\phi} = (s_1^{\phi}, \dots, s_p^{\phi})$, where p is the fibre dimension of E and $s_j^{\phi} \in C^{\infty}(U)$, $1 \leq j \leq p$. We define $\bar{\partial}s$, relative to the trivialization (U,ϕ), by

$$(\bar{\partial}s)^{\phi} = (\bar{\partial}s_1^{\phi}, \dots, \bar{\partial}s_p^{\phi})$$

We must show that $\bar{\partial}s$ is well defined, independent of choice of trivialization. But if $s^{\psi} = (s_1^{\psi}, \dots, s_p^{\psi})$, relative to the trivialization (V, ψ), then $s^{\psi} = g_{\psi\phi}s^{\phi}$ on the overlap $U \cap V$ and the transition function $g_{\psi\phi}$ is *holomorphic*. Hence

$$\begin{aligned} \bar{\partial}s^{\psi} &= (\bar{\partial}g_{\psi\phi})s^{\phi} + g_{\psi\phi}\bar{\partial}s^{\phi} \\ &= g_{\psi\phi}\bar{\partial}s^{\phi} \end{aligned}$$

and so $\bar{\partial}s$ is well defined. We define $\bar{\partial}: C^{p,q}(M,E) \to C^{p,q+1}(M,E)$ by

$$\bar{\partial}(\phi \otimes s) = \bar{\partial}\phi \otimes s + (-1)^{p+q}\phi \wedge \bar{\partial}s$$

where $\phi \in C^{p,q}(M)$ and $s \in C^{\infty}(E)$. We leave it to the reader to verify that $\bar{\partial}$ is well defined and that $\bar{\partial}^2 = 0$.

The proof that the above sequences are elliptic complexes is very similar to that indicated in Example 8 and we omit details.

We cannot construct ∂-sequences for E-valued forms, E a holomorphic vector bundle (unless E satisfies additional conditions – e.g. E is a trivial bundle). However, if we say that a complex vector bundle E is *antiholomorphic* when the bundle $\bar{E}$ is holomorphic, then, by conjugating forms, we obtain ∂-sequences of E-valued forms and these sequences are elliptic (if E has complex structure J, then $\bar{E}$ is the complex vector bundle which is equal to E as a real vector bundle but has complex structure –J).

Let E_1 and E_2 be complex vector bundles on M and let us suppose from now on that M is *compact* and *oriented* (the latter assumption is included because of our intended applications and is not needed for the existence and regularity theorem). Fix a Riemannian metric on M and denote the corresponding measure on M of volume 1 by $d\mu$. Choose fixed hermitian metrics $(,)_1$ and $(,)_2$ on E_1 and E_2 respectively (see sub-section B below or Ref. [31], these Proceedings). We may define hermitian inner products $((,))_j$ on the spaces $C^{\infty}(E_j)$, $j = 1, 2$, by setting

$$((s,t))_j = \int_M (s(x), t(x))_j \, d\mu(x), \quad s, t \in C^{\infty}(E_j)$$

The spaces $C^{\infty}(E_j)$ will not be complete with respect to the inner products $((,))_j$ except in trivial cases.

The proof of the following result may be found in Refs [8, 28, 29].

Proposition 4.6. Let M be a compact manifold, E and F be hermitian vector bundles on M and $D \in \mathrm{Diff}^k(E,F)$. Then there exists a unique $D^* \in \mathrm{Diff}^k(F,E)$ characterized by

$$((Ds,t))_2 = ((s, D^*t))_1$$

for all $s \in C^\infty(E_1)$ and $t \in C^\infty(E_2)$.

D^* is called the *(formal) adjoint* of D. The symbol of D^* is given in terms of the symbol of D by the relation

$$\sigma(D^*) = (-1)^k(\sigma(D))^*$$

where by $(\sigma(D))^*$ we mean that we are taking the usual hermitian adjoint on fibres of L(E, F).

Definition 4.7. Let M be a compact manifold and E be a hermitian vector bundle on M. We say that $D \in \mathrm{Diff}^{2k}(E,E)$ is a *strongly elliptic differential operator* if $(-1)^k \sigma(D)(\xi)$ is a strictly positive definite linear operator for all non-zero $\xi \in TM'$.

We leave as an exercise the proof of the following

Proposition 4.8. $D \in \mathrm{Diff}^k(E,F)$ is elliptic if and only if $D^*D \in \mathrm{Diff}^{2k}(E,E)$ is strongly elliptic.

Suppose that we are given a sequence

$$0 \longrightarrow C^\infty(E_0) \xrightarrow{D_1} C^\infty(E_1) \xrightarrow{D_2} \cdots \xrightarrow{D_n} C^\infty(E_n) \longrightarrow 0$$

of LPDOs of order k such that $D_{j+1}D_j = 0$ for every j. We shall denote this sequence by $\mathscr{E}$. Since $D_{j+1}D_j = 0$,

$$\mathrm{Image}\,(D_j) \subseteq \mathrm{Kernel}\,(D_{j+1}) \subseteq C^\infty(E_j)$$

Now Image (D_j) and Kernel (D_{j+1}) are complex vector subspaces of $C^\infty(E_j)$ and so we may define the *cohomology vector spaces* of the complex $\mathscr{E}$ by

$$H^0(M,\mathscr{E}) = \mathrm{Kernel}\,(D_1)$$

$$H^j(M,\mathscr{E}) = \mathrm{Kernel}\,(D_{j+1})/\mathrm{Image}\,(D_j),\ j \geqslant 1$$

Now suppose that we are given a Riemannian metric on M with associated measure $d\mu$ of volume 1 and hermitian inner products $(\ ,\ )_j$ on the bundles E_j together with the induced hermitian inner products $((\ ,\ ))_j$ on the spaces $C^\infty(E_j)$. By Proposition 4.6 we may define for every j

$$D_j^* \in \mathrm{Diff}^k(E_j, E_{j-1})$$

and since $(D_{j+1}D_j)^* = D_j^*D_{j+1}^*$, it follows that $D_j^*D_{j+1}^* = 0$.

We define the operators $\Delta_j \in \mathrm{Diff}^{2k}(E_j, E_j)$ by

$$\Delta_j = D_{j+1}^*D_{j+1} + D_jD_j^*$$

A little linear algebra, together with the expression for $\sigma(D^*)$ given by Proposition 4.6, gives us

Proposition 4.9. Let

$$0 \longrightarrow C^\infty(E_0) \xrightarrow{D_1} C^\infty(E_1) \xrightarrow{D_2} \cdots \xrightarrow{D_n} C^\infty(E_n) \longrightarrow 0$$

be an elliptic complex on M. Then the operators Δ_j are all strongly elliptic.

Notation. If $\mathscr{E}$ is an elliptic complex on M we set

$$\mathscr{H}^j(M,\mathscr{E}) = \text{Kernel}\,(\Delta_j),\ j \geqslant 0$$

Notice that $\mathscr{H}^j(M,\mathscr{E})$, unlike $H^j(M,\mathscr{E})$, depends on our choice of metrics on M and the bundles E_{j-1}, E_j, E_{j+1}. It follows from the identity

$$(\Delta_j s, t) = (D_{j+1}s, D_{j+1}t) + (D_j^* s, D_j^* t), \quad s,t \in C^\infty(E_j)$$

that

$$\mathscr{H}^j(M,\mathscr{E}) = \{s \in C^\infty(E_j) : D_{j+1}s = 0, \text{ and } D_j^* s = 0\}$$

Hence

$$\mathscr{H}^j(M,\mathscr{E}) = \text{Kernel}\,(D_{j+1}) \cap \text{Kernel}\,(D_j^*)$$

We may now state the fundamental existence and regularity theorem for elliptic complexes:

Theorem 4.10. Let $0 \longrightarrow C^\infty(E_0) \xrightarrow{D_1} \cdots \xrightarrow{D_n} C^\infty(E_n) \longrightarrow 0$ be an elliptic complex on the compact manifold M and suppose that we are given a Riemannian metric on M and hermitian metrics on the bundles E_j. Denoting the complex by $\mathscr{E}$, we have for $j \geqslant 0$:
(1) $\mathscr{H}^j(M,\mathscr{E})$ is a finite-dimensional subspace of $C^\infty(E_j)$.
(2) $C^\infty(E_j) = \mathscr{H}^j(M,\mathscr{E}) \oplus \text{Image}\,(D_j) \oplus \text{Image}\,(D_{j+1}^*) = \mathscr{H}^j(M,\mathscr{E}) \oplus \text{Image}\,\Delta_j$. ($\oplus$ denotes orthogonal direct sum). In particular, any $s \in C^\infty(E_j)$ has the unique harmonic decomposition:

$$s = D_j\alpha + D_{j+1}^*\beta + \phi$$

where $\alpha \in C^\infty(E_{j-1})$, $\beta \in C^\infty(E_{j+1})$ and $\phi \in \mathscr{H}^j(M,\mathscr{E})$.
(3) $\mathscr{H}^j(M,\mathscr{E}) \approx H^j(M,\mathscr{E})$. This canonical isomorphism is defined by mapping $\phi \in \mathscr{H}^j(M,\mathscr{E}) \subset \text{Kernel}\,(D_{j+1})$ to the corresponding class in $H^j(M,\mathscr{E})$.

Proof. It follows easily from Proposition 4.9 that it is sufficient to show that if $D \in \text{Diff}^{2k}(E,E)$ is strongly elliptic then

$$\dim(\text{Kernel}\,(D)) < \infty$$

$$C^\infty(E) = \text{Kernel}\,(D) \oplus \text{Image}\,(D)$$

We shall not prove this deep result about elliptic operators here and we refer the reader to Refs [8, 27, 28, 41] for proofs.

Let us look at some applications of this very powerful result.

Example 10. Let M be a compact oriented Riemannian manifold and let $\mathscr{D}$ denote the de Rham complex of M:

$$0 \longrightarrow C^\infty(M) \xrightarrow{d} C^\infty(\Lambda^1{}_c TM') \xrightarrow{d} \ldots \xrightarrow{d} C^\infty(\Lambda^n{}_c TM') \longrightarrow 0$$

The Riemannian metric on M is easily shown to induce hermitian metrics on the complex bundles $\Lambda^r{}_c TM'$. The operators

$$\Delta = \Delta_j = dd^* + d^*d : C^\infty(\Lambda^j{}_c TM') \longrightarrow C^\infty(\Lambda^j{}_c TM')$$

are called the *Laplace operators* of the Riemannian manifold M and elements of Kernel (Δ) are called *harmonic forms.* The vector spaces $H^j(M,\mathscr{E})$ are the usual *de Rham cohomology* groups $H^j(M,\mathbb{C})$ of M. The fundamental theorem tells us that:

(1) $\dim_{\mathbb{C}} H^j(M,\mathbb{C}) < \infty$, $j \geq 0$ (this can, of course, be proved directly).

(2) Any $\psi \in C^\infty(\Lambda^j{}_c TM')$ has the unique harmonic decomposition

$$\psi = d\alpha + d^*\beta + \phi$$

where ϕ is a harmonic j-form. The reader may easily verify that $d\psi = 0$ implies that $d^*\beta = 0$ and so

$$\psi - \phi = d\alpha$$

In other words, each cohomology class in $H^j(M,\mathbb{C})$ is *uniquely* represented by a *harmonic* j-form.

We therefore have the remarkable fact that on a *Riemannian* manifold the cohomology may be uniquely represented by *harmonic* forms. This in turn suggests links between the differential geometric structure of the manifold (as determined by the Riemannian metric) and the topological structure, and much work has been done in this area.

Example 11. Let M be a compact complex hermitian manifold and E be a hermitian holomorphic vector bundle on M. We let $\mathscr{D}_p(E)$ denote the elliptic complex:

$$0 \longrightarrow C^{p,0}(M,E) \xrightarrow{\bar\partial} C^{p,1}(M,E) \xrightarrow{\bar\partial} \ldots \xrightarrow{\bar\partial} C^{p,n}(M,E) \longrightarrow 0$$

The operators

$$\square = \square_j = \bar\partial\bar\partial^* + \bar\partial^*\bar\partial : C^{p,j}(M,E) \to C^{p,j}(M,E)$$

are strongly elliptic and second order. The vector spaces $H^j(M, \mathscr{D}_p(E))$ are the *Dolbeault cohomology* groups $H^j(M,\Omega^p(E))$ of M and E. The fundamental theorem implies that:

(1) $\dim_{\mathbb{C}} H^j(M,\Omega^p(E)) < \infty$, $j, p \geq 0$. In particular, $\dim_{\mathbb{C}} O(E) < \infty$ for all holomorphic vector bundles E on M.

(2) If we let $H^{p,j}(M,E)$ denote the space of harmonic forms

$$\mathscr{H}^j(M, \mathscr{D}_p(E)) = \{\phi \in C^{p,j}(M,E) : \square\phi = 0\}$$

then

$$H^{p,j}(M,E) \approx H^j(M,\Omega^p(E)), \quad p, j \geq 0$$

In sub-section C, we use this fact to give an easy proof of the Serre duality theorem.

If $E = \underline{\mathbb{C}}$, we recover the Dolbeault complexes $\mathscr{D}_p$:

$$0 \longrightarrow C^{p,0}(M) \xrightarrow{\bar{\partial}} C^{p,1}(M) \xrightarrow{\bar{\partial}} \dots \xrightarrow{\bar{\partial}} C^{p,n}(M) \longrightarrow 0$$

In this case

$$H^j(M, \mathscr{D}_p) = \{s \in C^{p,j}(M) : \bar{\partial}s = 0\}/\bar{\partial}C^{p,j-1}(M)$$

and the fundamental theorem implies that every $s \in C^{p,j}(M)$ has a unique harmonic decomposition

$$s = \bar{\partial}\alpha + \bar{\partial}^*\beta + \phi$$

where $\alpha \in C^{p,j-1}(M)$, $\beta \in C^{p,j+1}(M)$, and ϕ is harmonic. The operators $\Box = \bar{\partial}\bar{\partial}^* + \bar{\partial}^*\bar{\partial}$ are usually referred to as the complex *Laplace-Beltrami* operators of the hermitian manifold M.

Much of the remainder of this section will be working towards showing that there is a close connection between the de Rham and Dolbeault complexes when M is a Kähler manifold. In this way we shall find relationships between the complex structure, differential geometry and topology of such manifolds.

B. Contractions and linear algebra

Hermitian linear algebra and, in particular, contractions play an important role in complex differential geometry. We start this subsection by giving a brief résumé of "contractions".

Let E denote a finite-dimension vector space with complex structure J. We let E^* denote the dual space of E, $\bar{E}^*$ the conjugate dual space of E, and $\bar{E}$ the conjugate space of E, i.e. E with complex structure $-J$. We let $\wedge^r E$ denote the rth complex exterior power of E. We define a linear map

$$\theta : \wedge^p E \otimes \wedge^q E^* \to \wedge^{p-q} E, \; p \geqslant q$$

by requiring that θ is the transpose of "wedge product", i.e. for all $X \in \wedge^p E$, $\phi \in \wedge^q E^*$ and $\psi \in \wedge^{p-q} E^*$, we have

$$\langle \theta(X \otimes \phi), \psi \rangle = \langle X, \phi \wedge \psi \rangle$$

($\langle\,,\rangle$ denotes the pairing between a space and its dual). We similarly define

$$\theta : \wedge^p E \otimes \wedge^q E^* \to \wedge^{q-p} E^*$$

for $q \geqslant p$. We remark that if $p = q$, $\theta : \wedge^p E \otimes \wedge^p E^* \to \mathbb{C}$ is the usual dual pairing.

If $X \in \wedge^p E$ and $p \leqslant q$, we define

$$\theta_X : \wedge^q E^* \to \wedge^{q-p} E^*$$

by

$$\theta_X(\phi) = \theta(X \otimes \phi), \; \phi \in \wedge^q E^*$$

and we similarly define $\theta_\phi : \wedge^p E \to \wedge^{p-q} E$ when $p \geqslant q$ and $\phi \in \wedge^q E^*$. θ is an example of a *contraction* and θ_X, θ_ϕ are usually referred to as *interior products* (by X and ϕ respectively).

The properties of $\theta, \theta_X, \theta_\phi$ follow easily from the corresponding properties of wedge product. We leave as an exercise the verification that:

(1) $\theta_X\theta_Y = \theta_{Y\wedge X} = (-1)^{pq}\theta_Y\theta_X$, $X \in \Lambda^p E$, $Y \in \Lambda^q E$.

(2) If $X \in E$, the sequence of complex vector spaces and linear maps

$$\cdots \xrightarrow{\theta_X} \Lambda^{r+1}E^* \xrightarrow{\theta_X} \Lambda^r E^* \xrightarrow{\theta_X} \Lambda^{r-1}E^* \xrightarrow{\theta_X} \cdots$$

is *exact*.

If we have a tensor product of more than two vector spaces, say

$$\Lambda^p E \otimes \Lambda^q E^* \otimes \Lambda^r E^* \otimes \Lambda^s E$$

then

$$\theta^1_3 : \Lambda^p E \otimes \Lambda^q E^* \otimes \Lambda^r E^* \otimes \Lambda^s E \to \Lambda^q E^* \otimes \Lambda^{r-p}E^* \otimes \Lambda^s E,\ p \leqslant r$$

$$\to \Lambda^{p-r}E \otimes \Lambda^q E^* \otimes \Lambda^s E,\ p \geqslant r$$

is the map defined on generators by

$$\theta^1_3(X \otimes \phi \otimes \psi \otimes Y) = \phi \otimes \theta_X \psi \otimes Y,\ p \leqslant r$$

$$= \theta_\psi X \otimes \phi \otimes Y,\ p \geqslant r$$

More generally, the symbol θ^i_j means that the contraction is between the ith and j^{th} factors of a tensor product of exterior powers, where the ith factor (superscript) is an $\Lambda^* E$ and the jth factor (subscript) is an $\Lambda^* E^*$. The importance of the operators θ^i_j comes from the fact that most of the linear algebra used in differential geometry is built up by successive applications of these contraction operations (exercise!). We have not written down the formulae for contractions relative to a basis for two reasons: first, we do not need such formulae; second, except in simple cases, the co-ordinate expressions are *very* complex. This should not be taken to imply, however, that *all* linear algebra should be done "invariantly" as the use of orthonormal bases and diagonalization are often powerful and simplifying techniques.

Let E' denote the real dual $L_{\mathbb{R}}(E, \mathbb{R})$ of E. We form the complexifications

$${}_cE = E \otimes_{\mathbb{R}} \mathbb{C} \text{ and } {}_cE' = E' \otimes_{\mathbb{R}} \mathbb{C} = L_{\mathbb{R}}(E, \mathbb{C})$$

and we have the canonical isomorphisms

$$\Lambda^r {}_cE \approx \bigoplus_{p+q=r} \Lambda^p E \otimes \Lambda^q \bar{E}$$

$$\Lambda^r {}_cE' \approx \bigoplus_{p+q=r} \Lambda^p E^* \otimes \Lambda^q \bar{E}^*$$

We denote the set of vectors in $\Lambda^r {}_cE$ ($\Lambda^r {}_cE'$) corresponding to $\Lambda^p E \otimes \Lambda^q \bar{E}$ ($\Lambda^p E^* \otimes \Lambda^q \bar{E}^*$) by $\Lambda^{p,q}(E)$ ($\Lambda^{p,q}(E)'$). Elements of $\Lambda^{p,q}(E)$ ($\Lambda^{p,q}(E)'$) are called (p,q)-vectors ((p,q)-forms). The dual pairing $E \times E' \to \mathbb{R}$ induces a dual pairing

$$\Lambda^r {}_cE \times \Lambda^r {}_cE' \to \mathbb{C}$$

which in turn restricts to a dual pairing

$$\Lambda^{p,q}(E) \times \Lambda^{p,q}(E)' \to \mathbb{C},\ p + q = r$$

Conjugation in $\mathbb{C}$ induces conjugation in ${}_cE$ and ${}_cE'$ and so on $\Lambda^r{}_cE$ and $\Lambda^r{}_cE'$, $r \geqslant 0$. For example, if $X \otimes d \in {}_cE$, $X \in E$, $d \in \mathbb{C}$, we define

$$\overline{X \otimes d} = X \otimes \bar{d}$$

If $\phi \in E^*$ then $\bar{\phi} \in \bar{E}^*$ and conversely. More generally, we find that for all $p,q \geqslant 0$

$$\overline{\Lambda^{p,q}(E)} = \Lambda^{q,p}(E) \text{ and } \overline{\Lambda^{p,q}(E)'} = \Lambda^{q,p}(E)'$$

We say that $\phi \in \Lambda^r{}_cE'$ is a *real* form if $\bar{\phi} = \phi$. Notice that a (p,q)-form can be real only if $p = q$.

Let $L(E,\bar{E};\mathbb{C})$ denote the space of complex bilinear forms on $E \times \bar{E}$ (i.e. the space of real bilinear maps from $E \times E \to \mathbb{C}$ which are complex linear in the first variable and conjugate complex linear in the second variable). Now $L(E,\bar{E};\mathbb{C}) \approx E^* \otimes \bar{E}^*$. In the sequel we shall use this identificat[ion] to evaluate elements of $E^* \otimes \bar{E}^*$ on pairs of points of E.

Definition 4.11. An element $h \in E^* \otimes \bar{E}^*$ is said to be a *hermitian quadratic form* on E if h is conjugate symmetric:

$$h(e,f) = \overline{h(f,e)}, \quad \text{for all } e,f \in E$$

If, in addition, h is positive definite,

$$h(e,e) > 0, \text{ for all } e \neq 0$$

we shall say that h is a *hermitian metric* or *hermitian inner product* on E.

Remark. A hermitian metric h on E defines a complex inner product (,) on E by

$$(e,f) = h(e,f), \quad e, f \in E$$

and we shall regard the two concepts as equivalent.

Now $E^* \otimes \bar{E}^* \sim \Lambda^{1,1}(E)'$ and we define the *Kähler form* of the hermitian metric h to be the element $\hat{H} \in \Lambda^{1,1}(E)'$ defined by

$$\hat{H} = ih$$

The conjugate symmetry of h implies that $\hat{H}$ is a *real* (1,1)-form.

Since $E^* \otimes \bar{E}^* \approx L(E,\bar{E}^*)$ we may regard a hermitian metric h on E as a complex linear map $h: E \to \bar{E}^*$ satisfying the appropriate symmetry and positivity conditions (h must be an isomorphis[m] equal to its conjugate transpose and $\langle h(e),e\rangle$ must be strictly positive for all non-zero $e \in E$). The conjugate of h is the complex linear map $\bar{h}: \bar{E} \to E^*$ defined by $\bar{h}(\bar{e}) = \overline{h(e)}$, $e \in E$. Since ${}_cE \approx E \oplus \bar{E}$ and ${}_cE' \approx E^* \oplus \bar{E}^*$, we may take the direct sum of h and $\bar{h}$ to obtain a complex linear map

$$G = h + \bar{h}: {}_cE \to {}_cE'$$

G induces maps $\Lambda^rG: \Lambda^r{}_cE \to \Lambda^r{}_cE'$, $\Lambda^rG^{-1}: \Lambda^r{}_cE' \to \Lambda^r{}_cE$. For notational economy we shall denote these maps in the sequel by G and G^{-1} respectively. We remark that

$$G: \Lambda^{p,q}(E) \to \Lambda^{q,p}(E)'$$

The maps G and G^{-1} induce contractions on spaces of complex vectors and forms. For example, if $p \geqslant s$ and $q \geqslant r$, we may define

$$\theta(h): \Lambda^{p,q}(E) \otimes \Lambda^{r,s}(E) \to \Lambda^{p-s,q-r}(E)$$

by

$$\theta(h)(X \otimes Y) = \theta(X \otimes G(Y))$$

$$= G^{-1}\theta(G(X) \otimes Y)$$

where $X \in \Lambda^{p,q}(E)$, $Y \in \Lambda^{r,s}(E)$.

For $p \geqslant r$, $q \geqslant s$, we define

$$\bar{\theta}(h): \Lambda^{p,q}(E) \otimes \Lambda^{r,s}(E) \to \Lambda^{p-r,q-s}(E)$$

by

$$\bar{\theta}(h)(X \otimes Y) = \theta(X \otimes \overline{G(Y)}), \quad X \in \Lambda^{p,q}(E),\ Y \in \Lambda^{r,s}(E)$$

If $p = r$, $q = s$, we set

$$\bar{\theta}(h)(X \otimes Y) = (X,Y) \in \mathbb{C}$$

$(\, , \,)$ defines a hermitian metric on $\Lambda^{p,q}(E)$, and in the sequel we shall take this as the hermitian metric on $\Lambda^{p,q}(E)$ induced from the metric h on E. We may similarly define contraction operators $\theta(h^{-1})$, $\bar{\theta}(h^{-1})$, $\theta(h)^{i}_{j}$, etc., and hermitian metrics on $\Lambda^{p,q}(E)'$, $p,q \geqslant 0$.

Remark. Notice that all the above contractions are defined in terms of the dual pairings induced from $E \times E' \to \mathbb{R}$.

The complex structure J on E induces complex structures on E', ${}_cE$ and ${}_cE'$ which we continue to denote by J (recall though that the complex structure J on ${}_cE$ and ${}_cE'$ does *not* coincide with complex structure given by the complexification). It is easily verified that the map $G: {}_cE \to {}_cE'$ is conjugate complex linear with respect to the complex structure J and it follows that G is the complexification of a conjugate J-linear isomorphism

$$g: E \to E'$$

i.e. $gJ = -Jg$. g defines a real inner product $(\, , \,)$ on E by

$$(e,f) = \langle g(e), f \rangle, \quad e, f \in E$$

and $(\, , \,)$ is clearly J-invariant: $(Je, Jf) = (e,f)$, for all $e, f \in E$. The following formula gives the relation between the hermitian inner product defined on ${}_cE$ and the real inner product defined on E:

$$(X \otimes c, Y \otimes d) = c\bar{d}(X,Y)$$

where $X, Y \in E$, $c, d \in \mathbb{C}$.

Conversely, if we are given a real J-invariant inner product g on E, then g induces hermitian metrics on E and ${}_cE$ by reversal of the above arguments.

From now on we shall let g denote the real inner product on E associated to the hermitian metric h on E. Let $\dim_{\mathbb{C}} E = m$ and $\dim_{\mathbb{R}} E = n = 2m$. Since g is J-invariant we have well-defined "positive" vectors $\Lambda \in \Lambda^n E$ and $\Theta \in \Lambda^n E'$ of length 1 (see Ref. [31], these Proceedings). Indeed, if we choose a real basis $\{e_1, Je_1, e_2, \ldots, Je_m\}$ for E and denote the corresponding co-ordinate system on E by $(x_1, y_1, x_2, \ldots, y_m)$ then

$$\Lambda = (\sqrt{\det[g_{ij}]})^{-1} e_1 \wedge Je_1 \wedge e_2 \wedge \ldots \wedge Je_m$$

$$\Theta = (\sqrt{\det[g_{ij}]}) dx_1 \wedge dy_1 \wedge dx_2 \wedge \ldots \wedge dy_m$$

We call Λ the *orientation* form of g (or h) and Θ the *volume* form of g (or h).

Since $\Lambda^{2m}{}_c E = \Lambda^{m,m}(E)$ and $\Lambda^{2m}{}_c E' = \Lambda^{m,m}(E)'$, Λ and Θ define real (m, m)-forms in $\Lambda^{2m}{}_c E$ and $\Lambda^{2m}{}_c E'$ respectively. We continue to denote these forms by Λ and Θ.

As an exercise the reader may verify that

$\widehat{H}^m = m!\,\Theta$. In particular, $\widehat{H}^m \neq 0$.

Definition 4.12. Let h be a hermitian metric on E and let $\dim_{\mathbb{C}} E = m$, $\dim_{\mathbb{R}} E = n = 2m$. The *complex Hodge star operator* is the complex linear map

$$* : \Lambda^r{}_c E' \to \Lambda^{n-r}{}_c E'$$

defined for $r \geqslant 0$ by

$$*\phi = \theta(h^{-1})(\phi \otimes \Theta), \; \phi \in \Lambda^r{}_c E'$$

Using elementary properties of contractions and wedge products, one may easily prove the following facts about the star operator:

(1) $*1 = \Theta$.

(2) $(*\phi, \psi) = \langle \phi \wedge \overline{\psi}, \Lambda \rangle$, $\phi \in \Lambda^r{}_c E'$, $\psi \in \Lambda^{n-r}{}_c E'$.

(3) $*\phi = \theta(G(\phi) \otimes \Theta) = G^{-1}\theta(\phi \otimes \Lambda)$.

(4) $(\phi, \psi)\Theta = \phi \wedge * \overline{\psi}$, $\phi, \psi \in \Lambda^r{}_c E'$.

(5) $**\phi = (-1)^r \phi$, $\phi \in \Lambda^r{}_c E'$ (* is an "involution").

(6) $* : \Lambda^{p,q}(E)' \to \Lambda^{m-q,\, m-p}(E)'$, $p, q \geqslant 0$.

(7) $(*\phi, *\psi) = (\phi, \psi)$ (* is an "isometry").

(8) $\overline{*\phi} = *\overline{\phi}$ (* is a "real operator").

Remarks

(1) It follows from property (5) of the star operator that the star operator is a complex linear *isomorphism*.

(2) A real star operator may also be defined. Properties (1) to (7) continue to hold except that the sign in (5) depends on the dimension of E as well as on r.

Let F be another complex vector space with hermitian metric k. Now $k : F \to \overline{F}^*$ and so, taking conjugates of forms, k induces a conjugate complex linear map $\widetilde{k} : F \to F^*$. For $r \geqslant 0$, we define the conjugate complex linear map

$$\# : \Lambda^r{}_c E' \otimes F \to \Lambda^r{}_c E' \otimes F^*$$

by

$$\#(\phi \otimes f) = \overline{\phi} \otimes \widetilde{k}(f), \; \phi \in \Lambda^r{}_c E', \; f \in F$$

We remark that $\#\# = \mathrm{id}$.

If we extend the star operator to $\Lambda^r{}_c E' \otimes F$ as $* \otimes \mathrm{id}$, we clearly have

$$*\# = \#*$$

The hermitian metrics h and k induce a hermitian inner product on $\Lambda^r{}_c E' \otimes F$. We claim that this inner product is given in terms of $*$ and $\#$ by

$$(\phi, \psi)\,\Theta = \theta_4^2(\phi \wedge *\#\psi), \quad \phi, \psi \in \Lambda^r{}_c E' \otimes F$$

where

$$\theta_4^2 : \Lambda^r{}_c E' \otimes F \otimes \Lambda^{n-r}{}_c E' \otimes F^* \to \Lambda^r{}_c E' \otimes \Lambda^{n-r}{}_c E'$$

This fact follows from the definition of the tensor product metric on $\Lambda^r{}_c E' \otimes F$ together with property (4) of the star operator. In the sequel, we usually drop θ_4^2 from the above expression and just write $\phi \wedge *\#\psi$.

This concludes our survey of complex linear algebra. All the above definitions and constructions generalize immediately to complex vector bundles on (orientable) manifolds. In particular, suppose that M is a complex manifold with holomorphic tangent bundle TM. A *hermitian metric* on M is a section $h \in C^\infty(TM^* \otimes \overline{TM}^*)$ which restricts to a hermitian metric on the fibres of TM. M, together with a hermitian metric, is called a *hermitian manifold.* The *Kähler form* $\hat{H}$ of h is the section $ih \in C^{1,1}(M)$. Associated to the hermitian metric h on M we have a J-invariant *Riemannian metric* $g \in C^\infty(TM' \otimes TM')$ on M and hermitian metrics on all the complex exterior power bundles $\Lambda^{p,q}(M)$, $\Lambda^{p,q}(M)'$. From g we obtain well defined volume and orientation forms Θ and Λ on M and so we may define the complex *Hodge star operator*

$$* : C^{p,q}(M) \to C^{m-q,\,m-p}(M)$$

for $p,q \geq 0$ (m denotes the complex dimension of M).

If E is a complex vector bundle on M, a hermitian metric on E is a section $k \in C^\infty(E^* \otimes \bar{E}^*)$ which induces hermitian metrics on the fibres of E. E, together with a hermitian metric, is called a *hermitian vector bundle.* If M is a hermitian manifold and E a hermitian vector bundle we may define

$$* : C^{p,q}(M,E) \to C^{m-q,\,m-p}(M,E)$$

$$\# : C^{p,q}(M,E) \to C^{q,p}(M,E^*)$$

Finally, we remark that it is easily shown, using partitions of unity, that every complex manifold and complex vector bundle admits a hermitian metric.

C. Complex Laplace-Beltrami operator

In this subsection we shall obtain a rather more explicit description of the Laplace-Beltrami operators $\square$ defined in subsection A.

Let M be a compact complex manifold and suppose that we are given a hermitian metric h on M. As we showed at the end of the previous subsection, h induces hermitian metrics on the complex vector bundles $\Lambda^{p,q}(M)'$, $p,q \geq 0$, and a Riemannian metric g on M. We let Θ denote the volume form of g.

For p,q $\geqslant 0$, we may define a complex inner product $((\ ,\ ))$ on $C^{p,q}(M)$ by

$$((\phi, \psi)) = \int_M (\phi, \psi)\Theta, \quad \phi, \psi \in C^{p,q}(M)$$

where $(\ ,\ )$ denotes the hermitian metric on the bundle $\Lambda^{p,q}(M)$. Since $(\phi, \psi)\Theta = \phi \wedge * \bar{\psi}$, we have

$$((\phi, \psi)) = \int_M \phi \wedge * \bar{\bar{\psi}}, \quad \phi, \psi \in C^{p,q}(M)$$

Remark. Apart from the constant multiple $\int_M \Theta$, $((\ ,\ ))$ is the inner product defined in subsectic

We define the first-order LPDO

$$\vartheta : C^{p,q+1}(M) \to C^{p,q}(M)$$

by

$$\vartheta\phi = -(*\partial*)\phi, \ \phi \in C^{p,q+1}(M)$$

Proposition 4.13. The operator ϑ is the formal adjoint of $\bar{\partial}$.

Proof. We must show that for all $\phi \in C^{p,q}(M)$, $\psi \in C^{p,q+1}(M)$ we have

$$((\bar{\partial}\phi, \psi)) = ((\phi, \vartheta\psi))$$

Since $\phi \wedge * \bar{\psi} \in C^{m,m-1}(M)$ $(m = \dim_{\mathbb{C}} M)$, we have

$$\begin{aligned} d(\phi \wedge * \bar{\psi}) &= \bar{\partial}(\phi \wedge * \bar{\psi}) \\ &= \bar{\partial}\phi \wedge * \bar{\psi} + (-1)^{p+q} \phi \wedge \bar{\partial}(* \bar{\psi}) \\ &= \bar{\partial}\phi \wedge * \bar{\psi} - \phi \wedge * \overline{(*\partial*\psi)} \end{aligned}$$

where the last equality is obtained by using the involution property of the star operator. By Stoke's theorem we therefore have

$$0 = \int_M d(\phi \wedge * \bar{\psi}) = \int_M \bar{\partial}\phi \wedge * \bar{\psi} - \int_M \phi \wedge * \overline{(*\partial*\psi)}$$

$$= ((\bar{\partial}\phi, \psi)) - ((\phi, \vartheta\psi)). \text{ Q.E.D.}$$

Definition 4.14. Let M be a hermitian compact complex manifold. For $p,q \geqslant 0$ we define the complex *Laplace-Beltrami* operator to be the second-order LPDO

$$\Box : C^{p,q}(M) \to C^{p,q}(M)$$

defined by

$$\Box = \bar{\partial}\vartheta + \vartheta\bar{\partial}$$

We leave as an exercise the proof of

Proposition 4.15. $\Box$ is a strongly elliptic second-order LPDO. The symbol of $\Box$ is given by

$$\sigma(\Box)(\xi) = \frac{1}{2}\|\xi\|^2$$

where $\xi \in TM'$, $\|\xi\|$ is the norm of ξ (relative to the Riemannian metric g) and by the right-hand side of the above equation we mean multiplication by $\frac{1}{2}\|\xi\|^2$.

Remark. We may define the formal adjoint $\bar{\vartheta}$ of ∂ by $-*\bar{\partial}*$ and the corresponding conjugate complex Laplace-Beltrami operator $\bar{\Box} : C^{p,q}(M) \to C^{p,q}(M)$. $\bar{\Box}$ is strongly elliptic and has the same symbol as $\Box$.

We may repeat the above arguments for holomorphic bundle valued forms. Thus let us suppose that E is a holomorphic vector bundle on M. We give M a hermitian metric h as above and E a hermitian metric k. We define a complex inner product on $C^{p,q}(M,E)$ by setting

$$((s,t)) = \int_M (s,t)\Theta, \quad s,t \in C^{p,q}(M,E)$$

where (,) denotes the hermitian metric on the bundle $\Lambda^{p,q}(M,E)'$. We again have

$$((s,t)) = \int_M s \wedge \# * t$$

and, as in Proposition 4.13, it is easily verified that

$$\vartheta = -*\#\bar{\partial}\#* : C^{p,q}(M,E) \to C^{p,q}(M,E)$$

is the formal adjoint of $\bar{\partial}$. We define

$$\Box_E = \bar{\partial}\vartheta + \vartheta\bar{\partial} : C^{p,q}(M,E) \to C^{p,q}(M,E)$$

and $\Box_E$ is a strongly elliptic second-order LPDO with the same symbol as that of $\Box$. In the sequel we usually drop the subscript E and just write $\Box$.

Let us use the elliptic theory of subsection A to prove the following important result.

Theorem 4.16 (the Serre duality theorem). Let E be a holomorphic vector bundle on the compact complex manifold M. Then, letting $\dim_{\mathbb{C}} M = m$, we have isomorphisms

$$H^q(M,\Omega^p(E)) \cong H^{m-q}(M,\Omega^{m-p}(E^*)),\ p,q \geqslant 0$$

Proof. Recall from subsection A that

$$H^q(M,\Omega^p(E)) = \{u \in C^{p,q}(M,E) : \bar{\partial}u = 0\}/\bar{\partial}C^{p,q-1}(M,E)$$

Choose hermitian metrics on M and E. The fundamental existence and regularity theorem for elliptic operators (Theorem 4.10) implies that

$$\mathscr{H}^q(M, \mathscr{D}_p(E)) \approx H^q(M,\Omega^p(E))$$

$$\mathscr{H}^{m-q}(M, \mathscr{D}_{m-p}(E^*)) \approx H^{m-q}(M,\Omega^{m-p}(E^*))$$

(see Example 11 for notation). Now

$$\mathscr{H}^q(M, \mathscr{D}_p(E)) = \{\phi \in C^{p,q}(M,E) : \Box\phi = 0\}$$

$$\mathscr{H}^{m-q}(M, \mathscr{D}_{m-p}(E^*)) = \{\phi \in C^{m-p,m-q}(M,E^*) : \Box\phi = 0\}$$

But the isomorphism $*\#: C^{p,q}(M,E) \to C^{m-p,m-q}(M,E^*)$ is easily seen to *commute* with $\Box$ (this is where we use our explicit formula for $\Box$). Hence $\Box\phi = 0$ if and only if $\Box *\#\phi = 0$ and so

$$*\#(\mathscr{H}^q(M, \mathscr{D}_p(E))) = \mathscr{H}^{m-q}(M, \mathscr{D}_{m-p}(E^*)) \qquad \text{Q.E.D.}$$

Remarks

(1) Notice that the isomorphism between the Dolbeault cohomology groups $H^q(M,\Omega^p(E))$ and $H^{m-q}(M,\Omega^{m-p}(E^*))$ depends on our choice of metrics on M and E.

(2) If M is a compact Riemann surface and E is a holomorphic line bundle on M, then Serre-duality implies that $H^1(M,E) \cong O(TM^* \otimes E^*)$. This fact may be used to give an easy proof of the Riemann-Roch theorem – see Refs [23, 34].

(3) There are versions of Serre-duality for non-compact complex manifolds and we refer the reader to Ref. [34] for details as well as for an alternative proof of Theorem 4.16.

D. Connections

In this subsection we shall develop the basic theory of connections and complex differential geometry to the point where we can relate the de Rham and Dolbeault complexes and so prove the Hodge decomposition theorem for Kähler manifolds (subsection E).

Definition 4.17. Let E be a complex vector bundle on the differentiable manifold M. A (complex) *connection* on E is a complex linear map

$$\nabla : C^\infty(E) \to C^\infty({}_cTM' \otimes E)$$

which satisfies the condition

$$\nabla(fs) = df \otimes s + f \otimes \nabla s, \text{ for all } s \in C^\infty(E),\ f \in C^\infty(M)$$

Remarks

(1) Definition 4.17 is only one of very many equivalent definitions of a connection.

(2) As we have defined it, a connection is a first-order LPDO with symbol "tensor product".

(3) A *real* connection on a (real) vector bundle E on M is a real linear map $\nabla : C^\infty(E) \to C^\infty(TM' \otimes E)$ satisfying $\nabla(fs) = df \otimes s + f \otimes \nabla s$, for all $s \in C^\infty(E)$ and real valued C^∞ functions f on M. Most of the definitions and propositions we give for complex connections have obvious real counterparts and for this reason we shall usually only state the complex versions.

Example. Exterior differentiation defines a connection on $\underline{\mathbb{C}}$ by

$$d : C^\infty(\underline{\mathbb{C}}) \to C^\infty({}_cTM') = C^\infty({}_cTM' \otimes \underline{\mathbb{C}})$$

This connection is called the *canonical connection* of M.

Suppose that ∇ is a connection on E. For $X \in C^\infty({}_cTM)$, we may define a map

$$\nabla_X : C^\infty(E) \to C^\infty(E)$$

by

$$\nabla_X(s) = \theta_X \nabla s. \quad s \in C^\infty(E)$$

Definition 4.18. If ∇ is a connection on E, $s \in C^\infty(E)$ and $X \in C^\infty({}_cTM)$, then $\nabla_X s$ is called the *covariant derivative of s with respect to X.*

Remark. Notice that $\nabla_X s(x) = \theta(X(x) \otimes \nabla s(x))$, $x \in M$, and so $\nabla_X s(x)$ depends only on the value of X at x. One reason we introduce connections is so that we can differentiate sections of vector bundles and the covariant derivative is the generalization of the "directional derivative" to manifolds.

We leave the proof of the next proposition as an exercise.

Proposition 4.19. Let ∇ be a connection on E. Then

(1) $\nabla_X(fs + t) = (L_X f)s + f\nabla_X s + \nabla_X t$, for all $s, t \in C^\infty(E)$, $f \in C^\infty(M)$, $X \in C^\infty({}_cTM)$.

(2) $\nabla_{X+gY}(s) = \nabla_X s + g\nabla_Y s$, for all $X, Y \in C^\infty({}_cTM)$, $s \in C^\infty(E)$, $g \in C^\infty(M)$.

Conversely, if for each $X \in C^\infty({}_cTM)$ we are given a map $\nabla_X : C^\infty(E) \to C^\infty(E)$ satisfying (1) and (2) then there exists a unique connection ∇ on E such that

$$\nabla_X s = \theta_X \nabla s, \quad \text{for all } X \in C^\infty({}_cTM),\ s \in C^\infty(E)$$

We shall now briefly outline some constructions involving connections.

(a) Direct sum

Let ∇ and $\tilde{\nabla}$ denote connections on the complex bundles E and F respectively. We define the *direct sum* of ∇ and $\tilde{\nabla}$, "$\nabla \oplus \tilde{\nabla}$", to be the connection on $E \oplus F$ satisfying

$$(\nabla \oplus \tilde{\nabla})_X (s \oplus t) = \nabla_X s \oplus \tilde{\nabla}_X t$$

for all $s \in C^\infty(E)$, $t \in C^\infty(F)$ and $X \in C^\infty({}_cTM)$. It may easily be shown that $(\nabla \oplus \tilde{\nabla})_X$ is well defined for all $X \in C^\infty({}_cTM)$ and satisfies conditions (1) and (2) of Proposition 4.19. Hence, by the converse to Proposition 4.19, we have a well-defined connection on $E \oplus F$. Similar remarks hold for the remaining constructions that we give.

(b) Tensor product

Let ∇ be a connection on E. For $r > 0$, we define the r-fold tensor product of ∇, "$\otimes^r \nabla$", to be the connection on $\otimes^r E$ characterized by

$$(\otimes^r \nabla)_X (s_1 \otimes \ldots \otimes s_r) = \sum_{j=1}^{r} s_1 \otimes \ldots \otimes \nabla_X s_j \otimes \ldots \otimes s_r$$

for all $s_1, \ldots, s_r \in C^\infty(E)$ and $X \in C^\infty({}_c TM)$.

When $r = 0$, we define $\otimes^0 E = \mathbb{C}$ and take $\otimes^0 \nabla = d$ (the canonical connection on M).

In the sequel, we usually omit reference to r and write ∇ instead of $\otimes^r \nabla$, $r \geqslant 0$.

If we have connections ∇ and $\tilde{\nabla}$ on the bundles E and F, we define the tensor product connection, "$\nabla \otimes \tilde{\nabla}$", on $E \otimes F$ by

$$(\nabla \otimes \tilde{\nabla})_X (s \otimes t) = \nabla_X s \otimes t + s \otimes \tilde{\nabla}_X t$$

for all $s \in C^\infty(E)$, $t \in C^\infty(F)$ and $X \in C^\infty({}_c TM)$. In the sequel, we usually just write ∇ as opposed to $\nabla \otimes \tilde{\nabla}$.

(c) Exterior power

Let ∇ be a connection on E. For $r \geqslant 0$, we define the rth exterior power of ∇, "$\Lambda^r \nabla$", to be the connection on $\Lambda^r E$ characterized by

$$(\Lambda^r \nabla)_X (s_1 \wedge \ldots \wedge s_r) = \sum_{j=1}^{r} s_1 \wedge \ldots \wedge \nabla_X s_j \wedge \ldots \wedge s_r$$

for all $s_1, \ldots, s_r \in C^\infty(E)$ and $X \in C^\infty({}_c TM)$. Again we usually abbreviate $\Lambda^r \nabla$ to ∇.

(d) Dual connection

Let ∇ be a connection on E. We define the dual connection of ∇, "∇^*", to be the connectio on E^* characterized by

$$\langle \nabla^*_X \phi, s \rangle + \langle \phi, \nabla_X s \rangle = L_X \langle s, \phi \rangle$$

for all $s \in C^\infty(E)$, $\phi \in C^\infty(E^*)$ and $X \in C^\infty({}_c TM)$ ($\langle\, ,\rangle$ denotes the pairing between $C^\infty(E)$ and $C^\infty(E^*)$ induced from the "dual" pairing of E and E^*). Again, we usually drop the superscript * and write ∇.

If we are given a connection on E, the above constructions allow us to define connections on all the various tensor and exterior power bundles of E.

Our definitions ensure that the operators ∇_X are *derivations* (see Ref. [18]). More precisely we have

Proposition 4.20. Let ∇ be a connection on E.
(1) For all $X \in C^\infty({}_c TM)$, ∇_X commutes with wedge and tensor products. For example, the following diagram commutes for all $p, q \geq 0$:

$$\begin{array}{ccc} C^\infty(\Lambda^p E \otimes \Lambda^q E) & \xrightarrow{\wedge} & C^\infty(\Lambda^{p+q} E) \\ \downarrow \nabla_X & & \downarrow \nabla_X \\ C^\infty(\Lambda^p E \otimes \Lambda^q E) & \xrightarrow{\wedge} & C^\infty(\Lambda^{p+q} E) \end{array}$$

(2) For all $X \in C^\infty({}_c TM)$, ∇_X commutes with contractions. For example, the following diagram commutes for all $p \geq q \geq 0$:

$$\begin{array}{ccc} C^\infty(\Lambda^p E \otimes \Lambda^q E^*) & \xrightarrow{\theta} & C^\infty(\Lambda^{p-q} E) \\ \downarrow \nabla_X & & \downarrow \nabla_X \\ C^\infty(\Lambda^p E \otimes \Lambda^q E^*) & \xrightarrow{\theta} & C^\infty(\Lambda^{p-q} E) \end{array}$$

Proof. (1) is more or less immediate from the definition of the connections on tensor and exterior powers of E. (2) holds because of the way we defined the dual connection. Let us prove (2) in the special case $p = q = 1$. We must show that if $s \in C^\infty(E)$, $\phi \in C^\infty(E^*)$, then

$$\nabla_X \theta(s \otimes \phi) = \theta \nabla_X(s \otimes \phi)$$

Now $\theta(s \otimes \phi) = \langle s, \phi \rangle \in C^\infty(\mathbb{C})$ and ∇_X is defined to be L_X on $\mathbb{C}$. Hence $\nabla_X \theta(s \otimes \phi) = L_X \langle s, \phi \rangle$. On the other hand,

$$\begin{aligned} \theta \nabla_X(s \otimes \phi) &= \theta(\nabla_X s \otimes \phi + s \otimes \nabla_X \phi) \\ &= \langle \nabla_X s, \phi \rangle + \langle s, \nabla_X \phi \rangle \end{aligned}$$

and so the equality $\nabla_X \theta(s \otimes \phi) = \theta \nabla_X(s \otimes \phi)$ follows from the definition of the dual connection.

We let $\bar{E}$ and $\bar{E}^*$ respectively denote the conjugate and conjugate dual bundles of E. A complex connection ∇ on E induces complex connections $\bar{\nabla}$ and $\bar{\nabla}^*$ on $\bar{E}$ and $\bar{E}^*$ respectively. We define

$$\bar{\nabla}: C^\infty(\bar{E}) \to C^\infty({}_c TM' \otimes \bar{E})$$

by regarding a section s of $\bar{E}$ as a section of E (remember $E = \bar{E}$ as real bundles) and defining $\bar{\nabla} s = \nabla s$.

To define $\bar{\nabla}^*: C^\infty(\bar{E}^*) \to C^\infty({}_c TM' \otimes \bar{E}^*)$, we take conjugates of linear forms to define conjugate complex linear isomorphisms between $C^\infty(\bar{E}^*)$ and $C^\infty(E^*)$, $C^\infty({}_c TM' \otimes \bar{E}^*)$ and $C^\infty({}_c TM' \otimes E^*)$ and then set

$$\bar{\nabla}^* \phi = \overline{\nabla^* \bar{\phi}}, \ \phi \in C^\infty(\bar{E}^*)$$

It is easily shown that we have induced connections on exterior and tensor powers of $\bar{E}$ and $\bar{E}^*$ and that the analogue of Proposition 4.20 continues to hold.

We shall now introduce a little more structure into the theory.

Definition 4.21. Let E be a complex vector bundle on M with hermitian metric $k \in C^\infty(E^* \otimes \overline{E}^*)$. We shall say that a connection ∇ on E is a *metric* or *hermitian connection* if $\nabla k \equiv 0$. (By $\nabla k \equiv 0$ we mean that we have taken the induced connection on the bundle $E^* \otimes \overline{E}^*$ and that, relative to this connection, $\nabla k \equiv 0$.)

Notation. We shall let (E, k) denote a complex vector bundle E together with hermitian metric k.

Lemma 4.22. Let ∇ be a metric connection on (E, k). Then for all $X \in C^\infty({}_cTM)$ and $s \in C^\infty(E)$ we have

$$\nabla_X \theta(k \otimes s) = \theta(k \otimes \nabla_X s)$$

(∇_X "commutes with k").

Proof. $\nabla_X \theta(k \otimes s) = \theta(\nabla_X(k \otimes s))$, Proposition 4.17

$$= \theta(\nabla_X k \otimes s + k \otimes \nabla_X s)$$

$$= \theta(k \otimes \nabla_X s), \text{ since } \nabla k \equiv 0. \qquad \text{Q.E.D.}$$

From the definition of the induced connections on the tensor bundles of E follows

Lemma 4.23. If ∇ is a metric connection on (E, k), then the induced connections on $\otimes^r E$, $\Lambda^s E$, $\otimes^t E^*$, ... are metric connections with respect to the hermitian metrics induced by k on these bundles.

Lemmas 4.19, 4.20 imply

Proposition 4.24. Let ∇ be a metric connection on (E, k). Then for all $X \in C^\infty({}_cTM')$, ∇_X commutes with the (metric) contractions $\theta(k), \theta(k)$, etc. For example, the following diagram commutes for all $p \geqslant q \geqslant 0$:

$$\begin{array}{ccc} C^\infty(\Lambda^p E \otimes \Lambda^q E) & \xrightarrow{\overline{\theta}(k)} & C^\infty(\Lambda^{p-q} E) \\ \downarrow \nabla_X & & \downarrow \nabla_X \\ C^\infty(\Lambda^p E \otimes \Lambda^q E) & \xrightarrow{\overline{\theta}(k)} & C^\infty(\Lambda^{p-q} E) \end{array}$$

Example. Let ∇ be a metric connection on (E, k). Then if $s, t \in C^\infty(E)$,

$$L_X(s, t) = (\nabla_X s, t) + (s, \nabla_{\overline{X}} t), \quad \text{for all } X \in C^\infty({}_cTM)$$

Definition 4.25. A real connection on the tangent bundle TM of a differentiable manifold M is called a *(real) connection on M.*

A complex connection on the complex tangent bundle ${}_cTM$ of M is called a *(complex) connection on M.*

Remark. Any real connection on M induces a complex connection on M by complexifying.

Definition 4.26. Associated to any real (complex) connection on M we may define the *torsion form* of ∇ to be the tensor field $T \in C^\infty(\Lambda^2 TM' \otimes TM)(C^\infty(\Lambda^2{}_c TM' \otimes {}_c TM))$ characterized by

$$\langle T, X \wedge Y \rangle = \nabla_X Y - \nabla_Y X - [X, Y]$$

for all $X, Y \in C^\infty(TM)$ $(C^\infty({}_c TM))$. The pairing $\langle\ ,\ \rangle$ is that induced from the dual pairing of $\Lambda^2 TM$ and $\Lambda^2 TM'$.

Remark. Of course it must be shown that T is a well defined *tensor field.* This follows easily from the properties of the connection and Lie bracket and we refer the reader to Ref. [18] for more details.

Definition 4.27. If T is the torsion form of a real (complex) connection ∇ on M we define the reduced torsion of ∇, $T_R \in C^\infty(TM')$ $(C^\infty({}_c TM'))$, by

$$T_R = \theta T$$

Definition 4.28. A (real or complex) connection ∇ on M is said to be *symmetric* if the torsion form of ∇ vanishes identically.

For the proof of the following basic and well known result we refer to Ref. [18].

Theorem 4.29. Let g be a Riemannian metric on the differentiable manifold M. Then there exists a unique metric symmetric connection on M associated to g. This connection is called the *Levi-Cività* or *Riemannian* connection of g.

Remark. The complexification of the Levi-Cività connection defines a complex connection on M which is also torsion-free and metric with respect to the hermitian metric induced by g on ${}_c TM$. However, if M is a complex manifold, this connection will not always be the "right" connection for doing complex differential geometry on M.

We shall now do a little real differential geometry to show the relation between the Levi-Cività connection and the differential-topological structure of a differentiable manifold M.

Suppose that g is a Riemannian metric on M and that ∇ is a metric (not necessarily symmetric) connection on M. We take induced metrics and connections on all the tensor bundles of M. For $r \geqslant 0$, ∇ defines a map

$$\nabla : C^\infty(\Lambda^r TM') \to C^\infty(TM' \otimes \Lambda^r TM')$$

We define the operator

$$\tilde{d} : C^\infty(\Lambda^r TM') \to C^\infty(\Lambda^{r+1} TM')$$

by $\tilde{d} = \wedge \nabla$. $\tilde{d}$ is clearly a first-order differential operator and, using the definition of the exterior power connection on $\Lambda^r TM'$, it is not hard to see that $\tilde{d}$ satisfies the following characteristic property of exterior differentiation:

$$\tilde{d}(\phi \wedge \psi) = \tilde{d}\phi \wedge \psi + (-1)^p \phi \wedge \tilde{d}\psi, \text{ for all } \phi \in C^\infty(\Lambda^p TM'),\ \psi \in C^\infty(\Lambda^q TM')$$

However, as we shall shortly see, $\tilde{d}^2$ need not vanish.

We may also use ∇ to define the "adjoint" operator

$$\tilde{\delta}: C^{\infty}(\Lambda^{r+1}TM') \to C^{\infty}(\Lambda^r TM'),\ r \geq 0$$

by setting

$$\tilde{\delta}\phi = -\theta(g)\nabla\phi,\ \phi \in C^{\infty}(\Lambda^{r+1}TM')$$

The torsion form T of ∇ induces maps

$$\tilde{T}: C^{\infty}(\Lambda^r TM') \to C^{\infty}(\Lambda^{r+1}TM') \quad ; \quad \tilde{T}^{\#}: C^{\infty}(\Lambda^{r+1}TM') \to C^{\infty}(\Lambda^r TM')$$

defined by

$$\tilde{T}(\phi) = \Lambda(\theta_3^2(T \otimes \phi)),\ \phi \in C^{\infty}(\Lambda^r TM') \quad ; \quad \tilde{T}^{\#}(\phi) = \Lambda(\theta_3^1(T^{\#} \otimes \phi)),\ \phi \in C^{\infty}(\Lambda^{r+1}TM')$$

where $T^{\#} \in C^{\infty}(\Lambda^2 TM \otimes TM')$ is equal to $(\Lambda^2 g^{-1} \otimes g)T$ (regarding $g: TM \to TM'$). Both $\tilde{T}$ and $\tilde{T}^{\#}$ are linear over $C^{\infty}(M)$ and are therefore induced from vector bundle maps.

Proposition 4.30. Let ∇ be a metric connection on the Riemannian manifold M and let ∇ have torsion form T. Then

$$d = \tilde{d} + \tilde{T} \quad ; \quad \delta = \tilde{\delta} + \tilde{T}^{\#} - \theta_{g^{-1}(T_R)}$$

In particular, if ∇ is the Levi-Cività connection of M,

$$d = \tilde{d} \text{ and } \delta = \tilde{\delta}$$

Proof. We shall just prove a special case of the proposition. The only non-trivial part of the full proof is the use of Stokes' theorem and the uniqueness of the adjoint δ of d in the proof of the second identity. Let $\phi \in C^{\infty}(TM')$ and $X, Y \in C^{\infty}(TM)$. We have

$$\langle \tilde{d}\phi, X \wedge Y \rangle = \langle \nabla\phi, X \wedge Y \rangle = \langle \nabla\phi, X \otimes Y - Y \otimes X \rangle = \langle \nabla_X \phi, Y \rangle - \langle \nabla_Y \phi, X \rangle$$

Now, by definition of the dual connection,

$$\langle \nabla_X \phi, Y \rangle = -\langle \nabla_X Y, \phi \rangle + L_X \langle Y, \phi \rangle \text{ and } \langle \nabla_Y \phi, X \rangle = -\langle \nabla_Y X, \phi \rangle + L_Y \langle X, \phi \rangle$$

and so we have

$$\begin{aligned} \langle \nabla_X \phi, Y \rangle - \langle \nabla_Y \phi, X \rangle &= \langle \nabla_Y X - \nabla_X Y + [X,Y], \phi \rangle + L_X \langle Y, \phi \rangle - L_Y \langle X, \phi \rangle - \langle \phi, [X,Y] \rangle \\ &= -\langle T(X \wedge Y), \phi \rangle + \langle d\phi, X \wedge Y \rangle, \text{ definition of d and T} \\ &= \langle -\tilde{T}(\phi) + d\phi, X \wedge Y \rangle \qquad \text{Q.E.D.} \end{aligned}$$

Remarks

(1) It follows from Proposition 4.30 that if ∇ is the Levi-Cività connection then the Laplace operator $\Delta = \tilde{d}\tilde{\delta} + \tilde{\delta}\tilde{d}$. A little more work enables us to write Δ in terms of covariant differentiati[on] and curvature and so establish relationships between the topology and curvature of M using Hodge theory.

(2) We shall soon see that for *complex* hermitian manifolds the torsion forms of the naturally defined hermitian connection need not vanish unless the manifold is Kähler.

We have now said enough about the real case and it is time to start the study of complex analytic connections on holomorphic bundles.

Let E be a holomorphic vector bundle on the complex manifold M. The $\bar{\partial}$-operator defines a map

$$\bar{\partial} : C^\infty(E) \to C^\infty(\overline{TM}^* \otimes E)$$

Now a complex connection on E is a map

$$\nabla : C^\infty(E) \to C^\infty({}_c TM' \otimes E) \approx C^\infty((TM^* \oplus \overline{TM}^*) \otimes E)$$

and so we may think of $\bar{\partial}$ as defining one "half" of a complex connection on E. Similarly, if E is an antiholomorphic bundle, ∂ defines the holomorphic half of a complex connection on E.

Suppose that the holomorphic bundle E is given a hermitian metric k. Regarding k as a complex vector bundle isomorphism $k : E \to \bar{E}^*$, we may form the composition

$$(I \otimes k^{-1})\partial k : C^\infty(E) \to C^\infty(TM^* \otimes E)$$

which gives us a natural candidate for the holomorphic "half" of a complex connection on E. We leave it to the reader to verify that the map

$$\bar{\partial} + (I \otimes k^{-1})\partial k : C^\infty(E) \to C^\infty({}_c TM' \otimes E)$$

defines a complex connection on E.

Definition 4.31. Let E be a holomorphic vector bundle with hermitian metric k. The connection on E

$$\nabla : C^\infty(E) \to C^\infty({}_c TM' \otimes E)$$

defined by $\nabla = (I \otimes k^{-1})\partial k + \bar{\partial}$ is called the *hermitian connection of k.*

Remark. If E had been an antiholomorphic vector bundle with hermitian metric k, then we could have defined the hermitian connection of k by $\nabla = (I \otimes k^{-1})\bar{\partial} k + \partial$.

Proposition 4.32. Let (E, k) be a hermitian holomorphic vector bundle with hermitian connection ∇. The metric k induces metrics on the bundles $\bar{E}$, E^* and $\bar{E}^*$ and the corresponding hermitian connections on these bundles coincide with the conjugate, dual and conjugate dual connections defined earlier in this subsection.

Proof. The hermitian connections on $\bar{E}$, E^* and $\bar{E}^*$ are respectively defined by

$$\bar{\nabla} = (I \otimes \bar{k}^{-1})\bar{\partial}\bar{k} + \partial \;;\qquad \nabla^* = (I \otimes \bar{k})\partial \bar{k}^{-1} + \bar{\partial} \;;\qquad \bar{\nabla}^* = (I \otimes k)\bar{\partial} k^{-1} + \partial$$

To complete the proof of the proposition it suffices to show that ∇^* is indeed the dual connection of ∇ as the remaining results follow trivially by conjugation. The proof that ∇^* is the dual connection is most easily done by a local computation and we leave details to the reader.

From Proposition 4.32 follows

Proposition 4.33. Let (E, k) be a hermitian holomorphic vector bundle with hermitian connection k Then the hermitian connections associated to the induced hermitian metrics on the various tensor bundles of E coincide with the respective tensor powers of the connection ∇.

Proposition 4.34. Let (E, k) be a hermitian holomorphic vector bundle with hermitian connection ∇ Then ∇ is a metric connection.

Proof. The hermitian connection induced on $E^* \otimes \bar{E}^*$ is given by

$$\nabla = (I \otimes k)\bar{\partial}(I \otimes k^{-1}) + (\bar{k} \otimes I)\partial(\bar{k}^{-1} \otimes I)$$

We have to show $\nabla k \equiv 0$. But $(I \otimes k^{-1})\, k \in C^\infty(L(E, E))$ is just the identity map of E and so $\bar{\partial}(I \otimes k^{-1})k = 0$. Similarly, $\partial(\bar{k}^{-1} \otimes I)k = 0$ and so $\nabla k = 0$.

We may similarly prove

Proposition 4.35. Let (E, k) be a hermitian holomorphic vector bundle with hermitian connection ∇ Then, if we let J denote the complex structure of E $(J \in C^\infty(E^* \otimes E))$,

$$\nabla J \equiv 0$$

Let h be a hermitian metric on the complex manifold M (i.e. h is a hermitian metric on the holomorphic tangent bundle of M and so $h \in C^\infty(TM^* \otimes \overline{TM}^*)$). Associated to h we have connections on the holomorphic and antiholomorphic tangent bundles of M and so a connection ∇ on the complexified tangent bundle of M:

$$\nabla : C^\infty({}_cTM) \to C^\infty({}_cTM' \otimes {}_cTM)$$

∇ is called the *hermitian connection* of M associated to the metric h. The hermitian metric h on M induces metrics, and therefore metric connections, on the bundles $\Lambda^{p,q}(M)$, $\Lambda^{r,s}(M)'$, $p, q, r, s \geqslant 0$.

Remark. Associated to a hermitian metric h on M we have a riemannian metric g on M and so a (complexified) Levi-Cività connection on M. As we shall soon see, the Levi-Cività connection will *not* equal the hermitian connection unless h is a Kähler metric.

Proposition 4.36. Let h be a hermitian metric on the complex manifold M and let ∇ denote the associated hermitian connection. The torsion form of ∇ decomposes into a sum $S + \bar{S}$, where

$$S \in C^\infty(\Lambda^{2,0}(M)' \otimes TM) \text{ and } \bar{S} \in C^\infty(\Lambda^{0,2}(M)' \otimes \overline{TM})$$

and $\bar{S}$ is the conjugate of S.

Proof. A formal computation which uses the following facts about the Lie algebra of complex vector fields on M:
(1) If $X, Y \in C^\infty(TM)$ $(C^\infty(\overline{TM}))$, then $[X, Y] \in C^\infty(TM)$ $(C^\infty(\overline{TM}))$.
(2) If $X \in C^\infty(TM)$, $Y \in C^\infty(\overline{TM})$, then $[X, Y] = \theta_X \partial Y - \theta_Y \bar{\partial} X$.
We omit details.

As a straightforward corollary of Proposition 4.36 we have

Proposition 4.37. The torsion form T of a hermitian connection on M has the following properties:
(1) T is real.
(2) $T(X \wedge JY) = T(JX \wedge Y)$, for all $X, Y \in C^\infty({}_cTM)$.

Definition 4.38. The reduced torsion forms of a hermitian connection on M are the forms

$$S_R \in C^\infty(TM^*) \text{ and } \bar{S}_R \in C^\infty(\overline{TM}^*)$$

defined by $S_R = \theta S$ and $\bar{S}_R = \theta\bar{S}$ respectively.

From now on we assume that we have a fixed hermitian metric h on the complex manifold M and a hermitian holomorphic vector bundle (E, k) on M. We denote all the resulting metric connections on the bundles $\Lambda^{p,q}(M,E)$, $\Lambda^{p,q}(M,E)'$ by the same symbol ∇.

Since $\nabla: C^\infty(\Lambda^{p,q}(M,E)') \to C^\infty({}_cTM' \otimes \Lambda^{p,q}(M,E)')$ and ${}_cTM' = TM^* \oplus \overline{TM}^*$, ∇ may be written as a sum $\nabla' + \nabla''$ where

$$\nabla': C^\infty(\Lambda^{p,q}(M,E)') \to C^\infty(TM^* \otimes \Lambda^{p,q}(M,E)')$$

$$\nabla'': C^\infty(\Lambda^{p,q}(M,E)') \to C^\infty(\overline{TM}^* \otimes \Lambda^{p,q}(M,E)')$$

Just as we did for Riemannian manifolds, we may define maps

$$\tilde{\bar{\partial}}: C^{p,q}(M,E) \to C^{p,q+1}(M,E)$$

$$\tilde{\partial}: C^{p,q}(M,E) \to C^{p+1,q}(M,E)$$

by setting

$$\tilde{\bar{\partial}}\phi = \wedge\nabla''\phi, \quad \tilde{\partial}\phi = \nabla'\phi, \quad \phi \in C^{p,q}(M,E)$$

However, in the sequel we shall only be interested in the operator $\tilde{\partial}$ unless E is the trivial bundle when $C^{p,q}(M,\underline{\mathbb{C}}) = C^{p,q}(M)$.

A similar proof to that of the first part of Proposition 4.30 also proves

Proposition 4.39. The operator $\tilde{\bar{\partial}}: C^{p,q}(M,E) \to C^{p,q+1}(M,E)$ may be expressed in terms of $\bar{\partial}$ by the relation

$$\bar{\partial} = \tilde{\bar{\partial}} + \tilde{\bar{S}}$$

where $\tilde{\bar{S}}: C^{p,q}(M,E) \to C^{p,q+1}(M,E)$ is defined by $\bar{S}(\phi) = \wedge(\theta_3^2(\bar{S} \otimes \phi))$, $\phi \in C^{p,q}(M,E)$.

We also may express $\tilde{\partial}: C^{p,q}(M) \to C^{p+1,q}(M)$ in terms of ∂ by the relation

$$\partial = \tilde{\partial} + \tilde{S}$$

where $\tilde{S}(\phi) = \wedge(\theta_3^2(S \otimes \phi))$, $\phi \in C^{p,q}(M)$.

We recall that the metric h on M is said to be *Kähler* if

$$d\widehat{H} = 0$$

The hermitian manifold M is then called a *Kähler manifold.*

Proposition 4.40. The torsion forms S and $\bar{S}$ of a hermitian metric h on M are given by

$$S = -i(I \otimes h^{-1})\partial\widehat{H} \text{ and } \bar{S} = i(I \otimes h^{-1})\bar{\partial}\widehat{H}$$

In particular, h is a Kähler metric if and only if $S = \bar{S} = 0$.

Proof. Since ∇ is a metric connection, $\nabla\widehat{H} = 0$ (remember that $\widehat{H} = ih$ and that ∇ commutes with the operation of wedge product). Hence $\wedge\nabla\widehat{H} = 0$ and so, by Proposition 4.39, $0 = \widetilde{\partial}\widehat{H} = \partial\widehat{H} - \widetilde{S}(H$ Regarding h as an isomorphism from TM to $\overline{TM}^*$ and $I \otimes h$ as the isomorphism

$$I \otimes h: \Lambda^2 TM^* \otimes TM \to \Lambda^2 TM^* \otimes \overline{TM}^* \approx \Lambda^{2,1}(M)'$$

it is easily seen that $\widetilde{S}(\widehat{H}) = i(I \otimes h)S$. Therefore $\partial\widehat{H} = i(I \otimes h)S$ and so $S = -i(I \otimes h^{-1})\partial\widehat{H}$. Conjugating, we obtain the expression for $\bar{S}$.

Proposition 4.41. Let M be a complex manifold with hermitian metric h. The hermitian connection of h equals the (complexified) Levi-Cività connection of the Riemannian metric associated to h if and only if h is a Kähler metric.

Proof. Denote the hermitian connection of M by ∇ and the Riemannian metric on M associated to h by g. It is easily verified that ∇ restricts to a real connection $\widetilde{\nabla}$ on M and that $\widetilde{\nabla}$ is metric (with respect to g). The proposition now follows immediately from Proposition 4.40 together with the uniqueness of the Levi-Cività connection (see Ref. [18], page 158). Q.E.D.

Proposition 4.42. Let M be a Kähler manifold. Then
(1) For $p,q \geqslant 0$, the maps $\partial, \widetilde{\partial}: C^{p,q}(M) \to C^{p+1,q}(M)$ are equal.
(2) For $p,q \geqslant 0$, the maps $\bar{\partial}, \bar{\widetilde{\partial}}: C^{p,q}(M) \to C^{p,q+1}(M)$ are equal.
(3) If (E,k) is a hermitian holomorphic vector bundle on M then for $p, q \geqslant 0$ the maps $\bar{\partial}, \bar{\widetilde{\partial}}: C^{p,q}(M,E) \to C^{p,q+1}(M,E)$ are equal.

Remark. Notice that we can *deduce* (1) and (2) of Proposition 4.42 from the first part of Proposition 4.30 once we know Proposition 4.41.

Let us now turn to the study of the adjoint operators ϑ and $\bar{\vartheta}$. We define

$$\vartheta: C^{p,q+1}(M,E) \to C^{p,q}(M,E)$$

$$\bar{\vartheta}: C^{p+1,q}(M,E) \to C^{p,q}(M,E)$$

by

$$\vartheta\phi = -\theta(h^{-1})\nabla'\phi,\ \phi \in C^{p,q+1}(M,E)$$

$$\bar{\vartheta}\phi = -\theta(h^{-1})\nabla''\phi,\ \phi \in C^{p+1,q}(M,E)$$

Theorem 4.43. Let M be a Kähler manifold and (E, k) be a hermitian holomorphic vector bundle on M. Then

(1) For $p,q \geq 0$, the maps $\bar{\vartheta}, \tilde{\bar{\vartheta}}: C^{p,q+1}(M,E) \to C^{p,q}(M,E)$ are equal.

(2) For $p,q \geq 0$, the maps $\vartheta, \tilde{\vartheta}: C^{p+1,q}(M) \to C^{p,q}(M)$ are equal.

Proof. If E is the trivial bundle (with the standard metric!), (1) and (2) follow immediately from Proposition 4.30 together with Proposition 4.41. The general case is an easy corollary of this special case and we omit details.

Remark. For the non-Kähler case, $\tilde{\vartheta}$ may be expressed in terms of ϑ and torsion and reduced torsion forms of h (see Ref. [37]).

From now on we shall assume that M is a (compact) Kähler manifold with hermitian metric h.

For $p,q \geq 0$, the Kähler form $\hat{H}$ defines a map

$$L: C^{p,q}(M) \to C^{p+1,q+1}(M)$$

by

$$L(\phi) = \hat{H} \wedge \phi$$

Since $\hat{H}$ is a real form, L is a *real operator*, i.e. for all $\phi \in C^{p,q}(M)$ we have

$$\overline{L(\phi)} = L(\bar{\phi})$$

We define $\breve{H} \in C^\infty(\Lambda^{1,1}(M))$ by $\breve{H} = \overline{G^{-1}(\hat{H})}$ (see subsection B for notation – G is the complexification of the Riemannian metric associated to h). We define

$$L^*: C^{p+1,q+1}(M) \to C^{p,q}(M)$$

by setting

$$L^*(\phi) = \theta_{\breve{H}}\phi, \ \phi \in C^{p+1,q+1}(M)$$

We claim that L^* is the hermitian adjoint of L, i.e.

$$((L\phi, \psi)) = ((\phi, L^*\psi)), \ \phi \in C^{p,q}(M), \ \psi \in C^{p+1,q+1}(M)$$

Indeed (on fibres) we have

$$\begin{aligned}
(L\phi, \psi) &= (\hat{H} \wedge \phi, \psi) \\
&= \langle \hat{H} \wedge \phi, \overline{G^{-1}(\psi)} \rangle, \text{ definition of } (\ ,\ ) \\
&= \langle \phi, \theta(\hat{H} \otimes G^{-1}(\psi)) \rangle, \text{ definition of } \theta \\
&= \langle \phi, \overline{G^{-1}\theta(\overline{G^{-1}(H)} \otimes \psi)} \rangle, \text{ contractions ``commute'' with metric} \\
&= \langle \phi, \overline{G^{-1} L^* \psi} \rangle, \text{ definition of } L^* \\
&= (\phi, L^*\phi)
\end{aligned}$$

Before we give the main theorems of this section, we need a little more hermitian linear algebra.

Lemma 4.44. Let E be a complex vector space with hermitian metric h. Then for all $\phi, \psi \in E^*$ we have

$$(\phi, i\psi) = (\phi \wedge \overline{\psi}, \widehat{H})$$

Indeed this identity characterizes $\widehat{H}$.

Proof. Choose bases $\{dz_1, \ldots, dz_m\}$ and $\{d\overline{z}_1, \ldots, d\overline{z}_m\}$ for E^* and $\overline{E}^*$ respectively and compute.

Continuing with the assumptions of Lemma 4.44, we have

Lemma 4.45. Let $\phi \in E^*$, $\psi \in \Lambda^{p,q+1}(E)'$. Then

$$\theta(\theta_\phi \check{H} \otimes \psi) = i\theta(h^{-1})(\phi \otimes \psi)$$

Proof. It is clearly sufficient to prove $\theta_\phi \check{H} = iG^{-1}(\phi)$. Let $\overline{\gamma} \in \overline{E}^*$. Then

$$\begin{aligned}\langle \overline{\gamma}, \theta_\phi \check{H} \rangle &= \langle \phi \wedge \overline{\gamma}, \check{H} \rangle \\ &= (\phi \wedge \overline{\gamma}, \widehat{H}), \text{ definition of } (\ ,\ )\end{aligned}$$

On the other hand,

$$\begin{aligned}\langle \overline{\gamma}, iG^{-1}(\phi) \rangle &= \langle \overline{\gamma}, \overline{G^{-1}(-i\phi)} \rangle \\ &= (\overline{\gamma}, -i\overline{\phi}) \\ &= (\phi, i\gamma), \text{ conjugate symmetry of } (\ ,\ )\end{aligned}$$

and so the result follows from Lemma 4.44.

Continuing our assumptions on E, we define maps

$$\gamma_1, \gamma_2, \gamma_3 : E^* \otimes \Lambda^{p,q+1}(E)' \to \Lambda^{p,q}(E)'$$

by

$$\gamma_1(\phi \otimes \psi) = L^*(\phi \wedge \psi)$$

$$\gamma_2(\phi \otimes \psi) = \phi \wedge L^*\psi$$

$$\gamma_3(\phi \otimes \psi) = i\theta(h^{-1})(\phi \otimes \psi)$$

where $\phi \in E^*$ and $\psi \in \Lambda^{p,q+1}(E)'$.

Lemma 4.46. With the above notation,

$$\gamma_1 = \gamma_2 + \gamma_3$$

Proof. Let $Y \in \Lambda^{p,q}(E)$. Then

$$\begin{aligned}
\langle L^*(\phi \wedge \psi), Y\rangle &= \langle \theta_{\breve{H}}(\phi \wedge \psi), Y\rangle \\
&= \langle \phi \wedge \psi, \breve{H} \wedge Y\rangle \\
&= \langle \psi, \theta_\phi(\breve{H} \wedge Y)\rangle \\
&= \langle \psi, \theta_\phi(\breve{H}) \wedge Y + \breve{H} \wedge \theta_\phi(Y)\rangle \text{ (see subsection B)} \\
&= \langle \theta(\theta_\phi(\breve{H}) \otimes \psi), Y\rangle + \langle \theta_{\breve{H}}\psi, \theta_\phi Y\rangle \\
&= \langle i\theta(h^{-1})(\phi \otimes \psi), Y\rangle + \langle \phi \wedge L^*\psi, Y\rangle
\end{aligned}$$

by Lemma 4.45 and the definition of L^*. Since this equality holds for all $Y \in \Lambda^{p,q}(E)$, the result follows.

Theorem 4.47. Let M be a Kähler manifold. Then we have the following identities:
(1) $\partial L^* - L^*\partial = i\vartheta$.
(2) $\bar\partial L^* - L^*\bar\partial = -i\bar\vartheta$.
(3) $\vartheta L - L\vartheta = i\partial$.
(4) $\bar\vartheta L - L\bar\vartheta = -i\bar\partial$.

Proof. Let us denote the hermitian connection on M by ∇. We start by proving (1). Since $\partial = \tilde\partial$, we have

$$\begin{aligned}
\partial L^*\phi &= \wedge(\nabla'(\theta(\phi \otimes \breve{H}))), \ \phi \in C^{p,q}(M) \\
&= \wedge(\theta_2^3(\nabla'\phi \otimes \breve{H})), \text{ since } \nabla\breve{H} \equiv 0 \\
&= \gamma_2(\nabla'\phi), \text{ notation of Lemma 4.46} \\
&= \gamma_1(\nabla'\phi) - \gamma_3(\nabla'\phi), \text{ Lemma 4.46} \\
&= L^*\partial\phi + i\tilde\vartheta\phi \\
&= L^*\partial\phi + i\vartheta\phi, \text{ since } \vartheta = \tilde\vartheta \text{ on a Kähler manifold}
\end{aligned}$$

(2) follows from (1) by conjugating. Similarly (4) follows from (3) by conjugating. To complete the proof it is sufficient to verify (3). Let $\phi \in C^{p,q}(M)$, $\psi \in C^{p+1,q}(M)$. Then

$$(\vartheta L\phi, \psi) = (\phi, L^*\bar\partial\psi)$$

$$(L\vartheta\phi, \psi) = (\phi, \bar\partial L^*\psi)$$

Subtracting the second equation from the first, we have

$$\begin{aligned}
((\vartheta L - L\vartheta)\phi, \psi) &= (\phi, (L^*\bar\partial - \bar\partial L^*)\psi) \\
&= (\phi, -i\bar\vartheta\psi), \text{ by (2)} \\
&= (i\partial\phi, \psi), \text{ since } \bar\vartheta = \partial^*
\end{aligned}$$

Since this identity holds for all ϕ and ψ the result follows. Q.E.D.

Recall that if M is any hermitian manifold (not necessarily Kähler) we have an induced Riemannian structure on M and so associated Laplace operators $\Delta : C^\infty(\Lambda^r{}_c TM') \to C^\infty(\Lambda^r{}_c TM')$, $r \geqslant 0$.

Theorem 4.48. Let M be a Kähler manifold. Then we have the following identities:
(1) $\square = \bar\partial\vartheta + \vartheta\bar\partial = \bar\square = \partial\bar\vartheta + \bar\vartheta\partial = \frac{1}{2}\Delta$.
(2) $\partial\vartheta + \vartheta\partial = 0$, $\bar\vartheta\bar\partial + \bar\partial\bar\vartheta = 0$.

Proof. By Theorem 4.47 we have:

$$\square = -i(\bar\partial(\partial L^* - L^*\partial) + (\partial L^* - L^*\partial)\bar\partial)$$

$$= -i(\bar\partial\,\partial L^* - \bar\partial L^*\partial + \partial L^*\bar\partial - L^*\partial\bar\partial)$$

and $\bar\square = i(\partial(\bar\partial L^* - L^*\bar\partial) + (\bar\partial L^* - L^*\bar\partial)\partial)$

$$= -i(\bar\partial\,\partial L^* + \partial L^*\bar\partial - \bar\partial L^*\partial + L^*\partial\bar\partial)$$

Hence $\square = \bar\square$.

Now $\Delta = d\delta + \delta d = (\partial + \bar\partial)(\vartheta + \bar\vartheta) + (\vartheta + \bar\vartheta)(\partial + \bar\partial)$

$$= \square + \bar\square + (\partial\vartheta + \vartheta\partial) + (\bar\partial\bar\vartheta + \bar\vartheta\bar\partial)$$

To complete the proof of the theorem it is therefore sufficient to show that $\partial\vartheta + \vartheta\partial = 0$. But

$$i(\partial\vartheta + \vartheta\partial) = \partial(\partial L^* - L^*\partial) + (\partial L^* - L^*\partial)\partial = 0. \quad \text{Q.E.D.}$$

Remarks
(1) Theorem 4.48 is special to Kähler manifolds. It will enable us in subsection E to link the complex structure of M to the cohomology of M. If the manifold is not Kähler then the hermitian connection (which embodies the complex structure in the form of $\bar\partial$) will not equal the Riemanni connection (which contains the topological and geometric structure used in Hodge theory). If we denote the Riemannian connection associated to a hermitian metric on M by ∇, then M will be Kähler if and only if $\nabla J \equiv 0$, where J denotes the almost complex structure of M (see Ref. [19] for a proof of this fact).
(2) Theorems 4.47 and 4.48 have a generalization to holomorphic bundle valued forms. For example, if (E, k) is a hermitian holomorphic vector bundle on M we may define $L_E : C^{p,q}(M,E) \to C^{p+1,q+1}(M,E)$ and $L_E^* : C^{p+1,q+1}(M,E) \to C^{p,q}(M,E)$, $p,q \geqslant 0$, by tensoring L and L* with the identity map on E. We then find that $\tilde\partial L_E^* - L_E^*\tilde\partial = i\tilde\vartheta$ and $\bar\partial L_E - L_E\bar\partial = -i\tilde{\bar\vartheta}$. These identities are useful, for example, in calculations involving curvatures on M and E (see Ref. [37] for more details).

E. Introduction to the Hodge theory of Kähler manifolds

Let M be a hermitian manifold. Taking $E = \underline{\mathbb{C}}$ in Example 11 of subsection A we define the Dolbeault cohomology groups:

$$H^q(M,\Omega^p) = \{\phi \in C^{p,q}(M) : \bar\partial\phi = 0\}/\bar\partial C^{p,q-1}(M)$$

We set

$$H^{p,q}(M) = \{\phi \in C^{p,q}(M) : \square\phi = 0\}$$

Theorem 4.49 (Hodge, Kodaira, de Rham)
Let M be a compact Kähler manifold.

(1) $H^q(M,\Omega^p) \cong H^p(M,\Omega^q)$, $p,q \geqslant 0$.

(2) $H^r(M,\mathbb{C}) = \bigoplus_{p+q=r} H^p(M,\Omega^q)$, $r \geqslant 0$.

Proof. We already know from harmonic theory (Theorem 4.10, Example 11) that

$$H^{p,q}(M) = H^q(M,\Omega^p) \text{ and } H^{q,p}(M) = H^p(M,\Omega^q)$$

For (1) it is therefore sufficient to show that $H^{p,q}(M) = H^{q,p}(M)$. But on a Kähler manifold $\Box = \bar{\Box}$ (Theorem 4.48),and so $\Box\,\phi = 0$ if and only if $\Box\bar{\phi} = 0$. Hence, conjugation defines the required isomorphism between $H^{p,q}(M)$ and $H^{q,p}(M)$, proving (1).

In Example 10 of subsection A we showed that

$$H^r(M,\mathbb{C}) = \{\phi \in C^\infty(\Lambda^r{}_c TM') : \Delta\phi = 0\}$$

$$= \{\phi \in C^\infty(\Lambda^r{}_c TM') : \Box\phi = 0\}, \text{ since } \Box = \tfrac{1}{2}\Delta$$

Now any complex r-form ϕ on M may be written uniquely as a sum $\Sigma\phi_{p,q}$ of complex (p,q)-forms, $p+q = r$. Clearly $\Box\phi = 0$ if and only if $\Box\phi_{p,q} = 0$ for all $p+q = r$. Hence

$$H^r(M,\mathbb{C}) = \bigoplus_{p+q=r} H^{p,q}(M)$$

Harmonic theory completes the proof. Q.E.D.

For $p,q \geqslant 0$, we define

$$h^{p,q} = \dim_{\mathbb{C}} H^q(M,\Omega^p), \quad b_p = \dim_{\mathbb{C}} H^p(M,\mathbb{C})$$

We already know several facts about the numbers $h^{p,q}$. In particular, if $\dim_{\mathbb{C}} M = m$, we have:

(A) $h^{p,q} = h^{q,p}$, $p,q \geqslant 0$ (Theorem 4.49).

(B) $h^{p,q} = h^{m-p,m-q}$, $p,q \geqslant 0$ (Serre-duality).

The topological invariants b_r are given in terms of the $h^{p,q}$ by

(C) $b_r = \sum_{p+q=r} h^{p,q}$, $r \geqslant 0$.

From relations (A) and (C) follow immediately:

Corollary 4.49.1. On a compact Kähler manifold, b_{2r+1} is divisible by 2, $r \geqslant 0$.

Corollary 4.49.2. Let M be a compact Kähler manifold. Then, for $p \geqslant 0$,

$$b_{2p} \geqslant h^{p,p} \geqslant 1$$

Proof. $b_{2p} \geqslant h^{p,p}$ follows from relation (C). If we let ∇ denote the hermitian connection on M and $\widehat{H}$ the Kähler form of M, we have $\nabla\widehat{H} = 0$ (∇ is a metric connection) and so

$$\nabla'\widehat{H}^p = \nabla''\widehat{H}^p = 0, \ p \geqslant 0$$

Since $\bar\partial = \tilde{\bar\partial}$, $\vartheta = \tilde\vartheta$, it follows that $\bar\partial\widehat{H}^p = 0$ and $\vartheta\widehat{H}^p = 0$, $p \geqslant 0$. Therefore, $\square\widehat{H}^p = 0$. Since $\widehat{H}^m = m!\Theta$, $\widehat{H}^p \neq 0$, $1 \leqslant p \leqslant m$, and so $\dim_{\mathbb{C}} H^{p,p}(M) \geqslant 1$. Q.E.D.

Example. The only sphere that can admit a Kähler structure is S^2 (In fact, it is well known that if $n \neq 1,3$, then S^{2n} does not admit a complex structure. It is not yet known whether S^6 admits a complex structure.) See also Ref. [31], these Proceedings.

This is all we shall say about the topology of Kähler manifolds. For a much more thorough description of the cohomology of a Kähler manifold in terms of "primitive" elements we refer to Ref. [39].

We give one final corollary of Theorem 4.49 which again shows the relation between the complex and topological structure of Kähler manifolds.

Corollary 4.49.3. Every holomorphic p-form on a compact Kähler manifold is d-closed.

Proof. Let $\phi \in C^{p,0}(M)$ be holomorphic: $\bar\partial\phi = 0$. Now $\bar\partial\phi = 0$ implies $\square\phi = 0$ and so $\Delta\phi = 0$. Hence ϕ is harmonic. *A fortiori*, ϕ is d-closed. Q.E.D.

Remarks

(1) We refer to Ref. [20] for an example showing the necessity of the Kähler condition in Corollary 4.49.3.

(2) The proof of the Hodge decomposition theorem essentially rests on the identity $\Delta = \square/2$ (in fact this identity holds if and only if the manifold is Kähler). The proof that $\Delta = \square/2$ made us of the Lefschetz operator L (see Theorem 4.48). In the case of projective algebraic manifolds, the operator L may be interpreted geometrically (in homology) as the "intersection with a generic (transverse) hyperplane". Now Kähler manifolds need not be algebraic (see Refs [6,13]), and so the Lefschetz operator appears to lose its geometric significance when the manifold is Kähler but not algebraic. This perhaps gives an indication as to why some of the consequences of Theorem 4.49 – in particular that the odd-dimensional Betti numbers are even – are difficult to understand geometrically. In fact a Kähler manifold may, by an example of Hironaka, be complex analytically deformed into a non-Kähler complex manifold where, of course, the Lefsche operator is not defined at all but the topological structure is unchanged! We offer now a slightly different proof of the evenness of the odd-dimensional Betti numbers of a compact Kähler manifc First observe that to prove that b_{2r+1} is even, $r \geqslant 0$, it is sufficient to know that Δ maps (p,q)-for to (p,q)-forms since $\Delta\phi = 0$ if and only if $\Delta\bar\phi = 0$. Now on a Kähler manifold the Levi-Cività and hermitian connections coincide (Proposition 4.41). Denoting the Riemannian metric on M by g and either connection by ∇, we have the following expression for the Laplace operator:

$$-\Delta = \theta(g)\nabla(_\wedge\nabla) + {_\wedge\nabla}(\theta(g)\nabla)$$

Writing $\nabla = \nabla' + \nabla''$, we find, using the commutativity of ∇ with contractions and wedge produc that

$$-\Delta = \mathrm{trace}(\nabla'\nabla'' + \nabla''\nabla' + \nabla'^2 + \nabla''^2) + K$$

where K is a differential operator of order zero defined in terms of the curvature of the Riemannian metric g. Now trace ∇'^2 = trace $\nabla''^2 \equiv 0$ and so

$$-\Delta = \mathrm{trace}(\nabla' \nabla'') + \mathrm{trace}(\nabla' \nabla'') + K$$

Since trace $(\nabla' \nabla'')$ and trace $(\nabla' \nabla'')$ map (p, q)-forms to (p, q)-forms, we have shown that Δ is the sum of a second-order LPDO which maps (p,q)-forms to (p,q)-forms and a zero-order operator. Now the Levi-Cività connection on ${}_cTM$ restricts to hermitian connections on TM and $\overline{TM}$ (since we are assuming M Kähler) and so we find that the curvature form of ∇ splits as the sum of the curvature forms of the hermitian connections on TM and $\overline{TM}$. From this fact it follows easily that K maps (p, q)-forms to (p, q)-forms. Hence M has odd-dimensional Betti numbers even.

Denoting the real curvature form of the (uncomplexified) Levi-Cività connection by R, $R \in C^\infty(\Lambda^2 TM \otimes L(TM, TM))$, Bianchi's first identity implies that if M is Kähler then

$$\text{(A)} \qquad R(JX \wedge JY) = R(X \wedge Y)$$

for all $X, Y \in C^\infty(TM)$. Condition (A) says that if we take two, linearly independent, tangent vectors X and Y at a point $x \in M$ then the curvature associated to the plane $X \wedge Y$ is equal to that associated to the plane $J(X \wedge Y)$. Condition (A), of itself, does not immediately imply that Δ maps (p, q)-forms to (p, q)-forms (the Levi-Cività connection will not restrict to a connection on $\Lambda^{p,q}(M)'$ unless M is Kähler). However, since condition (A) is both "geometrical" and essential in the above proof of the evenness of odd-dimensional Betti numbers, it seems natural to ask whether:

(1) (A) implies that M is Kähler (with the assigned metric on M).

If (1) is false, which seems likely,

(2) (A) implies that odd-dimensional Betti numbers are even. A *direct* geometric proof would be of most interest here (see (3)).

(3′) If (A) holds, M is a complex deformation of a Kähler manifold.

(3″) If (A) holds, there exists a Kähler metric on M (same complex structure).

(4) Condition (A) is open (closed!) under complex deformations of M. (4) seems to me to be the most interesting problem. If condition (A) is open, but not closed, it might be rewarding to study Hironaka's example of a non-Kähler deformation in the framework of condition (A).

Finally the reader is warned not to take the above questions too seriously: they were inserted at the end of a long conference on a hot and humid day and it is quite possible that they are either bad questions or that the answers to them are already known.

REFERENCES

[1] ADAMS, J., GRIFFITHS, P., Topics in Algebraic and Analytic Geometry, Princeton University Press (1975).

[2] ANDREOTTI, A., HILL, C.D., E.E. Levi convexivity and the Hans Lewy problem I, II, Ann. Sc. Norm. Super. Pisa, Sci. Fis. Mat. **26** (1972) 299-324.

[3] ANDREOTTI, A., VESENTINI, E., Carleman estimates for the Laplace-Beltrami equation on complex manifolds, I.H.E.S. **25** (1965) 81-130.

[4] ATIYAH, M., SEGAL, G., SINGER, I.M., papers on the Atiyah-Singer index theorem in Ann. Math., especially I, II, III, starting **87** (1968) 531.

[5] BRICKELL, F., CLARK, R.S., Differentiable Manifolds, Van Nostrand Reinhold, London (1970).

[6] CORNALBA, M., "Complex tori and Jacobians", these Proceedings.

[7] DIEUDONNÉ, J., Foundations of Modern Analysis, Academic Press, New York (1960).

[8] EELLS, J., "Elliptic operators on manifolds", these Proceedings.

[9] FIELD, M.J., "Holomorphic function theory and complex manifolds", Global Analysis and its Applications I (Proc. Int. Course Trieste, 1972), IAEA, Vienna (1974) 83.

[10] FIELD, M.J., "Sheaf cohomology, structures on manifolds and vanishing theory", Global Analysis and its Applications II (Proc. Int. Course Trieste, 1972), IAEA, Vienna (1974) 167.

[11] FUKS, B.A., special chapters in Theory of Analytic Functions of Several Complex Variables **14**, AMS translations of mathematical monographs.

[12] GUNNING, C.R., ROSSI, H., Analytic Functions of Several Complex Variables, Prentice-Hall, Englewood Cliffs, N.J. (1965).

[13] DE LA HARPE, P., "Introduction to complex tori", these Proceedings.

[14] HILL, C.D., The Cauchy problem for $\bar{\partial}$, AMS Symposia in pure mathematics **23** (1973) 135-43.

[15] HIRZEBRUCH, F., Topological Methods in Algebraic Geometry, Springer (1966) (with appendices by SCHWARZENBERGER, R.L.E., BOREL, A.).

[16] HÖRMANDER, L., An Introduction to Complex Analysis in Several Variables, Van Nostrand, Princeton (1966).

[17] HÖRMANDER, L., Linear Partial Differential Operators, Springer, Berlin (1969).

[18] KOBAYASHI, S., NOMIZU, K., Foundations of Differential Geometry **I**, Interscience, New York (1963).

[19] KOBAYASHI, S., NOMIZU, K., Foundations of Differential Geometry **II**, Interscience, New York (1969).

[20] KODAIRA, K., MORROW, J., Complex Manifolds, Holt, Rinehart and Winston, New York (1971).

[21] LANG, S., Differential Manifolds, Addison-Wesley, Reading, Mass. (1972).

[22] MALGRANGE, B., Ideals of Differentiable Functions, Oxford University Press (1966) (Tata Institute, Studies in Mathematics, No.3).

[23] NARASIMHAN, M.S., "Vector bundles on compact Riemann surfaces", these Proceedings.

[24] NARASIMHAN, R., "Analytic spaces", presented at this Course.

[25] NARASIMHAN, R., Introduction to the Theory of Analytic Spaces, Springer Lecture Notes No. 25 (1966).

[26] NARASIMHAN, R., Several Complex Variables, University of Chicago Press (1971).

[27] NARASIMHAN, R., Analysis on Real and Complex Manifolds, Masson, Paris (1968).

[28] PALAIS, R.S., Seminar on the Atiyah-Singer Index Theorem, Ann. Study No. 57, Princeton University Press (1966).

[29] PALAIS, R.S., Foundations of Global Non-linear Analysis, Benjamin, New York (1968).

[30] HARVEY, F.R., Integral Formulae Connected by Dolbeault's Isomorphism, Rice Studies (1970) 77-97.

[31] ROBERTSON, S., "Elementary geometry of complex manifolds", these Proceedings.

[32] Seminaire Henri Cartan, 1952-54 and 1953-54, Benjamin, New York (1967).

[33] SERRE, J.P., Faisceaux algébriques cohérents, Ann. Math. **61** (1955).

[34] SERRE, J.P., Un théorème de dualité, Commun. Math. Helv. **29** (1955).

[35] SPIVAK, M., Calculus on Manifolds, Benjamin, New York (1965).

[36] TOUGERON, J.C., Idéaux de fonctions différentiables, Springer (1972).

[37] VESENTINI, E., On Levi convexivity and cohomology vanishing theorems, Tata Institute, Bombay (1967).

[38] VLADIMIROV, V.S., "Holomorphic functions of several complex variables, with non-negative imaginary part and some applications", these Proceedings.

[39] WEIL, A., Variétés Kähleriennes, Hermann, Paris (1958).

[40] WEIL, A., Foundations of algebraic geometry, AMS Coll. Publications Vol. 29, Providence, R.I. (1962).

[41] WELLS, R.O., Differential Analysis on Complex Manifolds, Prentice-Hall, Englewood Cliffs, N.J. (1973).

ASYMPTOTIC EVALUATION OF INTEGRALS, AND WIMAN-VALIRON THEORY

W.H.J. FUCHS
Department of Mathematics,
Cornell University,
Ithaca, New York,
United States of America

Abstract

ASYMPTOTIC EVALUATION OF INTEGRALS, AND WIMAN-VALIRON THEORY.

I. Asymptotic evaluation of integrals: The Laplace method; Faber's method; The saddle point method; Asymptotic evaluation of integrals with an oscillating integrand. II. Wiman-Valiron theory: Maximum term and central index; Logarithmic measure; The function g(x); The function $\nu(t)$; Estimates in terms of $\mu(r,f)$; Behaviour of an entire function near points where its maximum modulus is attained; Picard's theorem; An application to differential equations.

INTRODUCTION

The first part of these notes deals with the problem of approximate evaluation of integrals depending on a large parameter. The topic is of great importance in many branches of applied (and pure) mathematics. The notes give only a brief introduction to the elements of the subject; they cannot claim any originality. More detailed accounts of the subject and of related subjects can be found in:

DE BRUIJN, N.G., Asymptotic Methods in Analysis, North-Holland, Amsterdam (1958).
ERDÉLYI, A., Asymptotic Expansions, Dover, New York (1956).
MURRAY, J.D., Asymptotic Analysis, Clarendon Press, Oxford (1974).
OLVER, F.W., Asymptotics and Special Functions, Academic Press, New York (1974).
SIROVICH, L., Techniques of Asymptotic Analysis, Springer (1971).

The second part is concerned with the theory of Wiman-Valiron. This theory gives information about the behavior of an entire function f(z) near points where |f(z)| is close to the maximum modulus of f on the circle of radius |z| around the origin. The proofs of this theory have been simplified by Hayman, Kövari and others in recent times. The approach chosen here has not been published elsewhere. A fuller account of the theory will appear in a survey article by W.K. Hayman in the Bulletin of the Canadian Mathematical Society.

Part I
ASYMPTOTIC EVALUATION OF INTEGRALS

1. THE LAPLACE METHOD

<u>Aim</u>: To give simple, easily understood expressions for integrals involving a large parameter z .

Ex. $\int_a^b g(t)e^{zh(t)}dt$

z large, real.

Idea: Only parts of integrand which are near the places where integrand is close to its maximum give appreciable contributions. Near those points replace integrand by approximation which leads to easy integration.

Ex. $\int_0^\infty g(t)e^{-tz}dt$

Assume $g(t) = O((1+t)^A)$, g conts near 0, $g(0) \neq 0$.

Maximum of integrand is at 0 . Expect

$$\int_0^\infty g(t)e^{-tz}dt \approx \int_0^d g(t)e^{-tz}dt$$

$$\approx \int_0^\delta g(0)e^{-tz}dt = \frac{g(0)}{z}(1-e^{-\delta z})$$

$$\approx \frac{g(0)}{z}$$

Watson's Lemma. $F(z) = \int_0^\infty g(t)e^{-zt}dt$.

1. $g(t) = O((1+t)^A)$ $\qquad 0 \leq t < \infty$

2. Near $t = 0$

$$g(t) = a_0 t^{\lambda_0} + a_1 t^{\lambda_1} \ldots + a_p t^{\lambda_p} + O(t^{\lambda_p+\alpha})$$

$$(-1 < \lambda_0 < \lambda_1 \ldots < \lambda_p < \lambda_p + \alpha \ ; \ a_0 \neq 0 \ )$$

Conclusion

$$F(z) = \frac{\Gamma(\lambda_0+1)}{z^{\lambda_0+1}} a_0 + \frac{\Gamma(\lambda_1+1)}{z^{\lambda_1+1}} a_1 + \ldots + \frac{\Gamma(\lambda_p+1)}{z^{\lambda_p+1}} a_p + O(z^{-\lambda_p-\alpha-1}$$

<u>Proof.</u> $$D = F(z) - \frac{\Gamma(\lambda_0+1)}{z^{\lambda_0+1}} a_0 - \frac{\Gamma(\lambda_1+1)}{z^{\lambda_1+1}} a_1 \cdots - \frac{\Gamma(\lambda_p+1)}{z^{\lambda_p+1}} a_p$$

$$= \int_0^\infty g(t) e^{-zt} dt - \sum_{k=0}^{p} a_k \int_0^\infty t^{\lambda_k} e^{-zt} dt$$

since

$$\int_0^\infty t^m e^{-zt} dt = \frac{\Gamma(m+1)}{z^{m+1}}$$

$$D = \int_0^\infty \{g(t) - \sum_{k=0}^{p} a_k t^{\lambda_k}\} e^{-zt} dt$$

$$= \int_0^\delta + \int_\delta^\infty = I_1 + I_2$$

<u>In $0 \leq t \leq \delta$</u>

$$|g(t) - \sum_{k=0}^{p} a_k t^{\lambda_h}| < K(p) t^{\lambda_p+\alpha}$$

$$\therefore |I_1| < K(p) \int_0^\delta t^{\lambda_p+\alpha} e^{-zt} dt$$

$$< K(p) \int_0^\infty t^{\lambda_p+\alpha} e^{-zt} dt = \frac{K(p)\Gamma(\lambda_p+\alpha+1)}{z^{\lambda_p+\alpha+1}}$$

<u>In $\delta < t$</u>

$$|g(t) - \sum_{k=0}^{p} a_k t^{\lambda_k}| < C(1+t)^A + \max_{0 \leq k \leq p} (|a_k|)(1+t)^{\lambda_p}$$

$$< C(p)(1+t)^{\lambda_p+A}$$

$$|I_2| < C(p) \int_\delta^\infty (1+t)^{\lambda_p+A} e^{-1/2\, zt} \cdot e^{-1/2\, zt} dt$$

$$(1+t)^{\lambda_p+A} e^{-1/2\, zt} < (1+t)^{\lambda_p+A} e^{-t} < M(p) \qquad (0 < t;\ z > 2)$$

$$\therefore |I_2| < C_1(p) \int_\delta^\infty e^{-1/2\, zt} = C_1(p) e^{-1/2\, z\delta}$$

Therefore

$$D < |I_1| + |I_2| < \frac{K(p)\Gamma(\lambda_p+A+1)}{z^{\lambda_p+A+1}} + C_1(p)e^{-1/2\, z\delta}$$

$$< C_0(p)z^{-\lambda_p-A-1}$$

Q.e.d.

Asymptotic expansions

In applications of Watson's Lemma $g(t)$ can often be expanded in a power series near $t = 0$. Then the λ's are just the positive integers and in the expression of $\int_0^\infty g(t)e^{-zt}dt$ we may take $p = \lambda_p$ as large as we wish. However, if we let $p \to \infty$ for a given z it will turn out in most cases that the series $\sum_0^\infty \frac{\Gamma(\lambda_k+1)}{z^{\lambda_k+1}} a_k$ will not converge, because the error-term tends to ∞ as $p \to \infty$.

This does not mean that the series $\sum \frac{\Gamma(\lambda_k+1)}{z^{\lambda_k+1}} a_k$ is useless. By Watson's Lemma a fixed partial sum gives the value of $\int_0^\infty g(t)e^{-zt}dt$ with an error which tends to zero like $z^{-\lambda_p-\alpha-1}$ and by choosing the number of terms correctly for a given z one can find an approximation to the Laplace transform which may be very good.

Df. $\sum_0^\infty A_k z^{-\lambda_k}$ is an asymptotic expansion of $F(z)$ valid for real, positive z, if

(i) $\lambda_0 < \lambda_1 < \lambda_2 \dots < \lambda_k \to \infty$

(ii) $F(z) - \sum_0^p A_p z^{-\lambda_p} = O(z^{-\lambda_{p+1}}) \quad (z \to \infty)$

(Obvious generalisations to series involving also logarithm terms or other functions. Also to asymptotic expansions valid in an angle, $\alpha \leq \arg z \leq p$. In this case $H(z) = O(z^{-\lambda_{p+1}})$ means $|H(z)\cdot z^{\lambda_{p+1}}| < K$ for all $z = re^{i\theta}$ wit $\alpha \leq \theta \leq \beta$.)

Ex. 1. $\int_0^\infty \frac{e^{-t}}{t+z}\, dt$

For $z > 0, \quad t > 0$

$$\frac{1}{t+z} = \frac{1}{z} - \frac{t}{z^2} + \frac{t^2}{z^3} \cdots + (-1)^{p-1}\frac{t^{p-1}}{z^p} + (-1)^p \frac{t^p}{z^p(t+z)}$$

$$\int_0^\infty \frac{e^{-t}dt}{t+z} = \frac{\Gamma(1)}{z} - \frac{\Gamma(2)}{z^2} + \frac{\Gamma(3)}{z^3} \cdots + (-1)^p \frac{\Gamma(p)}{z^p} + R$$

$$|R| = \int_0^\infty \frac{e^{-t}t^p}{z^p(t+z)}\, dt$$

$$< \frac{1}{z^{p+1}} \int_0^\infty t^p e^{-t} dt = \frac{\Gamma(p+1)}{z^{p+1}}$$

$\int_0^\infty \frac{e^{-t}}{t+z}\, dt$ has the asymptotic expansion $\sum_{k=1}^{\infty} (-1)^{k+1} \frac{\Gamma(k)}{z^k}$.

Ex. 2. $F(z) = \int_z^\infty e^{-t^2} dt$

Use integration by parts:

$$\int_z^\infty e^{-t^2} dt = \int_z^\infty \frac{2te^{-t^2}}{2t}\, dt$$

$$= -\int_z^\infty \frac{1}{2t} \frac{d}{dt}(e^{-t^2}) dt$$

$$= \frac{1}{2z} e^{-z^2} - \frac{1}{2}\int_z^\infty t^{-2}e^{-t^2} dt$$

$$= \frac{1}{2z} e^{-z^2} + \frac{1}{2^2}\int_z^\infty t^{-3} \frac{d}{dt}(e^{-t^2}) dt$$

$$= \frac{1}{2z} e^{-z^2} - \frac{1}{2^2 z^3} e^{-z^2} + \frac{3}{2^2}\int_z^\infty t^{-4}e^{-t^2} dt$$

etc.

$$= \frac{1}{2z} e^{-z^2} - \frac{1}{2^2 z^3} e^{-z^2} + \frac{3}{2^3 z^5} e^{-z^2}$$

$$- \frac{1\cdot 3\cdot 5}{2^4 z^7} e^{-z^2} \cdots + \frac{1\cdot 3\cdot 5\cdots(2n-1)}{2^{n+1} z^{2n+1}} + R$$

$$|R| = \frac{1\cdot 3\cdots(2n-1)(2n+1)}{2^{n+1}} \int_z^{\infty} t^{-2n-2} e^{-t^2} dt$$

$$= -\frac{1\cdot 3\cdots(2n+1)}{2^{n+2}} \int_z^{\infty} t^{-2n-3} \frac{d}{dt}(e^{-t^2}) dt$$

$$< - \text{const.} \frac{e^{-z^2}}{z^{2n+3}} \int_z^{\infty} \frac{d}{dt} (e^{-t^2}) dt = \text{const.} \frac{e^{-t^2}}{z^{2n+3}}$$

$\therefore$ We have an asymptotic expansion.

[Easier way to estimate error:

If the expansion is

$$\int_z^{\infty} e^{-t^2} dt = u_0 + u_1 + u_2 \cdots + u_p + R_p$$

then sign of R_p is opposite of sign of u_p, same as sign of u_{p+1}.

But

$$R_p = u_{p+1} + R_{p+1}$$

Hence $|R_{p+1}| < u_{p+1}$; error is less than first neglected ter

2. LAPLACE METHOD

for the asymptotic evaluation of

$$F(z) = \int_a^b g(t) e^{zh(t)} dt$$

for large, positive values of z.

Theorem. Hypotheses:

1. $-\infty < a < b < \infty$.
2. $h(t)$ real-valued.

3. $h(t)$ has a maximum at $t = c$, $a < c < b$. Given $\delta > 0 \ \exists \ \epsilon = \epsilon(\delta) > 0$ such that

$$h(t) < h(c)-\epsilon \qquad (|t-c| > \delta)$$

4. Near $t = c$, $h(t)$ has continuous derivatives up to the 3rd order (inclusive) and $h''(c) < 0$, $h'(c) = 0$.

5. $g(t)$ continuous at c, $g(c) \neq 0$.

6. $|g(t)| < K \qquad (a \leq t \leq b)$.

<u>Conclusion</u>

$$F(z) \sim \sqrt{\frac{2\pi}{z|h''(c)|}}\; g(c)e^{zh(c)} \qquad (z \to \infty).$$

$[F(z) \sim G(z) \quad (z \to \infty)$ means

$F(z)/G(z) \to 1 \quad (z \to \infty)]$

Proof. By our assumptions on $h(t)$ and Taylor's Theorem with remainder there is a $\delta > 0$ such that

$$h'(t) = (t-c)h''(c) + O((t-c)^2) \qquad (|t-c| < \delta)$$

$$h(c)-h(t) = -\tfrac{1}{2}(t-c)^2h''(c) + O(|t-c|^3) \qquad (|t-c| < \delta)$$

It follows that

$$u(t) = \sqrt{h(c)-h(t)}$$

is a differentiable function in $c \leq t \leq c+\delta$ (one-sided derivatives at the endpoints) and that

$$\frac{du}{dt} = (-\frac{1}{\sqrt{2}}\, h''(c))^{\frac{1}{2}} + O[(t-c)] \qquad (c \leq t \leq c+\delta)$$

In particular $\frac{du}{dt}$ is positive and continuous in $c \leq t \leq c+\delta_1$ for some $\delta_1 > 0$. Therefore $u(t)$ has a continuously differentiable inverse function

$$t = \varphi(u)$$

in the interval $u(c) = 0 \leq u \leq \delta_2 = u(c+\delta_1)$.

Note that $\varphi'(0) = 1/u'(c) = \sqrt{2/(-h''(c))} = \sqrt{2/|h''(c)|}$.

Similarly

$$v(t) = \sqrt{h(c)-h(t)}$$

has a negative derivative in an interval

$$c-\delta_1' \leq t \leq c$$

and a continuously differentiable inverse function

$$t = \psi(v)$$

exists for $0 \leq v \leq \delta_1' = v(c-\delta_2')$; also

$$\psi'(0) = -\sqrt{2/|h''(c)|}$$

By change of variable we obtain

$$(2.1) \qquad \int_{c-\delta_1'}^{c+\delta_1} g(t)e^{zh(t)}dt = \int_0^{\delta_2'} g[\psi(v)]\cdot(-\psi'(v))e^{-zv^2+zh(c)}dv$$

$$+ \int_0^{\delta_2} g[\varphi(u)]\varphi'(u)e^{-zu^2+zh(c)}du$$

Given $\epsilon > 0$ we can choose δ_1 and δ_1', and therefore δ_2 and δ_2' so small that by the continuity of g near c

$$\left|g[\psi(v)](-\psi'(v))-g(c)\cdot\sqrt{2/|h''(c)|}\right| < \epsilon \qquad (0 < v < \delta_2')$$

$$\left|g[\varphi(u)]\varphi'(u)-g(c)\sqrt{2/|h''(c)|}\right| < \epsilon \qquad (0 < u < \delta_2)$$

Therefore, by (2.1)

$$(2.2) \qquad \left|\int_{c-\delta_1'}^{c+\delta_1} g(t)e^{zh(t)}dt - (\sqrt{2}\, g(c)e^{zh(c)}/\sqrt{|h''(c)|}) \int_{c-\delta_2'}^{c+\delta_2} e^{-zu^2}du\right|$$

$$< \epsilon\, e^{zh(c)} \int_{c-\delta_2'}^{c+\delta_2} e^{-zu^2}du < \epsilon e^{zh(c)} \int_{-\infty}^{\infty} e^{-zu^2}du$$

$$= \epsilon(\pi/z)^{1/2}e^{zh(c)}$$

By assumptions 3 and 6 we also have

$$|g(t)e^{zh(t)}| < Ke^{zh(c)-\epsilon z}$$

provided that δ_1 and δ_1' are chosen sufficiently small. Hence, by assumption 1

$$(2.3) \qquad \left|\int_a^b g(t)e^{zh(t)}dt - \int_{c-\delta_1'}^{c+\delta_1} g(t)e^{zh(t)}dt\right| < K(b-a)e^{zh(c)-\epsilon z}$$

By (2.2) and (2.3)

$$(2.4)\quad \left|\int_a^b g(t)e^{zh(t)}dt - \sqrt{2}\, g(c)e^{zh(c)}/\sqrt{|h''(c)|}\int_{-\delta_2'}^{\delta_2} e^{-zu^2}du\right|$$

$$< \varepsilon(\pi/z)^{1/2}e^{zh(c)} + K(b-a)e^{zh(c)-\varepsilon z}$$

Finally

$$\int_{-\delta'}^{\delta_2} e^{-zu^2}du = \int_0^{\delta_2} e^{-zu^2}du + \int_0^{\delta_2'} e^{-zu^2}du$$

$$= z^{-1/2}\int_0^{\delta_2 z^{1/2}} e^{-v^2}dv + z^{-1/2}\int_0^{\delta_2' z^{1/2}} e^{-v^2}dv$$

$$= 2z^{-1/2}\int_0^{\infty} e^{-u^2}du - z^{-1/2}\int_{\delta_2 z^{1/2}}^{\infty} e^{-v^2}dv - z^{-1/2}\int_{\delta_2' z^{1/2}}^{\infty} e^{-v^2}dv$$

and, by Ex. 2 of §1,

$$\left|\int_{\delta_2 z^{1/2}}^{\infty} e^{-v^2}dv\right| < \frac{1}{2\delta_2 z^{1/2}}\, e^{-\delta_2^2 z}$$

$$\left|\int_{\delta_2' z^{1/2}}^{\infty} e^{-v^2}dv\right| < \frac{1}{2\delta_2' z^{1/2}}\, e^{-\delta_2' z}$$

Also,

$$\int_0^{\infty} e^{-u^2}du = \sqrt{\pi}/2$$

Using these facts in (2.4) we obtain

$$\int_a^b g(t)e^{zh(t)}dt = \sqrt{\frac{2\pi}{|h''(c)|z}}\, g(c)e^{zh(c)} + E$$

$$|E| < \varepsilon(\pi/z)^{1/2}e^{zh(c)} + K(b-a)e^{zh(c)-\varepsilon z}$$

$$+ (g(c)e^{zh(c)}/\sqrt{2|h''(c)|})\{(\delta_2 z^{1/2})^{-1}e^{-\delta_2^2 z}$$

$$+(\delta_2' z^{1/2})^{-1}e^{-\delta_2' z}\}$$

Once $\varepsilon > 0$, and so δ_2 and δ_2' are chosen, we can find $Z(\varepsilon)$ so that $|E| < 2\varepsilon\sqrt{\frac{2\pi}{|h''(c)|z}}\cdot g(c)e^{zh(c)}$ for $z > Z(\varepsilon)$.

This proves the theorem.

Under stronger assumptions on $g(t)$ and $h(t)$ the integrands on the right-hand side of (2.1) can be expanded in powers of u (or v) and, after introducing the new variables $x = u^2$ and $y = v^2$, an asymptotic expansion can be found by using Watson's Lemma. The calculation of the functions φ and ψ and of their derivatives at c may be tiresome. It is usually more convenient to use an alternative method due to Faber. We state this method as a recipe and leave the justification as a problem to the reader.

3. FABER'S METHOD

Hypotheses. $F(z) = \int_a^b g(t)e^{zh(t)}dt$

1. $0 < a < b < \infty$

2. $h(t)$ has a maximum at $t = c$, $h''(c) < 0$.

3. Near $t = c$, g and h can be expanded in convergent power series in $(t-c)$.

4. Given $\delta > 0$ we can find $\epsilon > 0$ such that

$$h(t) < h(c)-\epsilon \qquad (|t-c| > \delta).$$

5. $|g(t)| < K$.

For simplicity of writing we suppose $c = 0$, the general case is easily reduced to this by a shift of origin.

Recipe. Expand $h(t)$ in powers of t:

$$h(t) = h(0) + \tfrac{1}{2}h''(0)t^2 + \ldots = h(0) + \tfrac{1}{2}h''(0)t^2 + h_1(t)$$

Rewrite integrand in the form

$$e^{zh(0)+\frac{1}{2}zh''(0)t^2} \sum_{k=0}^{\infty} c_k(z)t^k$$

by expanding $g(t)e^{zh_1(t)}$ in powers of t. The asymptotic expansion for $F(z)$ is obtained by integrating this power seri term-by-term from $t = -\infty$ to $t = \infty$ and collecting terms with the same power of z. The integrations are easily performed with the aid of the formulae

$$(2.5) \qquad \int_{-\infty}^{\infty} v^{2m+1}e^{-\alpha v^2}dv = 0 \qquad (m = 0,1,2\ldots,\alpha>0)$$

(integrand is an odd function),

$$\int_{-\infty}^{\infty} e^{-\alpha v^2}dv = (\pi/\alpha)^{1/2}$$

and hence, by differentiation under the integral sign,

$$(2.6) \quad \int_{-\infty}^{\infty} v^{2m} e^{-\alpha v^2} dv = \frac{1}{2} \cdot \frac{3}{2} \cdots \frac{(2m-1)}{2} \sqrt{\pi}\, \alpha^{-m-\frac{1}{2}}$$

$$(m = 1,2,3\ldots, \alpha > 0)$$

It is important to bear in mind that many integrals which do not exactly fall under the hypotheses that we have stated can still be treated by the same methods, either by reducing the integrals to the form discussed above by change of variables or by applying the idea of Laplace's method with suitable modifications.

<u>Ex.</u> $\Gamma(z+1) = \int_0^{\infty} e^{-t} t^z dt$

$t = zu$

$\Gamma(z+1) = z^{z+1} \int_0^{\infty} e^{z(\log u - u)} du$

$h(u) = \log u - u$

$h'(u) = \frac{1}{u} - 1 \qquad c = 1$

$h''(1) = -1$

$u - 1 = v$

$$e^{z(\log u-u)} = e^{-z} e^{[-v+\log(1+v)]z}$$

$$= e^{-z} e^{-\frac{v^2}{2}z + \frac{v^3}{3}z + \frac{v^4}{4}z \cdots}$$

$$= e^{-z-\frac{v^2 z}{2}} \left(1 + z\left(\frac{v^3}{3} - \frac{v^4}{4} \cdots\right) + \frac{1}{2}z^2\left(\frac{v^3}{3} - \frac{v^4}{4} \cdots\right)^2 + \frac{1}{6}z^3\left(\frac{v^3}{3} - \frac{v^4}{4} \cdots\right)^3 \cdots\right)$$

and obviously

$$\int_{-\infty}^{\infty} v^k e^{-\alpha v^2} dv = 0 \qquad (k \text{ odd})$$

Asymptotic expansion:

$$z^{z+1} e^{-z} \left\{ \int_{-\infty}^{\infty} e^{-\frac{1}{2}zv^2} dv - \frac{z}{4} \int_{-\infty}^{\infty} e^{-\frac{1}{2}zv^2} v^4 dv + \frac{1}{18} z^2 \int_{-\infty}^{\infty} e^{-\frac{1}{2}zv^2} v^6 dv + \ldots \right\}$$

Terms indicated by ... are of the form

$$\text{const. } z^{\ell} \int_{-\infty}^{\infty} e^{-\frac{1}{2}zv^2} v^k dv$$

where either $\ell = 1$ and $k \geq 6$ or $\ell = 2$, $k \geq 8$ or $\ell > 2$, $k \geq 3\ell$. In all cases these terms contribute in power $z^{-\nu}$ where $\nu \geq -\ell + \frac{k}{2} + \frac{1}{2} \geq 2$.

After evaluation of the integrals

$$\Gamma(z+1) = \sqrt{2\pi z}\, z^z e^{-z}\left(1 - \frac{1}{2}\cdot\frac{3}{2}\cdot\frac{2^2}{z^2}\cdot\frac{z}{4} + \frac{z^2}{18}\cdot\frac{1}{2}\cdot\frac{3}{2}\cdot\frac{5}{2}\cdot\frac{2^3}{z^3}\ \cdots\right)$$

$$= \sqrt{2\pi z}\, z^z e^{-z}\left(1 + \frac{1}{z}\{-\frac{5}{4} + \frac{15}{18}\}\ldots\right)$$

$$= \sqrt{2\pi z}\, z^z e^{-z}\left(1 + \frac{1}{12z}\ \ldots\right)$$

4. THE SADDLE POINT METHOD

(Combination of Laplace's Method with Cauchy's Theorem)

Assume now that

$$F(z) = \int_a^b g(t)e^{zh(t)}dt = \int_a^b G(t,z)dt$$

where $g(t)$ and $h(t)$ are entire functions of t.

We can now displace the path of integration by Cauchy's Theorem.

Desirable features of a path are:

1) The maximum of the absolute value of the integrand is as small as possible.

2) Absolute value of integrand drops off as rapidly as possib as one goes away from the maximum points along the path.

If $t = u+iv$, plot the surface

$$w = |G(t,z)|$$

above the u-v plane. Must connect the points $A \equiv (t=a, |G(a,z)|)$ and $B \equiv (t = b, |G(b,z)|)$ by a path that stays as low as possible. Usually A + B will be separated by several mountain ridges. Path has to go through the 'cols' or 'saddle-points' of these ridges. They are characterized by the fact that the tangent plane to the surface $w = |G(z,t)|$ is horizontal, i.e.

$$\frac{\partial}{\partial u}|G| = \frac{\partial}{\partial v}|G| = v$$

or

$$\frac{\partial}{\partial u}\log|G| = \frac{\partial}{\partial v}\log|G| = 0$$

But $$\frac{d}{dt}\log G(t,z) = \frac{\partial}{\partial u}\log|G| - i\frac{\partial}{\partial v}\log|G|$$

The cols are therefore given by those values of t for which

$$\frac{d}{dt}\log G(t,z) = 0$$

There will usually be a finite number of such t and in most cases there will be a single t_0 such that a path from a to b can be found which passes through t_0 and such that $|G|$ attains its largest value on the path at t_0.

Near a col $t = t_0$

$$\log G(t,z) = \log G(t_0,z) - \frac{(t-t_0)^2}{2}\frac{d^2}{dt^2}\log G(t_0,z)$$

$$+ \text{ higher-order terms.}$$

If $$t - t_0 = \rho e^{i\varphi} \qquad (\rho > 0)$$

$$\frac{d^2}{dt^2}\log G(t_0,z) = Ae^{i\alpha} \qquad \left(\text{assume } \frac{d^2}{dt^2}\log G(t_0,z) \neq 0 \text{ then } A > 0\right)$$

$$\log|G(t,z)| = R\log G(t,z)$$

$$= \log|G(t_0,z)| + \tfrac{1}{2}\rho^2 A\cos(2\varphi+\alpha)$$

$$+ \text{ terms of higher order in } \rho.$$

To make $\log|G(t,z)|$ as small as possible for given ρ, we must choose φ so that

$$\cos(2\varphi+\alpha) = -1$$

(1) $$\varphi = \frac{\pi-\alpha}{2} \quad \left(\text{or } -\frac{\pi+\alpha}{2}\right)$$

This leads to the <u>recipe</u>: To find asymptotic value for $\int_a^b G(t,z)dt$ (G holomorphic function of t), take the following steps:

1) Find saddle points $t_0, t_1 \ldots$ as roots of

$$\frac{d}{dt}\log G(t,z) = 0$$

2) Deform path of integration so that it passes through saddle point t_0 (and possibly other saddle points also).

3) Near t_0 choose as path of integration the straight line segment

$$t = t_0 + se^{i\varphi} \qquad (-\delta \leq s \leq \delta)$$

where φ is given by (1) above.

4) Evaluate contribution from neighborhood of t_0 by Laplace's method.

5) Check that contribution from parts of path away from the saddle point is negligible.

<u>Ex. 1</u>.

$$\frac{1}{n!} = \frac{1}{2\pi i} \oint e^t t^{-n-1} dt$$

Saddle point: $$\frac{\partial}{\partial t}(t - (n+1)\log t) = 1 - \frac{n+1}{t} = 0$$

$$t = n+1$$

$$\frac{\partial^2}{\partial t^2}(t - (n+1)\log t) = \frac{n+1}{t^2}$$

$$\alpha = 0$$

$$\varphi = \pm \frac{\pi}{2}$$

Choose as contour circle of radius $n+1$:

$$\frac{1}{n!} = \frac{(n+1)^{-n}}{2\pi} \int_{-\pi}^{\pi} e^{(n+1)(e^{i\varphi}-i\varphi)} e^{i\varphi} d\varphi$$

On path $|e^{(n+1)(e^{i\varphi}-i\varphi)}| = e^{(n+1)\cos\varphi} < e^{(n+1)}$

$\therefore$ Maximum at $\varphi = 0$; apply Laplace's method:

$$nh(\varphi) = n(e^{i\varphi} - i\varphi)$$

$$g(\varphi) = e^{e^{i\varphi}}$$

$$h'(\varphi) = ie^{i\varphi} - i$$

$$h''(\varphi) = -e^{i\varphi}$$

$$\frac{1}{n!} = \frac{1}{2\pi} e^{n+1} \cdot \sqrt{\frac{2\pi}{n}} (n+1)^{-n}$$

$$n! = \sqrt{2\pi n}\left(\frac{n}{e}\right)^n \frac{\left(1+\frac{1}{n}\right)^n}{e} \sim \sqrt{2\pi n}\left(\frac{n}{e}\right)^n$$

since $$\frac{\left(1+\frac{1}{n}\right)^n}{e} \to 1 \qquad (n \to \infty)$$

Ex. 2. $F(x) = \int_{-\infty}^{\infty} e^{i(tx - \frac{t^3}{3})} dt$ (x real)

Convergence of integral not obvious, but will be proved.

$$K(t) = i(tx - \frac{t^3}{3})$$

Saddle points: $K'(t) = i(x-t^2) = 0$

Case a) $x < 0$

$$t_1 = -i|x|^{1/2} \qquad t_2 = i|x|^{1/2}$$

$$K''(t) = -2t_i \qquad \arg K''(t_1) = \pi \ (= \alpha)$$

$$\arg K''(t_2) = 0$$

Correct direction for passing through $t_1: 2\varphi+\alpha = \pi; \ \varphi = 0$

Correct direction for passing through $t_2: \ \varphi = \frac{\pi}{2}$

$$K(t_1) = -|x|^{3/2} + \frac{|x|^{3/2}}{3} = -\frac{2|x|^{3/2}}{3}$$

$$K(t_2) = +\frac{2|x|^{3/2}}{3}$$

$$t = u+iv \quad \mathscr{R}K(t) = \mathscr{R}i((u+iv)x - \tfrac{1}{3}(u^3+3iu^2v-3uv^2-iv^3))$$

$$= -vx+u^2v-v^3/3$$

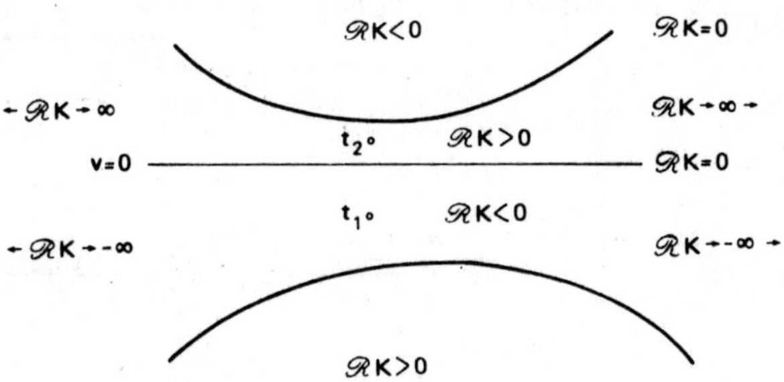

It is not possible to deform the line of integration into a curve passing through t_2 in the right direction without getting large contributions from $\pm\infty$.

Deform contour in $\int_{-N}^{M} e^{K(t)}dt$:

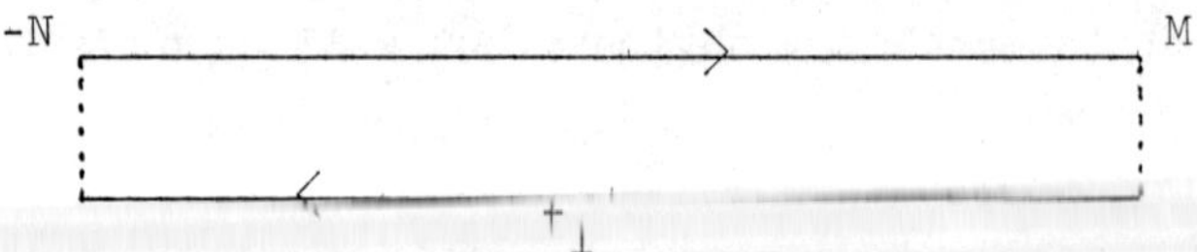

On vertical parts integrand $O(e^{+\frac{N^2}{2}v})$, $O(e^{\frac{M^2 v}{2}})$ resp.
$\therefore$ Contribution of vertical parts $\to 0$.

By Cauchy's Theorem

$$\int_{-\infty}^{\infty} e^{K(t)}dt = \lim_{N,M\to\infty} \int_{-N+t_1}^{M+t_1} e^{K(t)}dt$$

Let $\quad t = |x|^{1/2}(-i+u)$

$$\int_{-N+t_1}^{M+t_1} = |x|^{1/2} \int_{-N/|x|^{1/2}}^{M/|x|^{1/2}} e^{i|x|^{3/2}(i-u-\frac{1}{3}(-i+u)^3} du$$

$$\text{exponent} = -|x|^{3/2} + \frac{|x|^{3/2}}{3} - |x|^{3/2}u^2 - i|x|^{3/2}\frac{u^3}{3}$$

|Integrand| drops off rapidly from its value at $u = 0$. By method of steepest descent

$$F(x) \sim |x|^{\frac{1}{2}} \int_{-\infty}^{\infty} e^{-\frac{2}{3}|x|^{3/2}} \cdot e^{-|x|^{3/2}u^2} \sum_0^\infty \frac{1}{n!}\left(-\frac{i|x|^{3/2}u^3}{3}\right)^n du$$

Using (2.5) and (2.6),

$$F(x) \sim |x|^{-\frac{1}{4}} e^{-\frac{2}{3}|x|^{3/2}} \sqrt{\pi} \sum_0^\infty \frac{(-1)^n}{(2n)!} \frac{1}{9^n} \cdot \frac{1\cdot 3\cdots(6n-1)}{2\cdot 2\cdots 2} |x|^{-\frac{3n}{2}}$$

Ex. 3.

n^{th} Hermite polynomial defined by generating function

$$e^{-w^2+2xw} = \sum_0^\infty \frac{1}{n!} H_n(x) w^n$$

Take x real. Want asymptotic expression for large n of

$$\frac{1}{(n-1)!} H_{n-1}(x) = \frac{1}{2\pi i} \int e^{-w^2+2xw-n \log w} dw$$

Contour of integration loop encircling origin from -k to -k; $|\operatorname{Im} \log w| \leq \pi$.

Saddle points: $\frac{d}{dw}(-w^2+2xw-n \log w) = -2w+2x-\frac{n}{w} = 0$

$$\therefore w^2-xw+\frac{n}{2} = 0$$

$$w_+ = \frac{x}{2} + \sqrt{\frac{x^2}{4} - \frac{n}{2}}$$

$$w_- = \frac{x}{2} - \sqrt{\frac{x^2}{4} - \frac{n}{2}}$$

1) $2n \gg x^2$

$$w_+ = \frac{x}{2} + i\sqrt{\frac{n}{2} - \frac{x^2}{4}} \approx i\sqrt{\frac{n}{2}}$$

$$w_- = \frac{x}{2} - i\sqrt{\frac{n}{2} - \frac{x^2}{4}} \approx -i\sqrt{\frac{n}{2}}$$

At $\pm i\sqrt{\frac{n}{2}}$

$$\left(\frac{d}{dw}\right)^2 (-w^2 + 2xw - n \log w) = -2 + \frac{n}{w^2} < 0$$

A good direction for our contour near $\pm i\sqrt{\frac{n}{2}}$ is therefore parallel to the real axis $(\alpha = \pi)$.

Deform the contour of integration into a rectangle with corners at $\pm T \pm i\sqrt{\frac{n}{2}}$. Let $T \to \infty$. The contribution of the vertical sides tends to zero as $T \to \infty$,

$$\frac{1}{(n-1)!} H_{n-1}(w) = \frac{1}{2\pi i} \int_{-\infty-i\sqrt{\frac{n}{2}}}^{\infty-i\sqrt{\frac{n}{2}}} - \frac{1}{2\pi i} \int_{-\infty+i\sqrt{\frac{n}{2}}}^{\infty+i\sqrt{\frac{n}{2}}} e^{-w^2+2xw} w^{-n} dw$$

In the first integral set $w = \sqrt{\frac{n}{2}}\,(t-i)$:

$$J = \frac{1}{2\pi i}\int_{-\infty-i\sqrt{\frac{n}{2}}}^{\infty-i\sqrt{\frac{n}{2}}} e^{-w^2+2xw-n\log w}\,dw$$

$$= \frac{1}{2\pi i}\int_{-\infty}^{\infty} e^{-\frac{1}{2}nt^2+\sqrt{2n}\,xt+int+\frac{n}{2}-\sqrt{2n}\,xi-\frac{n}{2}\log\frac{n}{2}-n\log(t-i)}\,dt\sqrt{\frac{n}{2}}$$

The absolute value of the integrand, $\phi(t)$, say, is less than $|\phi(0)|\,e^{-\frac{1}{4}nt^2}(1+t^2)^{-\frac{n}{2}}$ for $|t| > \frac{4\sqrt{2x}}{n}$, so that only the immediate neighborhood of the origin is of importance. Here we apply Faber's method: Expand exponent in a power series:

$$-\,n\log(t-i) = \frac{n\pi}{2}i - nti - \frac{1}{2}nt^2 + \text{higher powers}$$

$$J = \frac{1}{2\pi}\int_{-\infty}^{\infty} e^{-nt^2+\sqrt{2n}xt+\frac{n}{2}+i(\sqrt{2n}x+(n-1)\frac{\pi}{2})-\frac{(n-1)}{2}\log\frac{n}{2}}(1+At^3+\dots)\,dt$$

Since we did not place the path of integration exactly through the saddle point, a linear term remains in the exponent. But this does not affect the essentials of the method. Integrati term-by-term gives as leading term:

$$J \sim \frac{1}{2\pi}\int_{-\infty}^{\infty} e^{-n(t-\sqrt{\frac{x}{2n}})^2+\frac{x^2}{2}+\frac{n}{2}+i(\sqrt{2n}\,x+(n-1)\frac{\pi}{2})}\left(\frac{2}{n}\right)^{\frac{n-1}{2}}dt$$

$$\sim \frac{1}{2\sqrt{n\pi}}\left(\frac{2}{n}\right)^{\frac{n-1}{2}} e^{\frac{x^2}{2}+\frac{n}{2}+i(\sqrt{2n}\,x+(n-1)\frac{\pi}{2})}$$

The value of $-\frac{1}{2\pi i}\int_{-\infty+i\sqrt{\frac{n}{2}}}^{\infty+i\sqrt{\frac{n}{2}}}$ is the conjugate complex of J.

Therefore

$$\frac{1}{(n-1)!}H_{n-1}(x) = 2\,\mathrm{Re}\,J \sim \frac{1}{\sqrt{n\pi}}\left(\frac{2}{n}\right)^{\frac{n-1}{2}} e^{\frac{x^2}{2}+\frac{n}{2}}\cos\left(\sqrt{2n}\,x+\frac{n-1}{2}\pi\right)$$

as $n \to \infty$

5. ASYMPTOTIC EVALUATION OF INTEGRALS WITH AN OSCILLATING INTEGRAND

[After lectures by A.W. Knapp]

(a) Monotone functions are harmless

Lemma 1. $F(x) \in C^1[a,b]$, $f(x) = F'(x)$. $m(x)$ monotone.

Then

$$\left|\int_a^b f(x)m(x)\,dx\right| \leq 4\|F\|_\infty \|m_\infty\|$$

$(\|g\|_\infty = \sup_{\alpha \leq x \leq b} |g(x)|)$.

Proof. $\int_a^b f(x)m(x)\,dx = F(x)m(x)\Big|_a^b - \int_a^b F(x)m'(x)\,dx$

$$\left|\int_a^b f(x)m(x)\,dx\right| \leq 2\|F\|_\infty \|m\|_\infty + \|F\|_\infty \left|\int_a^b m'(x)\,dx\right|$$

$$\leq 2\|F\|_\infty \|m\|_\infty + \|F\|_\infty \,|(m(b) - m(a))|$$

$$\leq 4\|F\|_\infty \|m\|_\infty$$

(b) Lemma 2 (Van der Corput's lemma)

A. $h(x)$ real-valued, $h(x) \in C^2[a,b]$ · $h'(x)$ monotone $|h'(x)| \geq k > 0$. Then

$$\left|\int_a^b e^{ih(x)}\,dx\right| \leq \frac{4}{k}$$

B. $m > 1$, $h(x) \in C^{m+1}[a,b]$, $|h^{(m)}(x)| \geq k > 0$.

Then

$$\left|\int_a^b e^{ih(x)}\,dx\right| \leq A_m\, k^{-\frac{1}{m}}$$

Proof of A. Since $h'(x)$ is differentiable and $\neq 0$, $h'(x)$ is of constant sign in $[a,b]$ and $h(x)$ is monotone.

$$y = h(x)$$

has a monotone, differentiable inverse function

$$x = M(y)$$

$$\frac{dx}{dy} = M'(y) = \frac{1}{h'(x)} \quad \text{is monotone.}$$

$$\left|\int_a^b e^{ih(x)}dx\right| = \left|\int_{a'}^{b'} e^{iy}M'(y)dy\right| \leq 4\|M'\|_\infty$$

by Lemma 1. $\|M'\|_\infty \leq \frac{1}{k}$, A is proved.

C. Suppose proved for $m < N$. Shall deduce result for $m = N$.

Without loss of generality $h^{(N)}(x) \geq k > 0$.

(a) Suppose first $h^{(N-1)}(x) > 0$. Then $(a < \gamma < b)$

$$h^{(N-1)}(\gamma) = h^{(N-1)}(a) + \int_a^\gamma h^{(N)}(x)dx$$

$$\geq k(\gamma - a)$$

$$\left|\int_a^b e^{ih(x)}dx\right| \leq \left|\int_a^\gamma\right| + \left|\int_\gamma^b\right| \qquad (a \leq \gamma \leq b)$$

$$\leq (\gamma - a) + A_{N-1}[k(\gamma - a)]^{-\frac{1}{N-1}}$$

(by the trivial estimate $|e^{ih(x)}| = 1$ and the hypothesis of induction).

If $b - a \leq k^{-\frac{1}{N}}$, choose $\gamma = b$ (2nd integral does not occ

If $b - a > k^{-\frac{1}{N}}$, choose $\gamma - a = k^{-1/N}$

In both cases

$$\left|\int_a^b e^{ih(x)}dx\right| \leq \frac{1+A_{N-1}}{k^{1/N}}$$

Similarly, if (b) $h^{(N-1)}(x) \leq 0$ $(a \leq x \leq b)$. [Use trivial estimate in $\gamma \leq x \leq b$].

(c) $h^{(N-1)}(x)$ changes sign in $[a,b]$. Since $h^{(N)}(x)$ is of constant sign, this can happen only once, at δ, say. Apply first parts of proof to $\int_a^\delta$, $\int_\delta^b$ separately. This proves result with $A_N = 2(1+A_{N-1})$.

To complete induction, must prove assertion true for $m = 2$. Note that $h'(x)$ is monotone and reduce to $m = 1$ by the argument used for reduction from N to $N - 1$.

Cor. $h(x)$ real-valued, $h \in C^2[a,b]$, h' monotone, $|h'(x)| \geq k > 0$ $(a \leq x \leq b)$, $m(x)$ monotone, $m \in C^1[a,b]$. Then

$$\left|\int_a^b e^{ih(x)} m(x)\,dx\right| \leq \frac{16\|m\|_\infty}{k}$$

[Lemma 1 and Lemma 2A. $F(x) = \int_a^x e^{ih(t)}dt$.]

Ex. $\int_0^1 e^{2\pi i\xi x+\frac{2\pi i}{x}} \frac{dx}{x}$ $\qquad (\xi > 0)$

$$h(x) = 2\pi(\xi x + \frac{1}{x})$$

$$h'(x) = 2\pi(\xi - \frac{1}{x^2}) \qquad |h'| > A\xi \quad \text{in} \quad |x-\xi^{-\frac{1}{2}}| > \frac{1}{2}\,\xi^{-\frac{1}{2}}$$

$$h''(x) = \frac{4\pi}{x^3} \qquad |h''| > A\xi^{3/2} \quad \text{in} \quad |x-\xi^{-1/2}| \leq \frac{1}{2}\,\xi^{-1/2}$$

[We use the letter A for a positive constant, not necessarily always the same]

By Lemma 2B

$$\left| \int_{\frac{1}{2}\xi^{-\frac{1}{2}}}^{\gamma} e^{ih(x)}dx\right| < \frac{A}{\xi^{3/4}} \qquad \frac{1}{2}\xi^{-1/2} < \gamma \leq \frac{3}{2}\xi^{-1/2}$$

By Lemma 1

$$\left|\int_{\frac{1}{2}\xi^{-1/2}}^{\frac{3}{2}\xi^{-1/2}} e^{ih(x)}\,\frac{dx}{x}\right| < A\xi^{-3/4}\cdot\xi^{1/2} = A\xi^{-1/4}$$

If $x_1 > \frac{3}{2}\xi^{-1/2}$, by Cor. of Lemma 2

$$\left| \int_{\frac{3}{2}\xi^{-1/2}}^{x_1} e^{ih(x)} \frac{dx}{x} \right| < \frac{A\xi^{1/2}}{A_1\xi} = A\xi^{-1/2}$$

$$\left| \int_{\epsilon}^{\frac{1}{2}\xi^{-1/2}} e^{[illegible]} \frac{dx}{x} \right| = \left| \int_{2\xi^{-1/2}}^{[illegible]} e^{[illegible]} \frac{du}{u} \right| [illegible]$$

$$< A\xi^{-1/2}$$

$$\therefore \int_0^1 e^{2\pi i(\xi x + \frac{1}{x})} \frac{dx}{x} = O(\xi^{-1/4}) \qquad \xi \to \infty$$

<u>Stationary phase</u>

<u>Theorem 1</u>. $h(x)$ real-valued, $\epsilon\ C^3\ [a,b]$

$$\left.\begin{array}{ll} \text{(i)} & 0 < \lambda \leq |h''(x)| \leq \beta\lambda \\ \text{(ii)} & |h'''(x)| \leq \mu \end{array}\right\} \quad a \leq x \leq b$$

$$\text{(iii)} \quad h'(c) = 0 \qquad\qquad a < c < b$$

Then

$$\int_a^b e^{ih(x)} dx = \sqrt{2\pi}\ e^{(\operatorname{sgn} h'')\frac{\pi i}{4} + ih(c)} |h''(c)|^{-1/2}$$

$$+ O(\lambda^{-4/5}\mu^{1/5})$$

$$+ O(\min \frac{\beta}{|h'(a)|},\ \lambda^{-1/2})$$

$$+ O(\min \frac{\beta}{|h'(b)|},\ \lambda^{-1/2})$$

The constants implied by the O - notation are absolute constan
independent of h .

$$\text{Pr.} \quad \int_a^b = \int_a^{c-\delta} + \int_{c-\delta}^{c+\delta} + \int_{c+\delta}^{b}$$

$$(a \leq c - \delta < c + \delta \leq b)$$

By condition (1) $h'(x)$ is monotone in $c \leq x \leq b$ and in
$a \leq x \leq c$ and

$$|h'(x)| \geq |h'(c+\delta)| \qquad (c+\delta \leq x \leq b)$$

$$|h'(x)| \geq |h'(c-\delta)| \qquad (a \leq x \leq c - \delta)$$

Also

$$|h'(c\pm\delta)| = |\int_c^{c\pm\delta} h''(x)dx| \geq \delta\lambda$$

By Lemma 2A

$$|\int_a^{c-\delta}| + |\int_{c+\delta}^{b}| < \frac{8}{\delta\lambda}$$

In $\quad c - \delta \leq x \leq c + \delta$

$$e^{ih(x)} = e^{ih(c) + \frac{i(x-c)^2}{2} h''(c) = iR(x)}$$

$$|R(x)| \leq \frac{|x-c|^3}{6} \mu$$

Since $\quad |e^{it} - 1| < |t| \quad ,$

$$e^{iR(x)} = 1 + R_1(x)$$

$$|R_1(x)| < \frac{|x-c|^3}{6} \mu$$

$$\therefore \int_{c-\delta}^{c+\delta} e^{ih(x)}dx = e^{ih(c)} \int_{c-\delta}^{c+\delta} e^{i\frac{(x-6)^2}{2} h''(c)} dx + R_2$$

$$|R_2| < 2 \int_0^{\delta} \frac{u^3}{6} \mu \, du = \frac{1}{12} \mu \, \delta^4$$

$$\int_{c-\delta}^{c+\delta} e^{i\frac{(x-6)^2}{2} h''(c)} dx = 2 \int_0^{\frac{\delta^2}{2}|h''(c)|} e^{\pm iu} \frac{du}{\sqrt{u}} \cdot \frac{1}{\sqrt{2|h''(c)|}}$$

$$\left(\frac{(x-c)^2}{2} |h''(c)| = u, \quad \pm = \text{sgn } h''(c)\right)$$

$$(*) \quad \begin{cases} = \sqrt{\frac{2}{|h''(c)|}} \int_0^{\infty} e^{\pm iu} \frac{du}{\sqrt{u}} + R_3 \\ |R_3| < \frac{4\sqrt{2}}{\delta\sqrt{|h''(c)|}} \cdot \sqrt{\frac{2}{|h''(c)|}} \qquad \text{(Lemma 1)} \end{cases}$$

$$|R_3| \leq \frac{8}{\delta|h''(c)|} \leq \frac{8}{\delta\lambda}$$

By contour integration 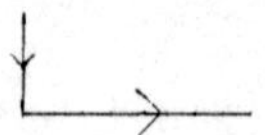

$$e^{\mp i\frac{\pi}{4}} \int_0^\infty \frac{e^{\pm iu}}{\sqrt{u}}\, du = \int_0^\infty \frac{e^{-u}}{\sqrt{u}}\, du = \sqrt{\pi}$$

$$\int_a^b e^{ih(x)}\, dx = \sqrt{\frac{2\pi}{|h''(c)|}}\, e^{\pm i\frac{\pi}{4}} + R_4 \quad (\pm = \operatorname{sgn} h''(c))$$

$$|R_4| < \frac{8}{\delta |h''(c)|} + \frac{1}{12}\mu\delta^4 + \frac{8}{\delta\lambda}$$

$$< \frac{16}{\delta\lambda} + \frac{1}{12}\mu\delta^4$$

$\delta = (\lambda\mu)^{-1/5}$:

$$|R_4| < 17\,\lambda^{-4/5}\,\mu^{1/5}$$

This result is obtained under the assumption that $\min(c-a, b-c) > (\lambda\mu)^{-1/5}$.

If $b-c < (\lambda\mu)^{-1/5}$, then we must subtract

$$e^{ih(c)} \int_b^{c+\delta} e^{\frac{i}{2}(x-c)^2 h''(c)}\, dx \quad \text{from the main term.}$$

By Lemma 2A this makes a change of at most

$$\frac{4}{(b-c)|h''(c)|} < \frac{4}{(b-c)\lambda}$$

But $$\beta(b-c)\lambda > \left|\int_c^b h''(x)\,dx\right| = |h'(b)|$$

Also, by Lemma 2B

$$\left|\int_b^{c+\delta} e^{\pm\frac{i}{2}(x-c)^2 h''(c)}\, dx\right| < \frac{A_2}{|h''(c)|^{1/2}} < A_2\,\lambda^{-1/2}$$

Similarly, if c is close to a .

This completes the proof of the theorem.

Theorem 2. $I = \int_a^b e^{\cdot h(x)} g(x)\,dx$

and $h(x)$ satisfies all hypotheses of Theorem 1 .

$g(x)$ monotone, $g(c) \neq 0$. $\omega(\delta) = \sup_{c-\delta \leq x \leq y \leq x+\delta \leq c+\delta} |g(y)-g(x)|$

Then for $a \leq c-\delta < c+\delta \leq b$

$$I = \sqrt{\frac{2\pi}{|h''(c)|}}\, e^{(\operatorname{sgn} h''(c))\frac{\pi i}{4} + ih(c)}\, g(c)$$

$$+ O\left(\frac{\|g\|_\infty}{\delta\lambda}\right)$$

$$+ O\{\min[(\lambda^{-1/2} + \mu\delta^4)\omega(\delta);\ \delta\omega(\delta)]\}$$

Proof. $I = \int_a^{c-\delta} + \int_{c-\delta}^{c+\delta} + \int_{c+\delta}^{b} e^{ih(x)} g(x)\,dx$

As in the proof of Th. 1

$$|h'(x)| \geq \delta\lambda \qquad (a \leq x \leq b,\ |x-c| \geq \delta)$$

and so, by the Corollary of Lemma 2

$$\left|\int_a^{c-\delta} + \int_{c+\delta}^{b} e^{ih(x)} g(x)\,dx\right| < \frac{32\|g\|_\infty}{\delta\lambda}$$

$$\int_{c-\delta}^{c+\delta} e^{ih(x)} g(x)\,dx = g(c) \int_{c-\delta}^{c+\delta} e^{ih(x)}\,dx + \int_{c-\delta}^{c+\delta} e^{ih(x)} (g(x)-g(c))\,dx$$

If $F(x) = \int_c^x e^{ih(t)}\,dt$,

then, as in the proof of Theorem 1

$$F(x) = \int_c^x e^{ih(c) + i\frac{(F-c)^2}{2} h''(c)}\,dt + O(\mu\delta^4) \quad (|x-c| \leq \delta)$$

and

$$\left|\int_0^x e^{ih(c) + i\frac{(t-c)^2}{2} h''(c)}\,dt\right| < A_2 \lambda^{-1/2} \qquad (a \leq x \leq b)$$

by Lemma 2B. By Lemma 1 with our choice of $F(x)$ and $m(x) = g(x) - g(c)$

$$\left| \int_{c-\delta}^{c+\delta} e^{ih(x)} (g(x)-g(0)) dx \right| < O(\lambda^{-1/2}\omega(\delta) + \mu\delta^{+4}\omega(\delta))$$

and, trivially, $\left| \int_{c-\delta}^{c+\delta} e^{ih(x)} (g(x)-g(0)) dx \right| < 2\delta\omega(\delta)$

Collecting all the information and using (*) (p.257) we obtain Theorem 2.

<u>Example.</u> $I = \int_0^1 e^{2\pi i(\xi x + \frac{1}{x})} \frac{dx}{x}$ ξ real, $\xi \to \infty$

We know

$$\left| \int_0^{\frac{1}{2}\xi^{-1/2}} + \int_{\frac{3}{2}\xi^{-1/2}}^{1} e^{2\pi i(\xi x + \frac{1}{x})} \frac{dx}{x} \right| < O(\xi^{-1/2})$$

Apply Th. 2 to $\int_{\frac{1}{2}\xi^{-1/2}}^{\frac{3}{2}\xi^{-1/2}}$:

$$\|g\|_\infty = \alpha_1 \xi^{1/2} \qquad c = \xi^{-1/2}$$

$$h''(x) = \frac{4\pi}{x^3}, \quad \lambda = \alpha_2 \xi^{3/2} \qquad h'''(x) = -\frac{12\pi}{x^4}, \quad \mu = \alpha_3 \xi^2$$

$$h''(c) = 4\pi\xi^{3/2} \quad h(c) = 4\pi\xi^{1/2}; \quad g(c) = \xi^{112}; \quad \omega(\delta) < \alpha_4 \cdot \frac{\delta}{(\xi^{-1/2})^2} =$$

$$I = \frac{1}{\sqrt{2}} e^{i(\frac{\pi}{4} + 4\pi\xi^{112})} \xi^{-\frac{1}{4}} + O(\frac{1}{\delta\xi}) + O\{\min(\delta^2\xi, \delta\xi^{-\frac{1}{2}} + \delta^5\xi^3)\} + O(\xi^{-\frac{1}{2}})$$

$= \text{leading term} + O(\xi^{-\frac{1}{3}})$, if we choose $\delta = \xi^{-\frac{2}{3}}$.

$$J = \int_0^1 e^{2\pi i(\xi x + \frac{1}{x})} \frac{dx}{x}$$

treated by the saddle point method:

1) $\xi < 0$; $\xi \to -\infty$

Saddle point at $\xi - \frac{1}{x^2} = 0$, $x = \pm i|\xi|^{-\frac{1}{2}}$

$x = |\xi|^{-\frac{1}{2}} t$

$$J = \int_0^{|\xi|^{\frac{1}{2}}} e^{2\pi i|\xi|^{+\frac{1}{2}}(-t+\frac{1}{t})} \frac{dt}{t}$$

$h(t) = 2\pi i(-t+\frac{1}{t})$

$h'(t) = 0 : \quad t = \pm i$

$h''(i) = 4\pi i\,(i)^{-3} = -4\pi$

$h''(-i) = 4\pi i(-i)^3 = 4\pi \qquad x = 0, \qquad \varphi = \frac{\pi}{2}$

If $t = re^{i\theta}$

$$Rh(t) = R\, 2\pi i(-re^{i\theta} + \frac{1}{r}e^{-i\theta})$$

$$= 2\pi(r + \frac{1}{r})\sin\theta$$

Integrand small in lower half-plane.

Change path of integration to

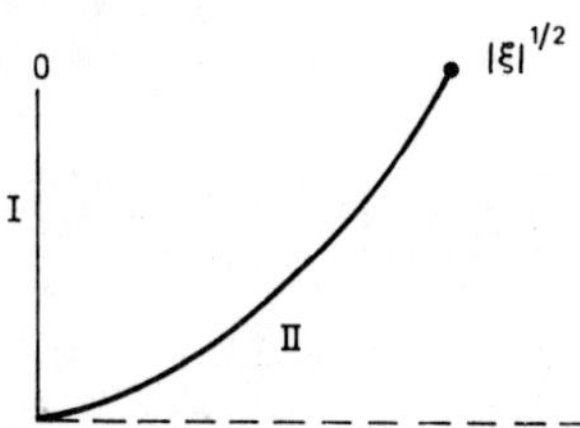

Saddle point method applies to $\int_I$

$$\therefore \quad \int_I = O\left(\frac{e^{2\pi i|\xi|^{+\frac{1}{2}}(+2i)}}{|\xi|^{1/4}}\right) = O\left(\frac{e^{-4\pi|\xi|^{1/2}}}{|\xi|^{1/4}}\right)$$

On II important contribution comes from points near $|\xi|^{1/2}$. $t = |\xi| e^{i\theta}$, then

$$\int_{II} = i\int_{-\frac{\pi}{2}}^{0} e^{-2\pi i|\xi| e^{i\theta} + 2\pi i e^{-i\theta}} d\theta$$

$$\text{Integrand} = e^{-2\pi i|\xi| + 2\pi|\xi|\theta} (1+\ldots)$$

(... indicates power series in θ, coefficients of $\theta^m = O(|\xi|^{\frac{m}{2}})$) . By Watson's Lemma

$$\int_{II} \sim \frac{ie^{-2\pi i|\xi|}}{2\pi|\xi|} + O(|\xi|^{-\frac{3}{2}})$$

2) $\xi > 0, \quad \xi \to \infty$

Saddle points at $\pm\xi^{-\frac{1}{2}}$

$$x = \xi^{-\frac{1}{2}} t$$

$$J = \int_0^{\xi^{\frac{1}{2}}} e^{2\pi i \xi^{\frac{1}{2}}(t+\frac{1}{t})} \frac{dt}{t}$$

$$h(t) = 2\pi i(t+\tfrac{1}{t}) . \qquad h''(1) = 4\pi i \cdot \alpha \cdot \tfrac{\pi}{2} \quad \varphi = + \tfrac{i}{4}$$

$$Rh(t) = Rh(re^{i\theta})$$

$$= R\, 2\pi i(re^{i\theta} + \tfrac{1}{r} e^{-i\theta})$$

$$= -2\pi \sin\theta(r-\tfrac{1}{r})$$

Deform contour to

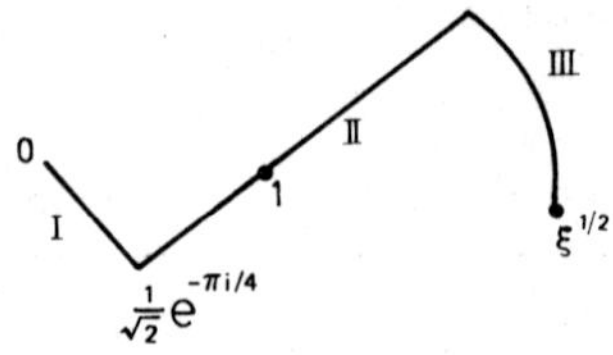

On I $|t| = r < \frac{1}{2r}$

$$e^{Rh(t)} \frac{1}{|t|} < e^{-\frac{\pi}{\sqrt{2}r}\xi^{\frac{1}{2}}} \frac{1}{r}$$

$$\left|\int_I e^{\xi^{\frac{1}{2}}h(t)} \frac{dt}{t}\right| < \int_{\sqrt{2}}^{\infty} e^{-\xi^{\frac{1}{2}}\frac{\pi}{\sqrt{2}}s} \frac{ds}{s} < \frac{1}{\pi\xi^{1/2}} e^{-\xi^{\frac{1}{2}}\pi}$$

On II $\frac{1}{t} = O(1)$ and $Rh(t)$ has a minimum at $t = 1$.

By Laplace's Method

$$\int_{II} \approx \sqrt{\frac{2\pi}{\xi^{1/2}.4\pi}}\; e^{\frac{\pi i}{4}+4\pi i\xi^{\frac{1}{2}}} = \frac{1}{\sqrt{2}\xi^{1/4}} e^{\frac{\pi i}{4}+4\pi\xi^{1/2}i}$$

On III

$$\left|\frac{e^{\xi^{\frac{1}{2}}h(t)}dt}{t}\right| < e^{-2\pi(\xi-1)\sin\theta}d\theta$$

$$< e^{-4(\xi-1)\theta}d\theta \qquad (\sin\theta > \tfrac{2}{\pi}\theta \quad (0<\theta<\tfrac{\pi}{2}))$$

Contribution of III:

$$\left|\int_{III}\right| < \int_0^{\infty} e^{-4(\xi-1)\theta}d\theta = \frac{1}{4(\xi-1)} = O(\tfrac{1}{\xi})$$

$$J \sim \frac{1}{\sqrt{2}\xi^{1/4}} e^{\frac{\pi i}{4}+4\pi i\xi^{1/2}}$$

Part II
WIMAN-VALIRON THEORY

1. MAXIMUM TERM AND CENTRAL INDEX

Let

$$f(z) = \sum_{n=0}^{\infty} c_n z^n$$

be a transcendental, entire function. Since

$$|c_n| r^n \to 0 \qquad (n \to \infty)$$

for every $r > 0$,

$$\mu(r,f) = \mu(r) = \sup_{n\geq 0} |c_n| r^n$$

is attained for some n.

Df. $\mu(r)$ is called the maximum term of $f(z)$. The largest index $n = \nu_0(r)$ such that

$$|c_{\nu_0(r)}| r^{\nu_0(r)} = \mu(r)$$

is called the central index.

The maximum modulus of $f(z)$ is defined by

$$M(r,f) = \sup_{|z|=r} |f(z)|$$

Exercise. Prove that $\nu_0(r) \to \infty \quad (r \to \infty)$.

An important part of Wiman-Valiron theory shows that the behavior of $f(z)$ at points where $|f(z)|$ is close to $M(r,f)$ is closely approximated by the behavior of a rather short sectic $\sum_{n=\nu_0(r)-k(r)}^{n=\nu_0(r)+k(r)} c_n z^n$. In particular it is possible to give good upper and lower bounds for $M(r,f)$ in terms of $\mu(r,f)$.

2. LOGARITHMIC MEASURE.

Definition: If E is a measurable subset of the interval $1 \leq r$, the logarithmic measure of E is

$$\ell(E) = \int_E \frac{dr}{r}$$

Many assertions of Wiman-Valiron hold only outside exceptional sets of finite logarithmic measure. A formula which is only valid outside such an exceptional set will be marked by two vertical lines to the right of the equation.

3. THE FUNCTION g(x)

Plot in a Cartesian coordinate system the points

$$P_n \equiv (n, -\log|c_n|) \quad (n = 0,1,2,\ldots)$$

Let $y = g_0(x)$ be the largest convex minorant of the set $\mathscr{P} = \{P_n\}_{n=0}^{\infty}$. That is to say that for any convex curve $y = h(x)$ with $h(x_1) > g_0(x_1)$ for some $x_1 > 0$ we can find a P_k with $-\log|c_k| < h(k)$. The curve $y = g_0(x)$ is a piecewis linear, convex curve whose corners are points of $\mathscr{P}$. Since for an entire function $-\frac{1}{n}\log|c_n| \to \infty$ as $n \to \infty$, the slopes of the segments forming $y = g_0(x)$ tend to ∞ as $x \to \infty$. We call $y = g_0(x)$ the <u>curve associated with</u> $f(z)$. Different entire functions can have the same associated curve. Among these

$$F_0(z) = \sum_{n=0}^{\infty} e^{-g_0(n)} z^n$$

has the largest coefficients and, in particular,

$$|c_n| \leq e^{-g_0(n)} \tag{3.1}$$

$$M(r,f) \leq \Sigma|c_n|r^n \leq \Sigma e^{-g_0(n)} r^n = M(r,F_0) \tag{3.2}$$

The associated curve allows one to find the maximal term and the central index by a geometrical construction. Let

$$r = e^t$$

Then

$$\log|c_n|r^n = nt + \log|c_n|$$

is the length of the intercept on the negative y-axis made by the line of slope t passing through $P_n \equiv (n, -\log|c_n|)$. It is now easy to see that

$$\log \mu(r,f) = \sup_n \,(nt + \log|c_n|)$$

is the intercept on the negative y-axis by the support line of slope t of the convex curve $y = g_0(x)$. The central index is the maximal x-coordinate of points $P \in \mathscr{P}$ lying on this support line.

It is clear from this interpretation that

$$\log \mu(r,f) = \log \mu(r,F_0) = \sup_{x>0}(xt - g_0(x)) \tag{3.3}$$

and that $f(z)$ and $F_0(z)$ have the same central index.

For ease of manipulation we replace the function $g_0(x)$ by a function $g(x)$ such that

$$g(x) \in C^2(0,\infty) \tag{3.4}$$

$$g''(x) > 0 \qquad (x > 0) \tag{3.5}$$

$$(3.6)\qquad g(x) - g_0(x) \to 0 \qquad (x \to \infty)$$

$$(3.7)\qquad g(x) - g_0(x) < 1 \qquad (x > 0)$$

It is not hard to see that these conditions can be satisfied. We put

$$F(z) = \sum_0^n e^{-g(n)} z^n$$

The geometrical interpretation makes it evident that $\nu_0(r,h)$ tends to ∞ for every transcendental entire function $h(z)$. It is therefore clear, by (3.6), that

$$\mu(r,F) \sim \mu(r,F_0) \qquad (r \to \infty)$$

Also, given $\varepsilon > 0$, we can find X_0 so that

$$\frac{1}{1+\varepsilon} e^{-g(x)} < e^{-g_0(x)} < (1+\varepsilon)e^{-g(x)} \qquad (x > X_0)$$

and therefore

$$(3.8)\qquad M(r,F_0) < (1+\varepsilon)M(r,F) + O(r^{X_0})$$

Since F is transcendental,

$$r^{X_0} = o(M(r,F)) \qquad (r \to \infty)$$

so that, by (3.8),

$$M(r,F_0) < (1+\varepsilon)M(r,F) \qquad (r > r_0(\varepsilon))$$

4. THE FUNCTION $\nu(t)$

Let

$$v(x) = -g(x) + tx$$

Then

$$v'(x) = -g'(x) + t$$

$$v''(x) = -g''(x) < 0$$

Therefore $v(x)$ increases to a maximum at $x = \nu(t)$ and decreases monotonely as x increases from 0 to ∞. Here $\nu(t)$ is the monotonely increasing function defined by

$$(4.1)\qquad g'(\nu(t)) = t$$

We shall show that the maximum of $v(x)$ at $\nu(t)$ is fairly pronounced, if t is outside an exceptional set of finite measure.

Lemma 4.1. $a > 0$, $\delta > 0$. Let E be the union of all intervals $[t,t']$ $(0 \leq t < t')$ such that

$$t' - t \leq [\nu(t') - \nu(t)]K(\nu(t')) \tag{4.2}$$

where

$$K(x) = \frac{a}{x \log(x+3)\{\log\log(x+3)\}^{1+\delta}}$$

Then E has finite measure.

Proof. Let $\{[t_k,t_k']\}$ be a finite or denumerable collection of non-overlapping intervals in which (4.2) holds. Since $\nu(t)$ is an increasing function of t, the intervals $[\nu(t_k),\nu(t_k')]$ are also non-overlapping and therefore

$$\sum_k (t_k'-t_k) < \Sigma[\nu(t_k') - \nu(t_k)]K(\nu(t_k'))$$

$$< \int_0^\infty K(x)dx = A(a,\delta) < \infty$$

The Lemma now follows immediately from

Lemma 4.2. Let A be the union of intervals I of the real axis. The I's may be open, half-open or closed, but none should reduce to a point. If the measure of any finite or denumerable set of non-overlapping intervals I is $\leq K$, then A can be enclosed in an open set of measure less than $(2+\epsilon)K$ for any $\epsilon > 0$.

Proof. 1) Assume all I open. If the measure of A is greater than $2K$, then it is possible to find a compact $F \subset A$ of measure $> 2K$. By the Heine-Borel Theorem we can find $I_1, I_2 \ldots I_m$ covering F. We may assume that no I_j is redundant, i.e. every I_j contains points not contained in any other I_k. We may also assume $I_j = (r_j,R_j)$ with $r_1 < r_2 < r_3 \ldots$ ($r_j = r_{j+1}$ would make the shorter of I_j, I_{j+1} redundant).

Since there is a point $\rho \in I_2$ not covered by any other I_j, we have

$$R_1 \leq \rho \leq r_3$$

so that I_1 and I_3 do not overlap. Similarly no two I's with odd index overlap and no two I's with even index overlap. Therefore the measure of F, $m(F)$, satisfies

$$2K < m(F) \leq \sum_{j \text{ odd}} m(I_j) + \sum_{j \text{ even}} m(I_j) \leq 2K$$

by our hypothesis. Contradiction.

2) If some of the I are not open, replace them by open intervals $I' \supset I$, $m(I') < (1+\epsilon)I$. The first part of the proof (with K replaced by $(1+\epsilon)K$) applies to the resulting set A'.

Remarks. 1) The idea of the Lemma is due to W. K. Hayman. 2) The factor $(2+\epsilon)$ can not be replaced by any number less than 2.

We can now prove

Proposition 4.1. Let

$$u(k,r) = e^{-g(k)}r^k = e^{-g(k)+tk}$$

where $k > 0$ need not be an integer. Define $\nu(t)$ by (4.1) and let

(4.3) $$h(x) = b/x \log(x+3)\{\log\log(x+3)\}^{1+\delta} \quad (b, \delta > 0)$$

Then

$$\| \quad u(k,r)/u[\nu(t),r] \leq \exp\{-\tfrac{1}{2}(k-\nu(t))^2 h(\nu(t))\}$$

$$(0 \leq k \leq 2\nu(t))$$

(Reminder: $\|$ means that the inequality is true for all r outside a set of finite logarithmic measure.)

Proof. Let E be the set described in Lemma 4.1. Since E is of finite measure, the set of r such that $\log r = t \in E$ is of finite logarithmic measure.
If $t \notin E$, then

(4.3) $$s-t > [\nu(s) - \nu(t)]K[\nu(s)] \quad (s > t)$$

and

(4.4) $$t-s < [\nu(t) - \nu(s)]K[\nu(t)] \quad (s < t)$$

By (4.3) $(\nu(s) = w)$ and (4.1)

(4.5) $$g'(w) - t > [w - \nu(t)]K(w) \quad (w > \nu(t))$$

In $\nu(t) \leq w \leq 2\nu(t)$

(4.6) $$K(w) > c\, K(\nu(t))$$

for a suitable $c = c(\delta)$, $0 < c < 1$. Therefore, by (4.5)

(4.7) $$g'(w) - t > [w-\nu(t)]c\, K(\nu(t)) \quad (\nu(t) < w \leq 2\nu(t))$$

Integration of (4.7) between the limits $\nu(t)$ and k $(\nu(t) < k \leq 2\nu(t))$ yields

$$- \log u(k,r) + \log u(\nu(t),r) > \frac{1}{2} c(k-\nu(t))^2 K(\nu(t)) \quad (\nu(t) < k \leq 2\nu(t)) \tag{4.8}$$

Similarly, by (4.4) and (4.1)

$$t - g'(w) > [\nu(t)-w]K(\nu(t)) \qquad (0 \leq w < \nu(t))$$

Integration from k to $\nu(t)$ $(0 \leq k < \nu(t))$ yields

$$\log u(\nu(t),r) - \log u(k,r) > + \frac{1}{2}[\nu(t)-k]^2 K(\nu(t)) \quad (0 \leq k < \nu(t)) \tag{4.9}$$

The proposition follows from (4.7) and (4.9), if the constant a in $K(x)$ is chosen so that

$$a > c \cdot a = b \tag{4.10}$$

Df. A value of r not belonging to the exceptional set of Proposition 4.1 will be called normal.

The normality of an r will depend on the choice of b and δ in $h(x)$, but this is of no importance to us.

5. ESTIMATES IN TERMS OF $\mu(r, f)$.

By (3.1) and (3.7) the general term of the power series for $f(z)$ is estimated by

$$|c_n| r^n \leq e^{-g_0(n)+tn} \leq e^{-g(n)+tn+1} \tag{5.1}$$

and for large n this can be improved to

$$|c_n| r^n \leq e^{-g_0(n)+tn} \leq e^{-g(n)+tn+o(1)} \tag{5.2}$$

because of (3.6).

If $r = e^t$ is normal, proposition 4.1 shows that

$$u(n,r) = e^{-g(n)+tn} < u(\nu(t),r)\exp\{-\frac{1}{2}(n-\nu(t)^2 h(\nu(t))\} \quad (0 \leq n \leq 2\nu(t)) \tag{5.3}$$

And, by (3.3) and the definition of $\nu(t)$,

$$\log u(\nu(t)) = \sup_{x>0}(-g(x)+tx) = \sup_{x>0}(-g_0(x)+tx+o(1)) \tag{5.4}$$

$$= \log \mu(r,f) + o(1)$$

[illegible] through normal values.

Next we obtain an estimate for $u(n,r)$ in the range $n > 2\nu(t)$. In this range $-g'(x)+t$ is a decreasing function of
Therefore, using (4.5), (4.6) and (4.10)

$$-g'(x) + t < -g'(2\nu(t)) + t \qquad (x > 2\nu(1))$$

$$< -\nu(t)K(2\nu(t))$$

$$< -\nu(t)cK(\nu(t)) = -\nu(t)h(\nu(t))$$

Integration from $2\nu(t)$ to n yields

$$-g(n) + tn < -g(2\nu(t)) + 2t\nu(t) - \alpha(n - 2\nu(t))$$

where $\alpha = \nu(t)h(\nu(t))$.

Hence

$$u(n,r) < u(2\nu(t),r)e^{-\alpha(n-2\nu(t))} \qquad (n > 2\nu(t))$$

$$< u(\nu(t),r)e^{-\frac{1}{2}(\nu(t))^2 h(\nu(t)) - \alpha(n-2\nu(t))} \tag{5.5}$$

where we have used (5.3).

We can now prove the main results of Wiman-Valiron Theory.

To save writing we abbreviate

$$\nu = \nu(t) \qquad\qquad h = h[\nu(t)]$$

It is important to bear in mind that h is defined as a functi
of ν and that $h \to 0$ as $\nu \to \infty$.

Theorem 5.1. Given $\varepsilon > 0$,

!! $$|\nu_0(r,f) - \nu| < \varepsilon h^{-\frac{1}{2}}$$

Proof. Since $f(z)$ and

$$F_0(z) = \Sigma\, e^{-g_0(n)} z^n$$

have the same maximum term and central index, it is enough to show that

$$e^{-g_0(n)+tn} < \mu(r,F_0) = \mu(r,f) \tag{5.6}$$

$$(|n-\nu| > \varepsilon\, h^{-\frac{1}{2}})$$

This follows from (5.1), (5.2), (5.3), (5.4) and (5.5).

Notation: $(r > r_0)$ reads: 'For all sufficiently large r'.

Theorem 5.2. Let

$$|z| = re^{s} = e^{t+s}, \qquad |s| < h^{\frac{1}{2}+\varepsilon}$$

$$h = h(\nu(t)) = b(\nu(t)\log(\nu(t)+3))^{-1}(\log\log(\nu+3))^{-1-\delta};$$

let k be an integer $0 \leq k \leq \nu = \nu(t)$

Then for normal r $(r > r_0)$

$$D = |f^{(\ell)}(z) - \sum_{|n-\nu|<k} \ell(n-1)\ldots(n-\ell+1)c_n z^{n-}|$$

$$< A2^{\ell}\nu\, \mu(r,f)e^{s\nu} e^{-\frac{1}{2}hk^2+|s|k} \cdot h^{-\frac{1}{2}}|z|^{-}$$

$$(\ell = 0,1,2\ldots)$$

where A is an absolute constant.

Proof. $|z|^{\ell} D < \sum_{|n-\nu|\geq k} n^{\ell}|c_n|e^{(t+s)n}$

$$< \sum_{|n-\nu|\geq k} n^{\ell} e^{-g(n)+(t+s)n+1} = \Sigma n^{\ell} u(n,r)e^{sn} \tag{5.8}$$

in the notation of (5.3).

By (5.5) and (5.4) for $r > r_0$

$$\ell \sum_{n>2\nu} n^{\ell} u(n,r) e^{sn} < 3\mu(r,f) e^{-\frac{1}{2}h\nu^2 + 2\alpha\nu} \sum_{n=2\nu}^{\infty} n^{\ell} e^{-(\alpha-s)n} \tag{5.9}$$

$$(\alpha = \nu h)$$

For $x > 2\nu$, and r sufficiently large

$$\frac{d}{dx}(x^{\ell} e^{-(\alpha-s)x}) = x^{\ell} e^{-(\alpha-s)x}\{\frac{\ell}{x} - (\alpha-s)\}$$

$$< 0$$

since in this range

$$\frac{\ell}{x} - (\alpha-s) < \frac{\ell}{2\nu} + h^{\frac{1}{2}+\varepsilon} - \nu h < 0 \qquad (r > r_0)$$

We can therefore estimate the sum in (5.9) by the remark that for any positive, monotonely decreasing function $\varphi(x)$

$$\sum_{n \geq a} \varphi(n) < \varphi(a) + \int_a^{\infty} \varphi(x)\,dx$$

('Integral test for series').

This yields

$$\sum_{n>2\nu} n^{\ell} e^{-(\alpha-s)n} < (2\nu)^{\ell} e^{-(\alpha-s)2\nu} + \int_{2\nu}^{\infty} x^{\ell} e^{-(\alpha-s)x} dx \; ; \tag{5.10}$$

$$I = \int_{2\nu}^{\infty} x^{\ell} e^{-(\alpha-s)x} dx < \frac{(2\nu)^{\ell} e^{-2(\alpha-s)\nu}}{\alpha - s} + \frac{\ell}{\alpha-s} \int_{2\nu}^{\infty} x^{\ell-1} e^{-(\alpha-s)x} dx$$

$$< \frac{(2\nu)^{\ell} e^{-2(\alpha-s)\nu}}{\alpha - s} + \frac{\ell}{2(\alpha-s)\nu} I$$

$$< \frac{(2\nu)^{\ell} e^{-2(\alpha-s)\nu}}{\alpha - s} + \frac{1}{2} I \qquad (r > r_0) \tag{5.11}$$

By (5.9), (5.10) and (5.11)

$$
\begin{aligned}
e \sum_{n>2\nu} n^{\ell} u(n,r) e^{sn} &< A_1 (2\nu)^{\ell} \mu(r,f) e^{s\nu} e^{-\frac{1}{2}h\nu^2+s\nu} \\
&< A_1 (2\nu)^{\ell} \mu(r,f) e^{s\nu} e^{-\frac{1}{2}hk^2+sk} \qquad (5.12) \\
&\qquad (0 \leq k \leq \nu)
\end{aligned}
$$

since

$$0 > -\frac{1}{2}h\nu^2 + s\nu = \min_{0 \leq x \leq \nu}(-\frac{1}{2}hx^2 + sx)$$

By (5.3) and (5.4),

$$(5.13) \qquad e \sum_{k \leq |n-\nu| \leq \nu} n^{\ell} u(n,r) e^{sn} < 6\mu(r,f) e^{s\nu} (2\nu)^{\ell} \sum_{j \geq k} e^{-\frac{1}{2}hj^2+|s|j} .$$

$$-\frac{1}{2}hx^2 + |s|x = -\frac{h}{2}(x - \frac{|s|}{h})^2 + \frac{|s|^2}{2h}$$

has a maximum at $|s|/h$ whose value is

$$|s|^2/2h = O(1) \qquad (r \to \infty)$$

We consider first the case that

$$k \leq h^{-\frac{1}{2}}$$

By splitting the sum on the right-hand side of (5.13) into two sums over monotonely decreasing terms, if necessary, we have

$$
\begin{aligned}
\sum_{j \geq k} e^{-\frac{1}{2}hj^2+|s|j} &< A_3\{\max_x e^{-\frac{1}{2}hx^2+|s|x} + \int_{-\infty}^{\infty} e^{-\frac{1}{2}hx^2+|s|x} dx \qquad (5.14) \\
&< A_4 h^{-\frac{1}{2}} < A_5 h^{-\frac{1}{2}} e^{-\frac{1}{2}hk^2+|s|k} \qquad (k \leq h^{-\frac{1}{2}})
\end{aligned}
$$

If $k > h^{-\frac{1}{2}}$, then the terms in the sum on the right-hand side of (5.13) are decreasing and by an integration by parts

$$\sum_{j \geq k} e^{-\frac{1}{2}hj^2+|s|j} < e^{-\frac{1}{2}hk^2+|s|k} + \int_k^\infty e^{-\frac{1}{2}hx^2+|s|x} dx$$

$$(5.15) \qquad < e^{-\frac{1}{2}hk^2+|s|k} + \frac{e^{-\frac{1}{2}hk^2+|s|k}}{hk-|s|}$$

$$< A_6 h^{-\frac{1}{2}} e^{-\frac{1}{2}hk^2+|s|k}$$

The Theorem follows from (5.8), (5.12), (5.13), (5.14) and (5.15).

Theorem 5.3

$$(5.16) \qquad \mu(r,f) \leq M(r,f)$$

$$\| \qquad M(r,f) < A\, h^{-\frac{1}{2}} \mu(r,f)$$

Proof. (5.13) is an immediate consequence of Cauchy's formula for the coefficients of a power series:

$$|c_n| r^n = \left|\frac{1}{2\pi} \int_{-\pi}^{\pi} f(re^{i\theta}) e^{\ in\theta} d\theta\right| \leq M(r,f)$$

The second assertion follows from Theorem 5.2 with $s = 0$, $k = 0$ applied to $\Sigma |c_n| z^n$.

Remark. By slightly more careful estimates the constant A in Th. 5.3 can be replaced by $\sqrt{2\pi}(1+O(1))$, as $r \to \infty$ through normal values.

6. BEHAVIOR OF AN ENTIRE FUNCTION NEAR POINTS WHERE ITS MAXIMUM MODULUS IS ATTAINED (following Hayman)

Theorem 6.1. If $f(z)$ is an entire function and if

$$|f(z_o)| \geq \frac{1}{2} M(|z_o|, f)$$

then in

$$|Z-z_o| < r\nu^{-\frac{1}{2}-\epsilon} \qquad (z > 0)$$

$$f(Z) = \left(\frac{Z}{z_o}\right)^{\nu} f(z_o)(1+o(1))$$

as $|z_o| \to \infty$ through normal values.

For the proof we require

Lemma 6.1. If $P(z)$ is a polynomial of degree m,

$$|P(z)| \leq M \qquad (|z| \leq r)$$

then

$$|P'(z)| \leq \frac{emMR^{m-1}}{r^m} \qquad (|z| \leq R; \quad R \geq r)$$

Proof. By the maximum principle applied to $z^{-m}P(z)$ in $r < |z| \leq \infty$

$$|z^{-m}P(z)| \leq Mr^{-m} \qquad (|z| \geq r)$$

Therefore, if $|z| = R \geq r$

$$|P'(z)| \leq \frac{1}{h} \max_{|\zeta-z|=h} |P(\zeta)| \leq \frac{M(R+h)^m}{hr^m}$$

$h = R/m$:

$$|P'(z)| \leq \frac{mR^{m-1}M}{r^m} \left(1 + \frac{1}{m}\right)^m < \frac{emMR^{m-1}}{r^m}$$

[The factor e can be eliminated by using results of S. Bernstein on the derivative of a polynomial.]

Lemma 6.2. $P(z)$ as in Lemma 6.1.
If $|z_o| = r$, $|z-z_o| < r/8m$, then

$$|P(z) - P(z_o)| < \frac{4mM}{r} |z-z_o|$$

Proof. By Lemma 6.1 in $|\zeta-z_o| < r/8m$

$$|P'(\zeta)| < \frac{emM\left(1 + \frac{1}{8m}\right)^m}{r} < \frac{4mM}{r}$$

and Lemma 6.2 follows by an obvious integration.

<u>Proof of Theorem 6.1</u>. Let

$$r = |z_o|, \quad |Z| = re^s$$

by the hypotheses of the Theorem

$$1-\nu^{-\frac{1}{2}-\epsilon} < e^s < 1+\nu^{-\frac{1}{2}-\epsilon}$$

so that

$$|s| < 2\nu^{-\frac{1}{2}-\epsilon} < h^{\frac{1}{2}+\frac{1}{2}\epsilon} \qquad (r > r_o)$$

Theorem 5.2 with $\ell = 0$

$$k = [h^{-\frac{1}{2}} \log^{\frac{2}{3}} \frac{1}{h}]$$

gives for $|Z| = re^s$

$$(6.1) \quad \Big| |f(Z) - \sum_{|n-\nu|<k} c_n Z^n| = o(\mu(r,f)e^{s\nu})$$

If q is the least integer satisfying

$$q-\nu > -k$$

we can write

$$\sum_{|n-\nu|<k} c_n Z^n = Z^q P(Z)$$

where $P(Z)$ is a polynomial of degree m,

$$m \leq 2k$$

Choosing $s = 0$ in (6.1) we see that

$$|P(z)| < r^{-q}M(r,f) + o(\mu(r,f)) \qquad (|z| \leq r)$$

$$< r^{-q}M(r,f)(1+o(1))$$

as $r \to \infty$ through normal values.
Therefore by Lemma 6.2

$$(6.2) \qquad |P(Z)-P(z_o)| < A_1 kM(r,f)r^{-q}.\nu^{-\frac{1}{2}-\epsilon} = o(M(r,f)r^{-q})$$

as $r \to \infty$ through normal values. By (6.1) and (6.2)

$$
\begin{aligned}
|Z^{-q}f(Z) - z_o^{-q}f(z_o)| &\leq |Z^{-q}f(Z)-P(Z)| \\
&\quad +|P(Z) - P(z_o)| \\
&\quad +|z_o^{-q}f(z_o)-P(z_o)| \\
&= o(M(r,f)|\tfrac{Z}{z_o}|^{\nu}\cdot|Z|^{-q}) \\
&\quad + o(M(r,f)r^{-q}) \\
&\quad + o(M(r,f)|z_o|^{-q})
\end{aligned}
$$

After multiplication by Z^q we have

$$f(Z) = (Z/z_o)^q\, f(z_o) + o(M(r,f)(|Z/z_o|^q + |Z/z_o|^{\nu}))$$

The Theorem follows if we observe that

$$|q-\nu+k| < 1$$

and so

$$
\begin{aligned}
(Z/z_o)^q &= (Z/z_o)^{\nu}\cdot(Z/z_o)^{q-\nu} \\
&= (Z/z_o)^{\nu}(1+o(1))
\end{aligned}
$$

since
$$
\begin{aligned}
(q-\nu)\log(Z/z_o) &= O(h^{-\frac{1}{2}}\log^{\frac{2}{3}}\tfrac{1}{h}\cdot\nu^{-\frac{1}{2}-\epsilon}) \\
&= o(1)
\end{aligned}
$$

<u>Theorem 6.2</u>. If

$$|f(z_o)| > \tfrac{1}{2}M(r,f)$$

and

$$|Z-z_o| < r\nu^{-\frac{1}{2}-\epsilon}$$

then for every non-negative integer ℓ

$$f^{(\ell)}(Z) = \nu^{\ell}(Z/z_o)^{\nu}\, f(z_o)z_o^{-\ell}(1+o(1))$$

as $|z_o| \to \infty$ through normal values.

<u>Proof</u>. We use induction on ℓ. The Theorem is true for $\ell = 0$, by Theorem 6.1. With q and $P(z)$ as in the proof of Theorem 6.1 we have

$$(\tfrac{d}{dz})^{\ell}\sum_{|n-\nu|<k} c_n z^n = (\tfrac{d}{dz})^{\ell}\, z^q P(z)$$

Repeated application of Lemma 6.1 yields [$|P(z)| < 2M(r,f)r^{-q}$!]

$$|P^{(\ell)}(z)| < A\frac{e^{\ell}m^{\ell}M(r,f)}{r^{\ell+q}} \qquad (|z| \leq r)$$

Therefore, by Leibnitz' theorem on the differentiation of produc

$$|(\frac{d}{dz})^{\ell} z^{q}P(z)| < (2e)^{\ell}q^{\ell}M(r,f)r^{-\ell} \qquad (|z| \leq r)$$

$$< (2e)^{\ell}\nu^{\ell}M(r,f)r^{-\ell}$$

Therefore

$$|R(z)| = |z^{-q+\ell}(\frac{d}{dz})^{\ell} z^{q}P(z)| < (2e)^{\ell}\nu^{\ell}M(r,f)r^{-q}(|z| \leq r)$$

Since $R(z)$ is a polynomial of degree $m < 2k$ we have by Lemma 6.1

$$(6.3) \qquad |R'(z)| < 2^{\ell+1} e^{\ell}\nu^{\ell}kM(r,f)r^{-q-m}(|z|^{m-1}+r^{m-1})$$

We choose now $k = [h^{-\frac{1}{2}} \log^{\frac{2}{3}} \frac{1}{h}]$.

With $|z_o| = r$,

$$|Z-z_o| < r\nu^{-\frac{1}{2}-\epsilon}$$

we obtain from Theorem 5.2 as $r \to \infty$ through normal values

$$(6.4)\begin{cases} |(\frac{Z}{z_o})^{\nu} \nu^{\ell+1}f(z_o)Z^{-\ell-1}-f^{(\ell+1)}(Z)| \\ < |(\frac{Z}{z_o})^{\nu} \nu^{\ell+1} f(z_o)Z^{-\ell-1} - \frac{d}{dZ} Z^{q-\ell} R(Z)| \\ + |f^{(\ell+1)}(Z) - \frac{d}{dZ}(Z^{q-\ell}R(Z)| \end{cases}$$

$$\begin{cases} < |(\frac{Z}{z_o})^{\nu} \nu^{\ell+1}f(z_o)Z^{-\ell-1} - (q-\ell)Z^{q-\ell}R(Z)\cdot Z^{-1}| \\ + |Z^{q-\ell}R'(Z)| \\ + o(\nu^{\ell+1}M(r,f)|Z/z_o|^{\nu}|Z|^{-\ell-1}) \end{cases}$$

Again by Theorem 5.2 and the hypothesis of induction

$$|(\frac{Z}{z_o})^{\nu} \nu^{\ell} f(z_o) Z^{-\ell} - Z^{q-\ell} R(Z)|$$

$$< |(\frac{Z}{z_o})^{\nu} \nu^{\ell} f(z_o) Z^{-\ell} - f^{(\ell)}(Z)|$$

$$+ |f^{(\ell)}(Z) - Z^{q-\ell} R(Z)| < o(\nu^{\ell} M(r,f) |\frac{Z}{z_o}|^{\nu} |Z|^{-\ell}$$

Using this together with

$$q-\ell = \nu(1+o(1))$$

we have

$$|(\frac{Z}{z_o})^{\nu} \nu^{\ell+1} f(z_o) Z^{-\ell-1} - (q-\ell) Z^{q-\ell-1} R(Z)|$$

(6.5)

$$= o(\nu^{\ell+1} M(r,f) |Z/z_o|^{\nu} |Z|^{-\ell-1})$$

The Theorem follows from (6.4), (6.3) and (6.5).

7. PICARD'S THEOREM. Every transcendental entire function assumes all complex values infinitely often, except perhaps for one exceptional value.

Proof. If the theorem is false, we can find an entire function $h(z)$ which assumes the values 0 and 1 only a finite number of times. Then

$$h(z) = p(z)e^{f(z)} = q(z)e^{g(z)} + 1$$

where f and g are entire functions, p and q are polynomials.

Case I. f and g are polynomials. By considering the maximum modulus of $h(z)$ on large circles we find that

$$f(z) = \alpha_m z^m + \text{lower powers}$$
$$g(z) = \alpha_m z^m + \text{lower powers} \qquad \alpha_m \neq 0$$

By choosing z such that $\alpha_m z^m$ is real and negative we arrive at the contradiction

$$o(1) = 1 + o(1)$$

as $|z| \to \infty$.

Case II. At least one of f and g is transcendental. Suppose, e.g., that $f(z)$ is transcendental. Then for normal r and $|z_0| = r$, $|f(z_0)| = M(r,f)$ we have, by Theorem 6.1,

$$f(z) = f(z_0)(z/z_0)^{\nu}(1+o(1)) \qquad |z-z_0| < r\nu^{-\frac{1}{2}-\epsilon}$$

By choosing $z = z_0 e^{i\theta}$, θ small, we see that near z_0 there are points on $|z| = r$ such that

$$\text{(7.1)} \qquad \operatorname{Re} f(z) > M(r,f)(1+o(1))$$

Since $M(r,f)$ grows more rapidly than any power of r as $r \to \infty$, it follows that $g(z)$ is also transcendental and that

$$M(r,g) > M(r,f)(1+o(1))$$

Interchanging the roles of f and g we find

$$M(r,g) = M(r,f)(1+o(1))$$

and at a point where (7.1) holds we must also have

$$\operatorname{Re} g(z) > M(r,g)(1+o(1))$$

Therefore Theorem 6.1 can be applied to f and g with the same z_0.

$$\text{(7.2)} \qquad \begin{aligned} f(z) &= f(z_0)(z/z_0)^{\nu}(1+o(1)) \qquad |z-z_0| < r\nu^{-\frac{1}{2}-\epsilon} \\ g(z) &= g(z_0)(z/z_0)^{\nu'}(1+o(1)) \qquad |z-z_0| < r\nu'^{-\frac{1}{2}-\epsilon} \end{aligned}$$

as $r \to \infty$ through values which are normal for both $f(z)$ and $g(z)$.

By considering $|h(z)|$ we see that $\operatorname{Re} f(z)$ and $\operatorname{Re} g(z)$ must assume large positive values at the same time. But by (7.2) this implies that $\operatorname{Re} f(z)$ and $\operatorname{Re} g(z)$ also take negative, numerically large values at the same time in the neighborhood of z_0. This leads to the contradiction

$$o(1) = 1 + o(1)$$

8. AN APPLICATION TO DIFFERENTIAL EQUATIONS

Let $p_1(z)$, $p_2(z)$ ... be holomorphic near z_0 and let at least one of the p's have a pole at z_0. We are interested in the behavior of the solutions $w(z)$ of the linear differential equation

$$(7.3) \qquad w^{(n)} + p_1 w^{(n-1)} + \dots + p_{n-1} w^{i} + p_n w = 0$$

near z_0.

Without loss of generality we may take $z_0 = \infty$ (otherwise introduce $z' = \frac{1}{z-z_0}$). Then all $p_j(z)$ will be holomorphic in $R_0 < |z| < \infty$ and

$$p_j(z) = \alpha_j z^{m_j} + \text{lower powers} \quad (|z| > R_0)$$

(If a $p_j(z)$ is $\equiv 0$, let $m_j = -\infty$)

By the general theory of linear differential equations there will be a complete set of linearly independent solutions of the form

$$w(t) = z^{\rho} \sum_{-\infty}^{\infty} c_k z^k$$

except for some exceptional cases when also powers of $\log z$ appear as factors before the Σ-sign. We omit discussion of the exceptional case.

We wish to find the possible orders of $w(z)$, i.e. the possible values of

$$\lambda = \overline{\lim_{r\to\infty}} \frac{\log\log M(r,w)}{\log r} \qquad (r > R_0)$$

By putting $w = z^{\rho} v$ we obtain a differential equation of the same type as (7.3) for $v(z)$. We may therefore assume right away that the solution of (7.3) is of the form

$$(7.4) \qquad w(z) = \sum_{-\infty}^{\infty} c_k z^k = \sum_{-\infty}^{-1} + f(z)$$

The entire function $f(z) = \sum_{n=0}^{\infty} c_n z^n$ satisfies

$$\lambda = \overline{\lim} \frac{\log\log M(r,w)}{\log r} = \overline{\lim} \frac{\log\log M(r,f)}{\log r}$$

and substitution of (7.4) into (7.3) gives

$$f^{(n)}(z) + p_1(z) f^{(n-1)}(z) \dots + p_n(z) f(z) = O(|z|^K)$$

as $z \to \infty$. By applying Theorem 6.2 with $Z = z_0$ ($|f(z_0)| > \frac{1}{2} M(r,f)$;

$|z_0| = r$ normal for f) we obtain

$$(7.5) \qquad 1 + \sum_{j=1}^{n} \alpha_j \nu^{-n+j} z_0^{j+m_j} (1 + o(1)) = O\left(\frac{|z_0|^K}{\nu^n M(r,f)}\right) = o(1)$$

Therefore $\nu < r^A$ $(r > r_0)$ where A is an absolute constant, since otherwise the 1 on the left-hand side of (7.5) cannot be balanced by the other terms.

Let $\nu(r) = r^{\gamma(r)}$ and let Γ be a limit point of the $\gamma(r)$ more precisely,

$$\Gamma = \lim \gamma(r_\nu)$$

where r_ν is a sequence tending to infinity whose members are normal for $f(z)$. Consideration of the largest terms in (7.5) shows that there must be at least two different values of j such that

$$m_j + j - j\Gamma = \sup_{0 \leq \ell \leq n} (m_\ell + \ell - \ell\Gamma), \quad \alpha_j \neq 0$$

$$(m_0 = 0)$$

This can be stated geometrically:

The only possible values of Γ are the slopes $\Gamma \geq 0$ of the least concave majorant of the set of points $(j, m_j + j)$ $(j = 0, 1, \ldots, n)$.

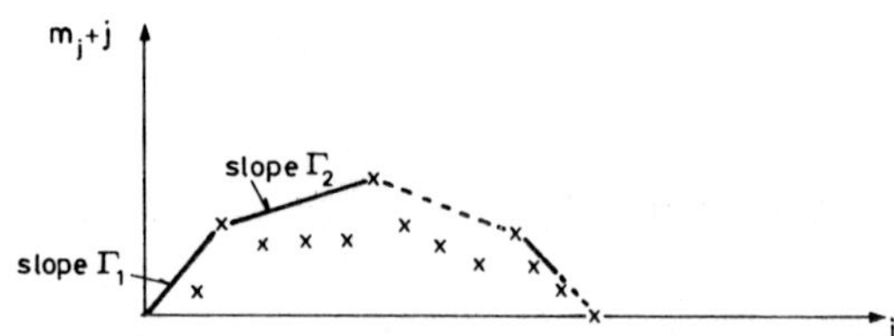

We show finally that these values of Γ are the only possible values that λ can assume.

We have, by (5.4),

$$\log \mu(r) + o(1) = -g(\nu(t)) + t\nu \qquad (t = \log r)$$

By (4.1) and the implicit function theorem, $\nu'(t)$ exists and

$$\frac{d}{dt}(-g(\nu(t)) + t\nu) = \nu$$

Therefore, by integration,

$$\log \mu(r) = \text{const.} + \int_0^t \nu(u)\,du + o(1)$$

Hence

$$\log \mu(r) < t\,\nu(t) + A$$

and

$$\log \mu(r) > \nu(t-1) + B$$

$$o(1) + \frac{\log\log \mu(r)}{\log r} < \frac{\log \nu(t)}{\log r} = \gamma(t) < \frac{\log\log \mu(re)}{\log r} + o(1)$$

Combined with Theorem 5.3 this shows that

$$\limsup_{r\to\infty} \frac{\log\log M(r)}{\log r} = \limsup_{r\to\infty} \frac{\log\log M(er)}{\log r} \geq \Gamma_0$$

where $\Gamma_0 = \limsup \gamma(r)$ as $r \to \infty$ through normal values. It also shows that $L = \limsup \frac{\log\log M(r)}{\log r}$ as $r \to \infty$ through normal values is $\leq \Gamma_0$. But for any $r > r_0$ the interval $[r, er]$ contains a normal value r' and so

$$\frac{\log\log M(r)}{\log r} < \frac{\log\log M(r')}{\log r' - 1} \qquad (r' \text{ normal})$$

Hence

$$\Gamma_0 \leq \limsup \frac{\log\log M(r')}{\log r'} < L \leq \Gamma_0$$

This proves our assertion.

CONFORMAL MAPPING
Summary*

C. POMMERENKE
Technische Universität Berlin,
Federal Republic of Germany

Abstract

CONFORMAL MAPPING: SUMMARY

This twelve-lecture course dealt with the theory of conformal mapping of simply connected plane domains with special emphasis on the boundary behaviour.

The course began with a quick review of the basic material, in particular the Riemann mapping theorem. Applications were given to hydrodynamics, electrodynamics and partial differential equations. The Schwarz-Christoffel formula was proved and was applied to deduce the mapping properties of elliptic integrals.

The second part dealt with the extension of the mapping function to the unit periphery. Necessary and sufficient geometric conditions for continuous and homeomorphic extension were given in terms of local connectivity and cut points.

The third and longest part dealt with the boundary behaviour of the derivative of the mapping function. Again the principal aim was to give, as far as possible, necessary and sufficient geometric conditions for various analytic properties. Some of the subjects considered were smooth boundary curves, the angular derivative, sets of finite linear measure, Plessner's theorem and the McMillan twist point theorem.

The course closed with a survey of the results about the family S of normalized univalent functions (without proofs). Linear-invariant families, the Loewner differential equation and quadratic differentials were briefly discussed.

BIBLIOGRAPHY

SOME BOOKS ON CONFORMAL MAPPING AND RELATED SUBJECTS

AHLFORS, L.V., Complex Analysis, 2nd Edn, McGraw-Hill, New York (1966).

AHLFORS, L.V., Conformal Invariants: Topics in Geometric Function Theory, McGraw-Hill, New York (1973).

BETZ, A., Konforme Abbildung, Springer-Verlag, Berlin (1964).

CARATHÉODORY, C., Conformal Representation, Cambridge Univ. Press (1963).

COLLINGWOOD, E.F., LOHWATER, A.J., The Theory of Cluster Sets, Cambridge Univ. Press (1966).

DUREN, P.L., Theory of H^p Spaces, Academic Press, New York (1970).

GOLUSIN, G.M., Geometrische Funktionentheorie, Deutscher Verlag d. Wiss., Berlin (1957).

HAYMAN, W.H., Multivalent Functions, Cambridge Univ. Press (1958).

JENKINS, J.A., Univalent Functions and Conformal Mappings, 2nd Edn, Springer-Verlag, Berlin (1965).

KOBER, H., A Dictionary of Conformal Representation, Dover Publications, New York (1957).

LAVRENTIEV, M.A., SHABAT, B.V., Methoden der komplexen Funktionentheorie, Deutscher Verlag d. Wiss., Berlin (1967).

* The lecture notes from this 12-lecture course are not published here since much of the material will be found in the literature. The third and longest part was taken from Chapter 10 of the author's book "Univalent Functions", listed in the Bibliography.

LEHTO, O., VIRTANEN, K.I., Quasikonforme Abbildungen, Springer-Verlag, Berlin (1965).

NEHARI, Z., Conformal Mapping, McGraw-Hill, New York (1952).

NEWMAN, M.H.A., Elements of the Topology of Plane Sets of Points, 2nd Edn, Cambridge Univ. Press (1964).

PÓLYA, G., LATTA, G., Complex Variables, Wiley, New York (1974).

POMMERENKE, C., Univalent Functions, Vandenhoeck & Ruprecht, Göttingen (1975).

TSUJI, M., Potential Theory in Modern Function Theory, Maruzen,Tokyo (1959).

IAEA-SMR-18/8

ELEMENTARY GEOMETRY OF COMPLEX MANIFOLDS

S.A. ROBERTSON
Department of Mathematics,
University of Southampton,
Southampton,
United Kingdom

Abstract

ELEMENTARY GEOMETRY OF COMPLEX MANIFOLDS.

This paper outlines the most elementary geometrical aspects of complex manifolds, stopping short of anything requiring serious use of sheaves or of the theory of connections. The main aim is to provide a clear account of basic definitions and examples, and to describe in detail some simple geometrical constructions. Most of the material is available in some form in one or other of the standard texts.

1. PSEUDOGROUPS OF TRANSFORMATIONS

Recall that a paracompact topological space X is said to be a (topological) m-manifold iff every point of X has an open neighbourhood U for which there is a homeomorphism $\xi: U \to U'$ onto an open subset U' of real linear m-space R^m. The homeomorphism ξ is called a **local co-ordinate system** or **chart** on X. Thus if $x \in U$, then $\xi(x) = p = (p_1, ..., p_m) \in R^m$, and the numbers $p_1, ..., p_m$ are called the **ξ-co-ordinates** of x.

Suppose now that $\xi: U \to U'$ and $\eta: V \to V'$ are charts on X, and that $W = U \cap V \neq \emptyset$. Then each point $x \in W$ has ξ-co-ordinates $p_1, ..., p_m$ and η-co-ordinates $q_1, ..., q_m$, which are related by the homeomorphism $\phi_{\xi\eta}: \xi(W) \to \eta(W)$ given by $\phi_{\xi\eta}(\xi(w)) = \eta(w)$, $w \in W$. Thus $q = \eta(w) = \eta(\xi^{-1}(p)) = \phi_{\xi\eta}(p)$.

By considering families of local co-ordinate systems for which the "change of co-ordinates" maps $\phi_{\xi\eta}$ are of some suitably restricted type, we can transfer structures on R^m to corresponding structures on X. We now describe in more detail how this is done.

A **transformation** in R^m is a homeomorphism $\phi: A \to B$ of one open subset A of R^m onto another B. A family $\mathscr{P}$ of such transformations is said to be a **pseudogroup** iff:

P1 $\phi \in \mathscr{P}, \psi \in \mathscr{P} \Rightarrow \psi \circ \phi \in \mathscr{P}$, whenever $\psi \circ \phi$ is defined;

P2 $\phi \in \mathscr{P} \Rightarrow \phi^{-1} \in \mathscr{P}$;

P3 A open in $R^m \Rightarrow 1_A \in \mathscr{P}$;

P4 A open in R^m, $\phi \in \mathscr{P} \Rightarrow \phi|A \in \mathscr{P}$;

P5 $\phi: A \to B$ a transformation in R^m, $A = \cup\{A_j : j \in J\}$, $\phi|A_j \in \mathscr{P}$, all $j \in J$, $\Rightarrow \phi \in \mathscr{P}$.

Examples of pseudogroups

(1) All transformations in R^m.

(2) All C^∞-diffeomorphisms between open subsets of R^m. Such a map is a transformation $\phi: A \to B$ which is C^∞, and whose inverse exists and is C^∞. Thus a C^∞-transformation is a C^∞-diffeomorphism iff, for all $a \in A$, $D\phi(a)$ is an automorphism of R^m, i.e. the Jacobian matrix $J\phi(a) = [\partial q_r/\partial p_s]_{p=a}$ is invertible, where $\phi(p) = q$.

(3) As (2), but with the condition $\det D\phi(a) \neq 0$ replaced by $\det D\phi(a) > 0$. That is, ϕ is **orientation-preserving.**

(4) The set of all invertible affine transformations in R^m. Thus $\phi: A \to B$ is in the set iff there is a map $\alpha: R^m \to R^m$ of the form $\alpha(p) = q$, where, for some $\mu \in R^m$ and some invertible $m \times m$ matrix $\Lambda = [\lambda_{rs}]$,

$$q_r = \sum_{s=1}^{m} \lambda_{rs} p_s + \mu_r$$

such that $\alpha \circ \iota_A = \iota_B \circ f$.

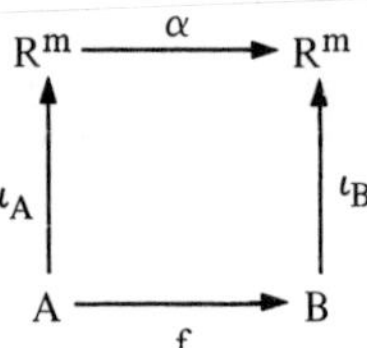

(Here ι_A, ι_B denote inclusions.)

(5) Let $\mathbb{C}$ denote the complex number field, and $\mathbb{C}^m = \mathbb{C} \times \mathbb{C} \times \ldots \times \mathbb{C}$ (m-factors) the complex linear m-space. Then the map $\rho: \mathbb{C}^m \to R^{2m}$, given by

$$\rho(z_1, \ldots, z_m) = (x_1, \ldots, x_m, y_1, \ldots, y_m)$$

where $z_s = x_s + y_s i$, is a homeomorphism (though *not* a linear map).

Thus if $\phi: A \to B$ is a transformation in R^m, then there is a unique homeomorphism $\widetilde{\phi}: \widetilde{A} \to \widetilde{B}$ between open subsets $\widetilde{A}$, $\widetilde{B}$ of $\mathbb{C}^m$, for which $\phi \circ (\rho|\widetilde{A}) = (\rho|\widetilde{B}) \circ \widetilde{\phi}$.

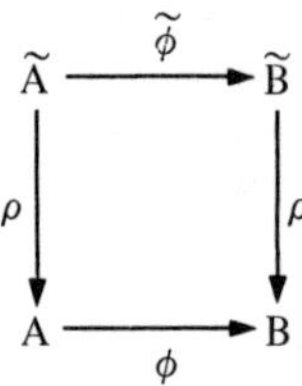

Then we can form a pseudogroup consisting of all ϕ such that both $\widetilde{\phi}$ and $(\widetilde{\phi})^{-1}$ are holomorphic.

The relations between the pseudogroups (2), (3) and (5) may be clarified by the following remarks about the various Jacobian matrices involved. We shall find that (5) $\subset$ (3) $\subset$ (2).

Let $\Delta = J\phi(x, y)$ denote the Jacobian matrix of a transformation $\phi: R^{2m} \to R^{2m}$ at $(x, y) \in R^m \times R^m = R^{2m}$, where $\widetilde{\phi}: \widetilde{A} \to \widetilde{B}$ is holomorphic. Then we can define the Jacobian matrix $\widetilde{\Delta}$ of $\widetilde{\phi}$ to be $J\widetilde{\phi}(z) = [\partial w_r/\partial z_s]$. Thus $\widetilde{\Delta}$ is complex $m \times m$ and Δ is real $2m \times 2m$. We want to show that $\det \Delta = |\det \widetilde{\Delta}|^2$, where $(x, y) = \rho(z)$.

Let $\phi(x, y) = (u, v)$. Since $\widetilde{\phi}$ is holomorphic, the Cauchy–Riemann equations

$$\frac{\partial u_r}{\partial x_s} = \frac{\partial v_r}{\partial y_s}, \quad \frac{\partial u_r}{\partial y_s} = -\frac{\partial v_r}{\partial x_s}$$

hold. Hence

$$\det \Delta = \begin{vmatrix} \frac{\partial u_r}{\partial x_s} & \frac{\partial u_r}{\partial y_s} \\ \frac{\partial v_r}{\partial x_s} & \frac{\partial v_r}{\partial y_s} \end{vmatrix} = \begin{vmatrix} \frac{\partial u_r}{\partial x_s} & -\frac{\partial v_r}{\partial x_s} \\ \frac{\partial v_r}{\partial x_s} & \frac{\partial u_r}{\partial x_s} \end{vmatrix}$$

$$= \begin{vmatrix} \frac{\partial u_r}{\partial x_s} + i\frac{\partial v_r}{\partial x_s} & i\frac{\partial u_r}{\partial x_s} - \frac{\partial v_r}{\partial x_s} \\ \frac{\partial v_r}{\partial x_s} & \frac{\partial u_r}{\partial x_s} \end{vmatrix} \qquad \begin{array}{l}\text{add } i \times \text{row } (r+m) \text{ to row } r \\ r = 1, \ldots, m\end{array}$$

$$= \begin{vmatrix} \frac{\partial u_r}{\partial x_s} + i\frac{\partial v_r}{\partial x_s} & 0 \\ \frac{\partial v_r}{\partial x_s} & \frac{\partial u_r}{\partial x_s} - i\frac{\partial v_r}{\partial x_s} \end{vmatrix} \qquad \begin{array}{l}\text{subtract } i \times \text{col } s \text{ from col}(s+m) \\ s = 1, \ldots, m\end{array}$$

$$= \begin{vmatrix} \frac{\partial w_r}{\partial z_s} & 0 \\ \frac{\partial v_r}{\partial x_s} & \overline{\frac{\partial w_r}{\partial z_s}} \end{vmatrix} = \det \begin{vmatrix} \widetilde{\Delta} & 0 \\ M & \overline{\widetilde{\Delta}} \end{vmatrix} = \det \widetilde{\Delta} \det \overline{\widetilde{\Delta}}$$

$$= \det \widetilde{\Delta} \, \overline{\det \widetilde{\Delta}} = |\det \widetilde{\Delta}|^2$$

It follows that if $(\widetilde{\phi})^{-1}$ is holomorphic, so that $\det \widetilde{\Delta} \neq 0$, all $z \in \widetilde{A}$, then for all $(x, y) \in A$, $\det \Delta > 0$. Thus every element of the "holomorphic" pseudogroup (5) is an element of the pseudogroup (3) of orientation-preserving C^∞-diffeomorphisms, in R^{2m}.

Exercises

(1) Show that any pseudogroup of transformations in R^m is a category whose objects are open subsets of R^m, and whose morphisms are the transformations (thus every morphism is an isomorphism, and the category is a groupoid).

(2) Deduce an inverse mapping theorem for holomorphic maps from the inverse mapping theorem for C^∞-maps.

(3) Compile a list of other examples of pseudogroups.

2. ATLASES AND STRUCTURES ON MANIFOLDS

Suppose that X is a topological m-manifold, and let $\mathscr{P}$ be a pseudogroup of transformations in R^m. A **$\mathscr{P}$-atlas** on X is a family $\mathscr{A} = \{\xi_j : U_j \to U_j' | j \in J\}$ of charts on X such that, for all

$j, k \in J$, the change of co-ordinates map $\phi_{jk} : \xi_j(W) \to \xi_k(W)$, $W = U_j \cap U_k \neq \emptyset$, is an element of $\mathscr{P}$, and $X = \cup\{U_j : j \in J\}$. We define an equivalence relation $\sim$ on the set of all $\mathscr{P}$-atlases on X by putting $\mathscr{A} \sim \mathscr{B}$ iff $\mathscr{A} \cup \mathscr{B}$ is a $\mathscr{P}$-atlas on X. That is, two $\mathscr{P}$-atlases $\mathscr{A}$ and $\mathscr{B}$ are equivalent iff the changes of co-ordinates between any overlapping co-ordinate systems from $\mathscr{A}$ and $\mathscr{B}$ are in the pseudogroup $\mathscr{P}$. The $\mathscr{P}$-atlases in any equivalence class are partially ordered by inclusion, and each equivalence class contains a unique maximal element $\mathscr{M}$ consisting of all the atlases in the class. Such a maximal $\mathscr{P}$-atlas is called a **$\mathscr{P}$-structure** on X, and the pair $(X, \mathscr{P})$ is called a **$\mathscr{P}$-manifold**.

Notice that any $\mathscr{P}$-atlas $\mathscr{A}$ on X determines a unique $\mathscr{P}$-structure on X equivalent to $\mathscr{A}$. It is therefore enough to work with a $\mathscr{P}$-atlas, which may consist of a small number of charts.

For each pseudogroup $\mathscr{P}$ in R^m, there are two basic questions:

(1) **Existence.** Given a topological m-manifold X, does there exist a $\mathscr{P}$-structure on X?

(2) **Classification.** Can we find a set of invariants with respect to a suitable notion of "$\mathscr{P}$-equivalence" which will enable us to decide whether two $\mathscr{P}$-manifolds $(X, \mathscr{M})$ and $(Y, \mathscr{N})$ are "$\mathscr{P}$-equivalent"?

The difficulty and interest of these questions depend on both $\mathscr{P}$ and X. We shall say a little about the situation for the holomorphic case, but first we try to clarify what we mean by $\mathscr{P}$-equivalence.

Suppose that $(X, \mathscr{M})$ and $(Y, \mathscr{N})$ are $\mathscr{P}$-manifolds of dimension m (i.e. the model space is R^m). A **$\mathscr{P}$-equivalence** from $(X, \mathscr{M})$ to $(Y, \mathscr{N})$ is a homeomorphism $\theta : X \to Y$ such that, for every chart $\xi : U \to U'$ in $\mathscr{M}$ and every chart $\eta : V \to V'$ in $\mathscr{N}$ such that $\theta(U) \cap V = \theta(W) \neq \emptyset$, the map $\theta_{\xi\eta} : \xi(W) \to \eta(\theta(W))$ given by $\theta_{\xi\eta}(\xi(w)) = \eta(\theta(w))$, $w \in W$, is in the pseudogroup $\mathscr{P}$.

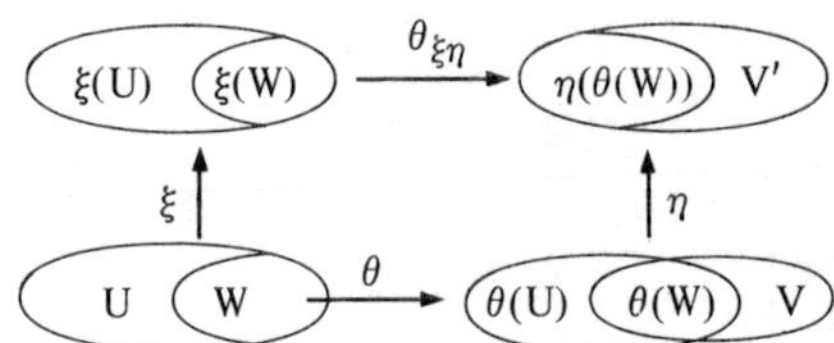

In the above definition of $\mathscr{P}$-equivalence of $\mathscr{P}$-manifolds we can again replace the structure $\mathscr{M}$ and $\mathscr{N}$ by atlases $\mathscr{A}$, $\mathscr{B}$ $\mathscr{P}$-equivalent to $\mathscr{M}$, $\mathscr{N}$ respectively.

It is important to realize that if $\mathscr{A}$ and $\mathscr{B}$ are $\mathscr{P}$-atlases on X, and $\mathscr{M}$, $\mathscr{N}$ are the $\mathscr{P}$-structures determined by $\mathscr{A}$, $\mathscr{B}$, then $\mathscr{A}$ is $\mathscr{P}$-equivalent to $\mathscr{B}$ iff $\mathscr{M} = \mathscr{N}$, i.e. iff $1_X : X \to X$ is a $\mathscr{P}$-equivalence from $(X, \mathscr{A}) = (X, \mathscr{M})$ to $(X, \mathscr{B}) = (X, \mathscr{N})$. We may yet have a $\mathscr{P}$-equivalence $\theta : X \to X$ between distinct $\mathscr{P}$-manifolds $(X, \mathscr{M})$, $(X, \mathscr{N})$.

The standard example, for the case $m = 1$, with $\mathscr{P}$ the pseudogroup of C^∞ diffeomorphisms in R, is as follows. Let $X = R$, and let $\mathscr{A}$, $\mathscr{B}$ consist respectively of the single charts $\xi : R \to R$, $\eta : R \to R$, where $\xi(x) = x$ and $\eta(x) = x^3$. Let $\mathscr{M}$, $\mathscr{N}$ be the associated C^∞-structures on R. Then $\mathscr{M} \neq \mathscr{N}$, so $\mathscr{A}$ is not C^∞-equivalent to $\mathscr{B}$. However, there is a C^∞-diffeomorphism (i.e. a C^∞-equivalence) from $(X, \mathscr{M})$ to $(X, \mathscr{N})$. In fact, $\theta : X \to X$, given by $\theta(x) = x^{1/3}$, is such a homeomorphism.

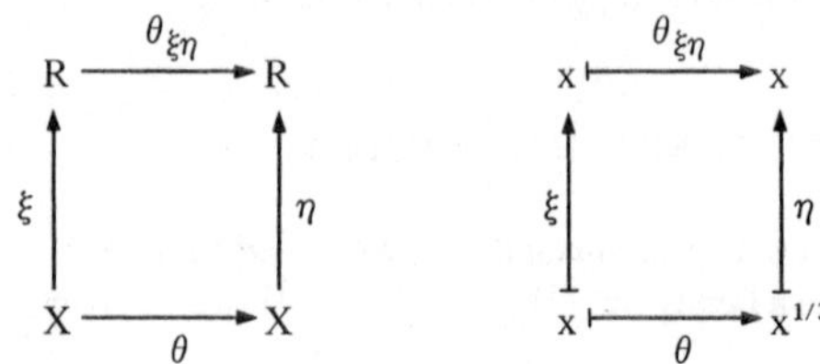

3. COMPLEX MANIFOLDS

A **complex m-manifold** is a $\mathscr{P}$-manifold $(X, \mathscr{M})$, where $\mathscr{P}$ is the pseudogroup of holomorphic transformations in $\mathbb{C}^m = R^{2m}$ with holomorphic inverses. Here we identify $\mathbb{C}^m$ with R^{2m}, as topological spaces, under the homeomorphism ρ of §1. Thus the complex structure of X is specified by the choice of an atlas $\mathscr{A}$ in which all changes of local co-ordinates are holomorphic.

Example 1. Let $X = \mathbb{C}^m$, $\mathscr{A} = \{1_{\mathbb{C}^m} : \mathbb{C}^m \to \mathbb{C}^m\}$.

Example 2. Let X be the two-dimensional sphere $S^2 = \{x \in R^3 : x_1^2 + x_2^2 + x_3^2 = 1\}$. For $\mathscr{A}$ we take the atlas consisting of just two charts $\xi : U \to \mathbb{C}$, $\eta : V \to \mathbb{C}$, where U is the complement $S^2 \setminus (0, 0, 1)$ of the north pole, and V the complement $S^2 \setminus (0, 0, -1)$ of the south pole, with

$$\xi(a, b, c) = \frac{a}{1-c} + i\frac{b}{1-c} \qquad (c \neq 1)$$

and

$$\eta(a, b, c) = \frac{a}{1+c} - i\frac{b}{1+c} \qquad (c \neq -1)$$

Thus for $w \in U \cap V$, $\xi(w)\eta(w) = 1$. So $\mathscr{A}$ determines a complex structure $\mathscr{M}$ on S^2. Of course $(S^2, \mathscr{M})$ is the **Riemann sphere.**

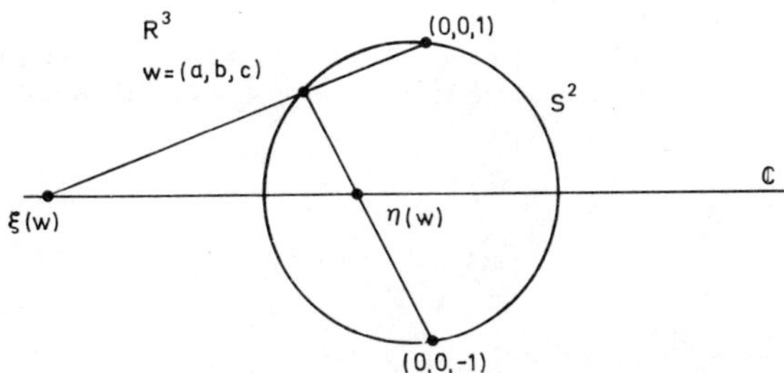

Example 3. Let $X = \{z \in \mathbb{C} : \operatorname{im} z = y > 0\}$ = upper half plane, $\mathscr{A} = \{1_X : X \to \mathbb{C}\}$, the atlas consisting of one chart, the inclusion of X in $\mathbb{C}$.

Example 4. $X = \{z \in \mathbb{C} : |z| < 1\}$, and again $\mathscr{A} = \{1_X : X \to \mathbb{C}\}$, the inclusion of the open unit disc in $\mathbb{C}$.

Taking m = 1 in Example 1, we obtain in all four examples of complex 1-manifolds, each connected and simply-connected. Of course we can translate the classical facts into the statement that every connected, simply-connected complex 1-manifold is holomorphically equivalent to one of these four, and in fact the last two examples are holomorphically equivalent. (An explicit equivalence from the upper half-plane to the open unit disc is given by the map sending z to $(i - z)/(i + z)$.)

Exercise. Find all the holomorphic self-equivalences of $\mathbb{C}$, S^2, and the upper half-plane, with respect to these complex structures.

Example 5. Let a and b be linearly independent vectors in R^2, and define $\gamma: R^2 \to S^1 \times S^1$ by $\gamma(\lambda a + \mu b) = (e^{2\pi i\lambda}, e^{2\pi i\mu})$, $\lambda, \mu \in R$. Then we can construct a complex structure $\mathscr{S}_{a,b}$ on the torus $T = S^1 \times S^1$ as follows. If U′ is an open subset of R^2, then $\gamma|U'$ is a homeomorphism onto its image $U = \gamma(U')$ iff for all $x \in U'$ and for any integers m and n not both zero, $x + ma + nb \notin U'$. Suppose then that $\gamma| : U' \to U$ is a homeomorphism, and let $\xi = (\gamma|U')^{-1}$. Then ξ is a chart on T, and the set of all such charts is the complex structure $\mathscr{S}_{a,b}$ $(R^2 = \mathbb{C})$.

Exercise. For any $\xi, \eta \in \mathscr{S}_{a,b}$, determine the change of local coordinates map $\phi_{\xi\eta}$ and check that it is holomorphic.

Exercise. Let $a = (1, 0) = c$ and suppose that b and d have positive imaginary part ($b_2 > 0$, $d_2 > 0$ Suppose further that

$$d = \frac{pb + q}{rb + s}$$

where p, q, r, s are integers and $ps - rq = 1$. Show that $(T, \mathscr{S}_{a,b})$ is complex equivalent to $(T, \mathscr{S}_{c,d}$

Example 6. Let H denote the upper half-plane $\{z \in \mathbb{C} : \text{im } z > 0\}$, and define an equivalence relation $\sim$ on $H \times \mathbb{C}$ by putting $(z, w) \sim (z', w')$ iff $z = z'$ and for some integers m, n, $w' = w + m +$ Consider the topological space $X = (H \times \mathbb{C}) / \sim$, and let $\theta : H \times \mathbb{C} \to X$ be the quotient map. Then we can define a complex structure on X as in Example 5, this time taking those open subsets U′ of $H \times \mathbb{C} \subset \mathbb{C}^2$ for which $\theta|U'$ is a homeomorphism onto $U = \theta(U')$. The set of charts $\xi: U \to U'$, $\xi = (\theta|U')^{-1}$ so obtained is the required complex structure $\mathscr{S}$ on X. Thus $(X, \mathscr{S})$ is a complex manifold of (complex) dimension 2.

If we denote the $\sim$-class of (z, w) by [z, w], then there is a map $\Gamma: X \to H$ given by $\Gamma[z, w] = z$ and we obtain the relation $\Gamma \circ \theta = \pi_1$, where $\pi_1 : H \times \mathbb{C} \to H$ is projection to the first factor.

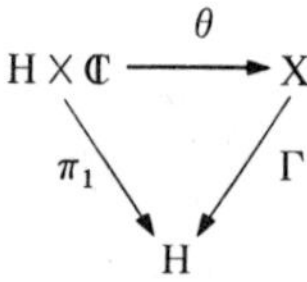

Exercise. Try to relate the fibre $\Gamma^{-1}(z)$ with the complex torus $(T, \mathscr{S}_{a,b})$ of Example 5 with $a = (1, 0)$ and $b = z$.

This example suggests the notion of **deformation** of complex manifolds. We can think of X as a family of complex tori parametrized by the points of H.

Example 7. Complex tori

The construction described in Example 5 can be generalized as follows. Take a basis $\{a_1, b_1, \ldots, a_m, b_m\}$ for R^{2m} and define a map $\gamma: R^{2m} \to (S^1)^{2m} = S^1 \times \ldots \times S^1$ (2m factors) by

$$\gamma\left(\sum_{s=1}^{m} \lambda_s a_s + \mu_s b_s\right) = (e^{2\pi i\lambda_1}, \ldots, e^{2\pi i\lambda_m})$$

and proceed to construct charts $\xi: U \to U' \subset R^{2m} = \mathbb{C}^m$ as before.

Example 8. Complex projective m-space

Define an equivalence relation $\sim$ on $\mathbb{C}_*^{m+1} = \mathbb{C}^{m+1} \backslash \{0\}$ by putting $z \sim w$ iff there is a complex number α ($\neq 0$) such that $z = \alpha w$. The quotient space $P_m(\mathbb{C}) = \mathbb{C}_*^{m+1}/\sim$ is a topological 2m-manifold, with a canonical complex structure defined by the following atlas. Let $U_r = \{[z] \in P_m(\mathbb{C}) : z_r \neq 0\}$, $r = 1, ..., m+1$, where $[z]$ denotes the $\sim$ class of $z \in \mathbb{C}_*^{m+1}$. Define $\xi_r : U_r \to \mathbb{C}^m$ by

$$\xi_r([z]) = z_r^{-1}(z_1, ..., z_{r-1}, z_{r+1}, ..., z_{m+1})$$

It is trivial to check that $\mathscr{A} = \{\xi_1, ..., \xi_{m+1}\}$ is a complex atlas on $P_m(\mathbb{C})$, and therefore determines a complex structure on $P_m(\mathbb{C})$. We call the complex manifold so obtained **complex projective** m-**space.**

Example 9. Grassmann manifolds

Let G(r, s) denote the set of all r-dimensional complex linear subspaces of $\mathbb{C}^{r+s} = \mathbb{C}^m$, where $1 \leqslant r \leqslant m-1$. We impose a complex structure on G(r, s) as follows.

Let $\lambda = (\lambda_1, ..., \lambda_r)$ be a sequence of integers such that $1 \leqslant \lambda_1 < \lambda_2 < ... < \lambda_r \leqslant m$, and let $\lambda^* = (\lambda_1^*, ..., \lambda_s^*)$ be the complementary sequence, with $1 \leqslant \lambda_1^* < ... < \lambda_s^* \leqslant m$. Thus each number n between 1 and m occurs exactly once either in λ or in λ^*.

Now consider the standard basis $e_1, ..., e_m$ for $\mathbb{C}^m$, and let A_λ, A_λ^* denote the linear subspaces generated by $\{e_p : p \in \lambda\}$ and $\{e_q : q \in \lambda^*\}$ respectively. Thus $\mathbb{C}^m = A_\lambda \oplus A_\lambda^*$, and there is a natural projection $\theta_\lambda : \mathbb{C}^m \to A_\lambda$. Consider the subset U_λ of G(r, s) consisting of all r-dimensional linear subspaces S of $\mathbb{C}^m$ such that $\theta_\lambda | S$ is an isomorphism onto the r-dimensional linear subspace A_λ. Any such r-plane S is given by a uniquely determined set of linear equations of the form

$$z_q = \sum_{h=1}^{r} \sigma_{hk} z_p, \qquad h = 1, ..., s$$

where $q = \lambda_h^*$ and $p = \lambda_k$. That is, $z \in S$ iff $z_1, ..., z_m$ satisfy these equations.

Define $\xi_\lambda : U_\lambda \to \mathbb{C}^{rs}$ by putting $\xi_\lambda(S) = w$, where $w_{r(h-1)+k} = \sigma_{hk}$. The $\binom{m}{r}$ sets U_λ cover G(r, s), and each ξ_λ is a bijection. Thus we can induce a topology on G(r, s), and check that the set of all the ξ_λ is a complex atlas on G(r, s). The complex manifold G(r, s) is called a **Grassmann** manifold. Note that dim G(r, s) = rs.

Exercise. Work out the details of exactly how the topology of G(r, s) is specified.

Exercise. Check that the change of co-ordinates is holomorphic in a simple case, say for the charts ξ_λ, ξ_μ on G(2, 2), where $\lambda = (1, 2)$, and $\mu = (2, 4)$.

Exercise. Develop the notion of Grassmann manifold from the point of view of homogeneous spaces. Observe that the standard action of the general linear group $GL(m, \mathbb{C})$ on $\mathbb{C}^m$ induces an action on G(r, s). Prove that this action is transitive, and determine the isotropy group Γ of A_λ, $\lambda = (1, 2, ..., r)$. Hence construct a homeomorphism ϕ from G(r, s) to the homogeneous space $GL(m, \mathbb{C})/\Gamma$.

Note that $G(1, m-1) = P_{m-1}(\mathbb{C})$.

Example 10. Hopf manifolds

Let $\alpha_1, ..., \alpha_m \in \mathbb{C}_*$, with $|\alpha_r| \neq 1$, $r = 1, ..., m$. Then there is a homeomorphism

$$d : S^{2m-1} \times R \to \mathbb{C}_*^m$$

given by $d((z_1, ..., z_m), t) = (\alpha_1^t z_1, ..., \alpha_m^t z_m)$, where S^{2m-1} is the unit sphere $\{z \in \mathbb{C}^m : |z_1|^2 + ... + |z_m|^2 = 1\}$ in $\mathbb{C}^m$. Now the additive group R of real numbers acts on $\mathbb{C}_*^m$ by $t \cdot z = (\alpha_1^t z_1, ..., \alpha_m^t z_m)$

and hence so does the additive group Z of integers. In fact, the action $\zeta : Z \times \mathbb{C}^m_* \to \mathbb{C}^m_*$ is induced under d by the natural action ϕ of Z on $S^{2m-1} \times R$, $\phi(n,(z,t)) = (z, t + n)$.

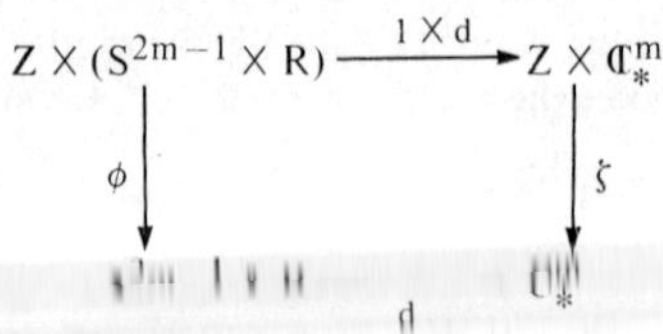

Hence $\mathbb{C}^m_*/Z$ is homeomorphic to $S^{2m-1} \times S^1$. But Z acts properly discontinuously without fixed points on $\mathbb{C}^m_*$. So we can induce a complex structure on $\mathbb{C}^m_*/Z$ and hence on $S^{2m-1} \times S^1$, as in Examples 5, 6 or 7.

These are only some of the best-known complex manifolds. Among the many other methods by which complex manifolds may be obtained, we mention the fact that if X and Y are complex manifolds, then so is $X \times Y$. Thus, for example, $P_m(\mathbb{C}) \times P_m(\mathbb{C})$ has a natural complex structure.

4. REAL AND COMPLEX LINEAR SPACES

Let E be a complex linear space. Then we can define a $\mathbb{C}$-linear endomorphism $J : E \to E$ by $J(v) = iv$. It follows that $J \circ J = J^2 = -1_E$.

Conversely, suppose that F is a real linear space. Then we can give F a complex structure if there is an R-linear endomorphism $\tilde{J} : F \to F$ for which $\tilde{J} \circ \tilde{J} = -1_F$. To do this, we define scalar multiplication by complex numbers by

$$(a + ib)(w) = aw + b\tilde{J}(w)$$

Exercise. Show that the map $\tilde{J} : F \to F$ is $\mathbb{C}$-linear with respect to this given complex structure, and coincides with the $\mathbb{C}$-endomorphism $w \mapsto iw$.

If $\{e_1, ..., e_m\}$ is a basis for V over $\mathbb{C}$, then $\{e_1, ..., e_m, J(e_1), ..., J(e_m)\}$ is a basis for V over R. Hence V must be of even dimension over R, if $\dim_R V < \infty$.

For instance, if $E = \mathbb{C}^m$, then $J : \mathbb{C}^m \to \mathbb{C}^m$ is given by $J(z_1, ..., z_m) = (iz_1, ..., iz_m)$ $= (ix_1 - y_1, ..., ix_m - y_m)$. Conversely, if $F = R^{2m}$, we define $\tilde{J} : R^{2m} \to R^{2m}$ by $\tilde{J}(x_1, ..., x_m, y_1, ..., y_m)$ $= (-y_1, ..., -y_m, x_1, ..., x_m)$ and recover the space $\mathbb{C}^m$, generated by the standard basis $e_1, ..., e_m \in$ where

$$e_1 = (1, 0, ..., 0)$$

$$e_2 = (0, 1, ..., 0)$$

....

$$e_m = (\underbrace{0, 0, ..., 1}_{m}, \underbrace{0, ..., 0}_{m})$$

This basis over $\mathbb{C}$ for $\mathbb{C}^m$ extends to a basis $\{e_1, ..., e_m, J(e_1), ..., J(e_m)\}$ for R^{2m} over R, where

$$J(e_1) = (\underbrace{0, ..., 0}_{m}, \underbrace{1, 0, ..., 0}_{m})$$

$$J(e_2) = (0, ..., 0, 0, 1, ... 0)$$

$$J(e_m) = (0, ..., 0, 0, ... 0, 1)$$

Another way of constructing a complex linear space from a real linear space V is to form the tensor product $V_{\mathbb{C}} = V \otimes_R \mathbb{C}$ of V with $\mathbb{C}$ over the real numbers. This space is called the **complexification** of V, and in case $\dim_R V = m$, we have $\dim_{\mathbb{C}} V_{\mathbb{C}} = m$ also. More explicitly, we put $V_{\mathbb{C}} = V \times V$, and define multiplication of $(z, w) \in V_{\mathbb{C}}$ by $a + ib \in \mathbb{C}$ in writing

$$(a + ib)(z, w) = (az - bw, bz + aw)$$

It is natural to write the elements of $V_{\mathbb{C}}$ in the form $z + iw$, where $z, w \in V$. (Thus (z, w) and $z + iw$ have the same meaning, as in the usual notational conventions for complex numbers, where $x = x + i0 = (x, 0)$, $iy = 0 + iy = (0, y)$.) We define **conjugation** in $V_{\mathbb{C}}$ by saying that $\overline{(z, w)} = (z, -w)$, i.e. $\overline{z + iw} = z - iw$.

Suppose now that V is a real linear space on which a real linear endomorphism J with $J^2 = -1_V$ is defined. Then we can use J to induce another such endomorphism $\tilde{J}$ on $V_{\mathbb{C}}$ with $\tilde{J}(x, y) = J(x) + iJ(y)$. It follows that $\tilde{J}$ is a $\mathbb{C}$-linear endomorphism of $V_{\mathbb{C}}$, with $\tilde{J}^2 = -1_{V_{\mathbb{C}}}$. Since $\tilde{J}^2 = -1_{V_{\mathbb{C}}}$, the eigenvalues of $\tilde{J}$ are $\pm i$. Hence we can split $V_{\mathbb{C}}$ into a direct sum $V' \oplus V''$, where V' is the eigenspace of i, and V'' the eigenspace of $-i$. In fact,

$$\tilde{J}(z + iw) = i(z + iw)$$

$$\Leftrightarrow J(z) + iJ(w) = iz - w$$

$$\Leftrightarrow \quad z = J(w) \text{ and } w = -J(z)$$

$$\Leftrightarrow z + iw = z - iJ(z)$$

Thus

$$V' = \{z - iJ(z) : z \in V\}$$

Similarly,

$$V'' = \{z + iJ(z) : z \in V\}$$

Exercise. Show that conjugation maps V' onto V'', and conversely, and is a linear isomorphism over R between V' and V''.

Exercise. Show that we can regard $\mathbb{C}^m$ as $R^m_{\mathbb{C}}$. Also examine $R^{2m}_{\mathbb{C}}$ and determine the subspaces $(R^{2m})'$, $(R^{2m})''$.

5. HERMITIAN INNER PRODUCTS

Let $V, J : V \to V$ be as above. An inner product $\gamma : V \times V \to R$ is said to be **Hermitian** iff, for all $x, y \in V$,

$$\gamma(J(x), J(y)) = \gamma(x, y)$$

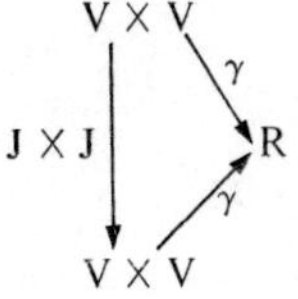

i.e. $\gamma \circ (J \times J) = \gamma$.

It follows that for any vector $z \in V$, $J(z)$ is perpendicular to z with respect to γ, since

$$\gamma(z, J(z)) = \gamma(J(z), J^2(z)) = \gamma(J(z), -z)$$

$$= -\gamma(J(z), z) = -\gamma(z, J(z))$$

Taking this argument a little further, consider the case where $\dim_R V = 2m$, and let z be a unit vector in V, $\gamma(z,z) = 1$. Then $\gamma(J(z), J(z)) = \gamma(z,z) = 1$ also, and so z, J(z) form a pair of orthonormal vectors, generating a 2-dimensional linear subspace of V. The orthogonal complement W of this 2-plane in V with respect to γ is then a real linear space of real dimension 2(m-1), and $J(W) = W$. It follows inductively, therefore, that V admits an orthonormal basis of the form $\{z_1, \ldots, z_m, J(z_1), \ldots, J(z_m)\}$.

Exercise. Let J be the endomorphism $J(x,y) = (-y, x)$ on $R^{2m} = R^m \times R^m$, and let $\gamma: R^{2m} \times R^{2m}$ be the standard inner product on R^{2m},

$$\gamma(z, w) = \gamma((x,y),(u,v)) = \sum_{r=1}^{m} (x_r u_r + y_r v_r)$$

Then γ is Hermitian with respect to J.

We can use such a Hermitian inner product on a real linear space V with respect to J to construct a complex symmetric bilinear form $\tilde{\gamma}$ on $V_{\mathbb{C}}$, given by

$$\tilde{\gamma}(z, w) = \tilde{\gamma}(x + iy, u + iv)$$

$$= \gamma(x,u) + i\gamma(y,u) + i\gamma(x,v) - \gamma(u,v)$$

We note:

(1) For all $z, w \in V_{\mathbb{C}}$, $\tilde{\gamma}(\bar{z}, \bar{w}) = \overline{\tilde{\gamma}(z, w)}$.

Proof. $\tilde{\gamma}(\bar{z}, \bar{w}) = \tilde{\gamma}(x - iy, u - iv)$

$$= \gamma(x,u) - \gamma(y,v) - i\gamma(x,v) - i\gamma(y,u)$$

$$= \overline{\tilde{\gamma}(z, w)}$$

(2) For all $z \in V_{\mathbb{C}}$, $\tilde{\gamma}(z, \bar{z}) \in R$, and for $z \neq 0$, $\tilde{\gamma}(z, \bar{z}) > 0$.

Proof. $\tilde{\gamma}(z, \bar{z}) = \tilde{\gamma}(x + iy, x - iy)$

$$= \gamma(x,x) + i\gamma(y,x) - i\gamma(x,y) + \gamma(y,y)$$

$$= \gamma(x,x) + \gamma(y,y)$$

(3) For all $z \in V'$, $w \in V''$, $\tilde{\gamma}(z, \bar{w}) = 0$ (i.e. for all $z, w \in V'$, $\tilde{\gamma}(z, w) = 0$).

Proof. $z \in V' \Leftrightarrow z = x - iJ(x)$
$w \in V'' \Leftrightarrow w = y + iJ(y)$

So $\tilde{\gamma}(z,\bar{w}) = \tilde{\gamma}(x - iJ(x), y - iJ(y))$

$$= \gamma(x,y) - \gamma(J(x),J(y)) - i\gamma(J(x),y) - i\gamma(x,J(y))$$

$$= \gamma(x,y) - \gamma(x,y) - i\gamma(J^2(x),J(y)) - i\gamma(x,J(y))$$

$$= + i\gamma(x,J(y)) - i\gamma(x,J(y)) = 0$$

Exercise. Show that any complex symmetric bilinear form $\tilde{\gamma}$ on $V_{\mathbb{C}}$ for which (1), (2) and (3) are true is obtained from a Hermitian inner product γ on V with respect to J in the way described above.

To any Hermitian inner product γ in V with respect to J, and for any $\mathbb{C}$-basis $\{\epsilon_1, ..., \epsilon_m\}$ for V′, we can construct a complex Hermitian matrix $H = [h_{r\bar{s}}]$ as follows. Observe that conjugation preserves $\mathbb{C}$-independence of vectors in $V_{\mathbb{C}}$, so the set $\{\bar{\epsilon}_1, ..., \bar{\epsilon}_m\}$ is a basis for V″. We define $h_{r\bar{s}} \in \mathbb{C}$ by

$$\tilde{\gamma}(\epsilon_r, \bar{\epsilon}_s) = h_{r\bar{s}}$$

then $h_{r\bar{s}} = \bar{h}_{s\bar{r}}$, so H is a Hermitian matrix in the sense that $\bar{H}^\dagger = H$.

6. HERMITIAN INNER PRODUCTS AND MULTILINEAR ALGEBRA

Let V and J be as in the previous section, and denote the dual L(V, R) of V by V*. Then we can define an R-linear endomorphism $J^* : V^* \to V^*$ by $J^*(\phi) = \phi \circ J$, and it follows at once that $(J^*)^2 = -1_{V^*}$.

We can also form the complex linear spaces $(V_{\mathbb{C}})^* = L(V_{\mathbb{C}}, \mathbb{C})$ and $(V^*)_{\mathbb{C}}$. It is an easy exercise to construct a $\mathbb{C}$-linear isomorphism from $(V_{\mathbb{C}})^*$ to $(V^*)_{\mathbb{C}}$, and so we identify these two spaces, writing $V^*_{\mathbb{C}}$ for either.

As before, $V^*_{\mathbb{C}}$ can be expressed as the direct sum of the eigenspaces $V^{*\prime}$, $V^{*\prime\prime}$ of the eigenvalues i, −i of J^*. It follows that the tensor space

$$T^r_s(V_{\mathbb{C}}) = \underbrace{V_{\mathbb{C}} \otimes ... \otimes V_{\mathbb{C}}}_{\text{r factors}} \otimes \underbrace{V^*_{\mathbb{C}} \otimes ... \otimes V^*_{\mathbb{C}}}_{\text{s factors}}$$

may be decomposed into a direct sum of spaces of the form

$$A_1 \otimes ... \otimes A_r \otimes B_1 \otimes ... \otimes B_s$$

where each A_h is either V′ or V″, and each B_k is either $V^{*\prime}$ or $V^{*\prime\prime}$.

Similar considerations apply to exterior algebra, and we are concerned particularly with the exterior algebra $\Lambda V^*_{\mathbb{C}}$. First note that $\Lambda V^{*\prime}$ and $\Lambda V^{*\prime\prime}$ are subalgebras of $\Lambda V^*_{\mathbb{C}}$, and for each pair of non-negative integers, let $\Lambda^{r,s} V^*_{\mathbb{C}}$ denote the set of all elements of $\Lambda V^*_{\mathbb{C}}$ generated by elements of the type $\alpha \wedge \beta$, where $\alpha \in \Lambda^r V^{*\prime}$ and $\beta \in \Lambda^s V^{*\prime\prime}$. Thus $\Lambda^{r,s} V^*_{\mathbb{C}}$ is a complex linear subspace of $\Lambda V^*_{\mathbb{C}}$, and we can write

$$\Lambda V^*_{\mathbb{C}} = \sum_{k=0}^{2m} \Lambda^k V^*_{\mathbb{C}} = \sum_{k=0}^{2m} \left(\sum_{r+s=k} \Lambda^{r,s} V^*_{\mathbb{C}} \right)$$

Note also that the conjugation map of $V_{\mathbb{C}}$ to itself, which is R-linear on V′ and maps V′ R-isomorphically onto V″ (and V″ to V′), induces an R-linear isomorphism from $\Lambda^{r,s} V^*_{\mathbb{C}}$ to $\Lambda^{s,r} V^*_{\mathbb{C}}$.

Now suppose that γ is a Hermitian inner product on V with respect to the endomorphism J ($J^2 = -1_V$ as before). Then we construct an element $\omega \in \Lambda^2 V^*$, that is to say a skew-symmetric bilinear map $\omega : V \times V \to R$, by $\omega(x,y) = \gamma(x, J(y))$, $x, y \in V$. The bilinearity of ω is immediate. Its skew-symmetry follows from the observation that $\omega(x,y) = \gamma(x, J(y)) = \gamma(J(y), x)$ $= \gamma(J(y), -J^2(x)) = \gamma(y, -J(x)) = -\gamma(y, J(x)) = -\omega(y, x)$.

Next we extend ω to an element $\tilde{\omega}$ of $\Lambda^2 V^*_{\mathbb{C}}$ just as we did with γ itself.

Exercise. Show that $\tilde{\omega} \in \Lambda^{1,1} V^*_{\mathbb{C}}$.

To complete this section, we express $\tilde{\omega}$ in terms of a suitable basis for $\Lambda^{1,1} V^*_{\mathbb{C}}$, as follows.

Let $\{\epsilon_1, \ldots, \epsilon_m\}$ be a $\mathbb{C}$-basis for V′, and $\{\bar{\epsilon}_1, \ldots, \bar{\epsilon}_m\}$ be the conjugate basis for V″. Then there are dual bases $\{\epsilon_1^*, \ldots, \epsilon_m^*\}$ for $V^{*\prime}$ and $\{\bar{\epsilon}_1^*, \ldots, \bar{\epsilon}_m^*\}$ for $V^{*\prime\prime}$ (where $(\bar{\epsilon}_j)^* = \overline{(\epsilon_j^*)}$). Hence there are bases

$$\{\epsilon_{k_1}^* \wedge \ldots \wedge \epsilon_{k_r}^* : 1 \leqslant k_1 < \ldots < k_r \leqslant m\}$$

for $\Lambda^r V^{*\prime}$ and

$$\{\bar{\epsilon}_{\ell_1}^* \wedge \ldots \wedge \bar{\epsilon}_{\ell_s}^* : 1 \leqslant \ell_1 < \ldots < \ell_s \leqslant m\}$$

for $\Lambda^s V^{*\prime\prime}$, and a basis

$$\{\epsilon_{k_1}^* \wedge \ldots \wedge \epsilon_{k_r}^* \wedge \bar{\epsilon}_{\ell_1}^* \wedge \ldots \wedge \bar{\epsilon}_{\ell_s}^*\} \text{ for } \Lambda^{r,s} V^*_{\mathbb{C}}.$$

Exercise. Show that

$$\tilde{\omega} = -2i \sum_{r,s=1}^{m} h_{r\bar{s}}\, \epsilon_r^* \wedge \bar{\epsilon}_s^*$$

where $[h_{r\bar{s}}]$ is the Hermitian matrix defined above for $\tilde{\gamma}$.

7. VECTOR BUNDLES

We recall some definitions and constructions concerning vector bundles. We have application to the tangent bundle (and associated vector bundles) of a real or complex manifold in mind. However, we examine first the more general situation where the base is an arbitrary topological space and the fibre is a real or complex linear space. In particular, the notions of **almost-complex structure** and **Hermitian structure** can be presented in this general setting in a very simple way.

Let $p : B \to X$ be a continuous map of B onto X, where B and X are topological spaces. Suppose that E is a real or complex linear space of finite dimension m (thus E is a topological space in a natural way). Then p is said to be a **vector bundle** with **total space** B, **base** X, **fibre** E (and p is also referred to as the **projection**) iff the following two conditions are satisfied:

(1) For some open covering $\mathscr{U} = \{U_\alpha : \alpha \in A\}$ of X, for all $\alpha \in A$ there is a homeomorphism $\phi_\alpha : U_\alpha \times E \to p^{-1}(U_\alpha)$ such that the diagram

$$\begin{array}{ccc} U_\alpha \times E & \xrightarrow{\phi_\alpha} & p^{-1}(U_\alpha) \\ & {\scriptstyle \pi_1}\searrow \quad \swarrow{\scriptstyle p|U_\alpha} & \\ & U_\alpha & \end{array}$$

commutes, where π_1 denotes projection to the first factor. Thus ϕ_α is fibre-preserving.

(2) For all $\alpha,\beta \in A$ with $W = U_\alpha \cap U_\beta \neq \emptyset$, there is a continuous map $g_{\alpha\beta} : W \to GL(E)$ such that, for all $w \in W$, the diagram

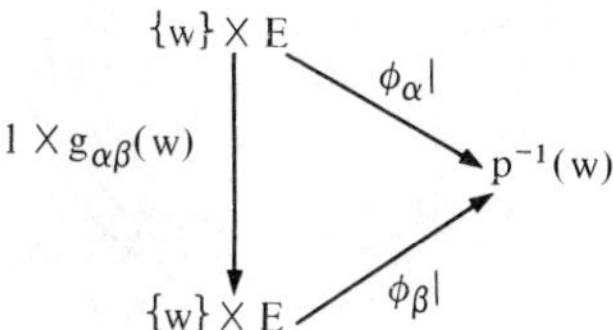

commutes. (Here GL(E) denotes the group of linear automorphisms of E.)

Alternatively, we can define a vector bundle $p : B \to X$ in terms of the **transition functions** $g_{\alpha\beta}$ as follows. Suppose that $\mathscr{U} = \{U_\alpha : \alpha \in A\}$ is an open covering of X, and for each α,β with $W = U_\alpha \cap U_\beta \neq \emptyset$ there is a continuous map $g_{\alpha\beta} : W \to GL(E)$ such that $g_{\alpha\beta}(w) \circ g_{\beta\gamma}(w) = g_{\alpha\gamma}(w)$, $w \in W$, and $g_{\alpha\alpha}(w) = 1_E$. Consider the topological sum (disjoint union) of the products $U_\alpha \times E$, which we can identify conveniently with the subspace

$$\mathscr{B} = \{(w,v,\alpha) : w \in U_\alpha,\ v \in E,\ \alpha \in A\}$$

of $X \times E \times A$ (giving A the discrete topology), and define an equivalence relation $\sim$ on $\mathscr{B}$ by putting $(w,v,\alpha) \sim (z,u,\beta)$ iff $w = z$ and $u = g_{\alpha\beta}(w)(v)$. Denote the $\sim$ class of (w,v,α) by $[w,v,\alpha]$, and define $p : B \to X$ by $p([w,v,\alpha]) = w$, where $B = \mathscr{B}/\!\sim$. Then p is a vector bundle in the previous sense, with $\phi_\alpha : U_\alpha \times E \to p^{-1}(U_\alpha)$ given by $\phi_\alpha(v,w) = [w,v,\alpha]$.

If $p : B \to X$ and $q : C \to Y$ are vector bundles (both over $\mathbb{C}$ or both over R), then a **vector bundle map** from p to q is a commutative diagram with f

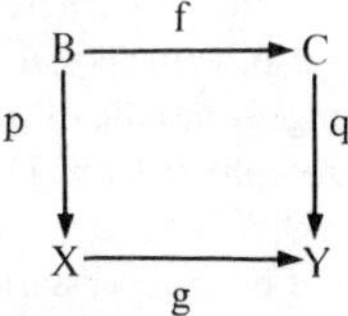

and g continuous, for which f is linear on fibres (i.e. $\forall\, x \in X$, f maps $E_x = p^{-1}(x)$ linearly into $F_{g(x)} = q^{-1}(g(x))$).

This is enough to establish the category of vector bundles (with fixed field of scalars), and so we have notions of bundle isomorphism, subbundle, and so on, automatically defined.

Examples

(1) Let $X \times E = B$, $p: B \to X$ the projection to the first factor. Then p is an E-bundle, the **trivial** or **product** E-bundle on X.

(2) (**Möbius band**). Let $X = S^1 = \{e^{i\theta} \in \mathbb{C} : \theta \in R\}$, and let $U = S^1 \setminus \{-1\}$, $V = S^1 \setminus \{1\}$. Thus $U \cap V = W = P \cup Q$ where $P = \{z \in S^1 : \text{im } z > 0\}$ and $Q = \{z \in S^1 : \text{im } z < 0\}$. Define $g_{uv}: W \to GL(E)$ to be constant on each of the components P, Q of W, with $g_{uv}|P = 1_E$ and $g_{uv}|Q = -1_E$.

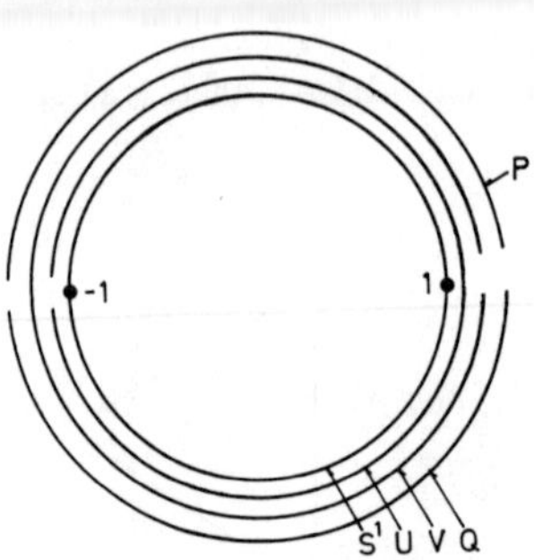

For E = R, the resulting R-bundle over S^1 is called the **Möbius band**, and this bundle is not vector bundle isomorphic to the cylinder $S^1 \times R$.

Exercise. Investigate whether the E-bundle defined above is isomorphic to $S^1 \times E$, for any real or complex E.

(3) **Tangent bundles.** Suppose that X is a C^∞ m-dimensional manifold with C^∞-structure $\mathscr{M}$ determined by a C^∞-atlas $\mathscr{A}$. We construct a vector bundle on X with fibre R^m as follows. For each $\xi_\alpha, \xi_\beta \in \mathscr{A}$, define $g_{\alpha\beta}: U_\alpha \cap U_\beta \to GL(m, R)$ by $g_{\alpha\beta} = D\phi_{\alpha\beta} \circ \xi_\alpha$. Then these transition functions $g_{\alpha\beta}$ determine the bundle required. We denote the total space by TX, and the projection by $\pi_X: TX \to X$. This bundle is called the **tangent bundle** of X.

Exercises

(1) For each $\alpha \in A$, let $\zeta_\alpha: \pi_X^{-1}(U_\alpha) \to U'_\alpha \times R^m \subset R^m \times R^m = R^{2m}$ be given by $\zeta_\alpha(v) = (u', t)$, where $(u, t) = \phi_\alpha^{-1}(v)$ and $u' = \xi_\alpha(u)$. Show that $\mathscr{B} = \{\zeta_\alpha : \alpha \in A\}$ is a C^∞-atlas on TX, and hence that TX has a natural structure as a C^∞ manifold of dimension 2m.

(2) Copy the construction of TX for the case where $\mathscr{A}$ is a complex atlas on X, and show that TX has a natural complex structure also, with fibre $\mathbb{C}^m$. Thus TX is a complex 2m-manifold ($\dim_{\mathbb{C}} X = m$), called the **holomorphic tangent bundle** of X.

(3) Construct a vector bundle isomorphism from TS^1 to $S^1 \times R$.

The tangent bundle of a C^∞ manifold is not necessarily isomorphic to a product bundle. For example, $TS^m \approx S^m \times R^m$ iff m = 1,3 or 7. A C^∞-manifold X for which TX is isomorphic to $X \times R^m$ is said to be **parallelizable.**

The following constructions have been noted earlier for linear spaces. Each can be **mobilized** over the base of a vector bundle in a natural way by applying the construction to each fibre. We indicate how this is done in one case, and leave the others as exercises.

Suppose that $p: B \to X$ is a real vector bundle with fibre E. We can form the dual bundle $p^*: B^* \to X$ as follows. If $\{g_{\alpha\beta}\}$ is the set of transition functions defining p with respect to some open covering $\mathscr{U} = \{U_\alpha : \alpha \in A\}$ of X, then we define transition functions $g^*_{\alpha\beta}: U_\alpha \cap U_\beta \to GL(E^*)$ by $g^*_{\alpha\beta}(w)(\lambda) = \lambda \circ g_{\alpha\beta}(w)$ for $w \in U_\alpha \cap U_\beta$, $\lambda \in E^* = L(E, R)$.

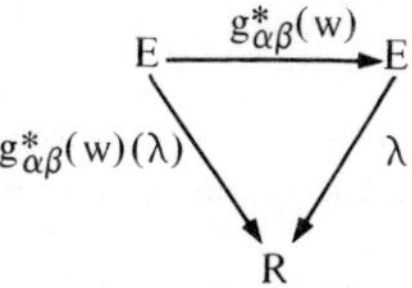

In like fashion, we can construct the **complexification** $p_{\mathbb{C}}$ of a real vector bundle $p: B \to X$, with fibre $E_{\mathbb{C}}$.

If E is a complex linear space, then of course we can define the dual bundle $p^*: B^* \to X$ as above, with fibre $E^* = L(E, \mathbb{C})$. If $p: B \to X$ and $q: C \to Y$ are both real or both complex, with fibres E and F respectively, then we can define the **Whitney sum** $p \oplus q$ with fibre $E \times F$ and the **tensor product** $p \otimes q$ with fibre $E \otimes F$. Hence the tensor bundles $T^r_s p$ of p are well defined, as are the bundles $\Lambda^r p$ of exterior forms, as before.

A **section** of a vector bundle $p: B \to X$ is a (continuous) map $\sigma: X \to B$ such that $p \circ \sigma = 1_X$. Every vector bundle has a section called the **zero section** which assigns to each $x \in X$ the zero element of the fibre $E_x = p^{-1}(x)$ at x. However, a vector bundle need not have any **nowhere zero** sections. For instance, the Möbius band, a real line-bundle on S^1, has no nowhere zero section (see Fig. 1).

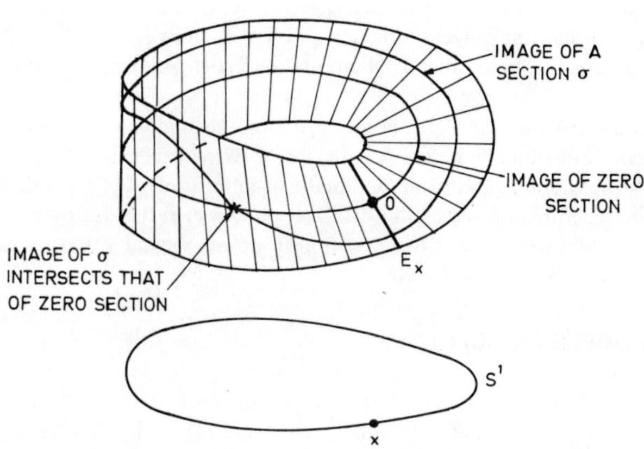

FIG.1. The Möbius band.

Exercises

(1) Show that a vector bundle $p: B \to X$ admits a nowhere zero section only if it has a trivial line subbundle ($\approx R \times X$ in real case). Is the converse true?

(2) Show that a vector bundle $p: B \to X$ is isomorphic to a product, $B = X \times E$, iff p admits m (everywhere) linearly independent sections, where $m = \dim E$.

Examples

(1) A **vector field** on a C^∞-manifold X is a section of TX. A **holomorphic** vector field on a complex manifold X is a holomorphic section of TX. (We have not written out a formal definition of either a C^∞-map between two C^∞-manifolds or a holomorphic map between two complex manifolds. However, this is easy to do, using suitable atlases on domain and codomain, and follows closely the definitions of C^∞-diffeomorphism and holomorphic equivalence given above.)

(2) A C^∞-**tensor field** on a C^∞ manifold X (contravariant order r, covariant order s) is a C^∞-section of $T^r_s X$. A C^∞-section of $\Lambda^r X$, the vector bundle associated with TX with fibre $\Lambda^r R^m$, is a C^∞ **r-form** on X.

(3) If we start with the tangent bundle $\pi_X : TX \to X$ of a C^∞-manifold, and construct the vector bundle whose fibre is the linear space of endomorphisms of $E = R^m$ (so the fibre is $\text{End}\, R^m = L(R^m, R^m)$), then a section J of this bundle is an **almost-complex structure on** X, provided $J^2 = -1$ (i.e. $\forall x \in X$, $J(x)^2 = -1_{E_x}$, $E_x = T_x X$ = tangent space to X at x). Thus an almost-complex structure on a C^∞-manifold is a continuous (or, usually, C^∞) map J which assigns to each $x \in X$ an endomorphism $J(x)$ of the tangent space $T_x X$ to x at X into itself, with $J(x)^2 = -1_{T_x X}$, $J(x)$ varying continuously (or C^∞-smoothly) with x over X.

If J is an almost-complex structure on a C^∞-manifold X, then we can extend J to the complexification $T_{\mathbb{C}} X$, and split this bundle into the Whitney sum $T'X \oplus T''X$ of subbundles of fibre dimension m, whose fibres are the eigenspaces of $\tilde{J}$.

(4) In (2) and (3), we have used the smooth structure of the manifold X only to give the appropriate sections a corresponding smoothness. Thus, for instance, we can define the notion of almost-complex structure in any vector bundle.

Exercise. Let $p : B \to X$ be a Whitney square, that is $p = q \oplus q$, for some vector bundle $q : C \to X$ on X with fibre F, say. Thus p has fibre $F \oplus F$. Define J on B by $J(x)(u, v) = (-v, u)$. Then J is an almost-complex structure on B (or in p). As a special case, take $X = T^*M$, where M is some C^∞ manifold, and consider the tangent bundle $\pi_X : TX \to X$, i.e. $\pi_X : T(T^*M) \to T^*M$. Show that π_X is a Whitney square, and hence that T^*M admits an almost-complex structure.

(5) Hermitian structures in vector bundles give another simple example of the notion of section. If we start with a bundle $p : B \to X$ with fibre E, we consider the situation where J is an almost-complex structure in p, and from the bundle of bilinear maps of $E_x \times E_x$ into R, $x \in X$. Then a Hermitian structure in p is a section γ of this bundle such that for each $x \in X$, $\gamma(x)$ is a Hermitian inner product with respect to J. Obviously we can extend γ to $\tilde{\gamma}$ on $p_{\mathbb{C}}$, and construct the associated 2-form $\tilde{\omega}$ as already described.

8. ALMOST-COMPLEX MANIFOLDS

Let $\mathcal{M} = \{\xi_\alpha : \alpha \in A\}$ be a complex structure on a manifold X of topological dimension 2m. Then each chart $\xi_\alpha : U_\alpha \to U'_\alpha$ in $\mathcal{M}$ determines a chart $\zeta_\alpha : TU_\alpha \to U'_\alpha \times \mathbb{C}^m \subset \mathbb{C}^m \times \mathbb{C}^m = \mathbb{C}^{2m}$ on the tangent bundle TX of X, the family $\mathcal{A} = \{\zeta_\alpha : \alpha \in A\}$ being a complex atlas on TX. The charts ζ_α induce a complex linear structure on each fibre (tangent space) $T_w X$, for which ζ_α maps $T_w X$ complex-isomorphically onto $\{\xi_\alpha(w)\} \times \mathbb{C}^m = \mathbb{C}^m$, all $\alpha \in A$.

Thus there is a well defined map $J : TX \to TX$ which sends a tangent vector $h \in T_w X$ to $ih \in T_w X$. That is, J is a section of $\text{End}_{\mathbb{C}} TX$ with $J^2 = -1_{TX}$. So J is an almost-complex structure on the manifold X.

It is natural to ask the converse question. Suppose that X is a C^∞-manifold, and J is a (say, C^∞) almost-complex structure on X. Does there exist a complex structure on X for which J coincides with the almost-complex structure defined above? As the name "almost-complex" suggests, such a complex structure exists in some cases but not in others.

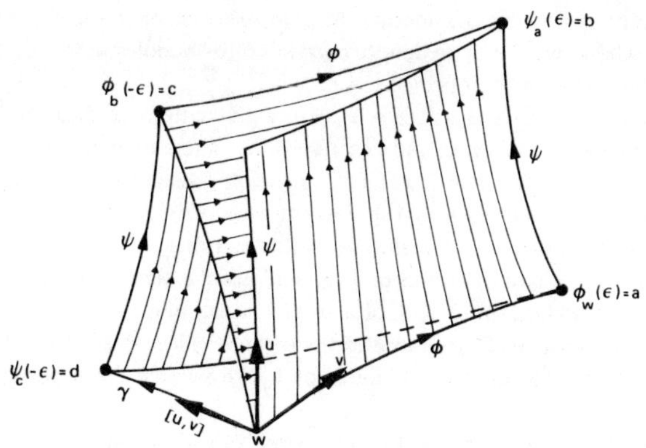

FIG.2. The Lie bracket.

We shall try to illustrate and explain the content of a theorem, due essentially to A. Newlander and L. Nirenberg, which gives necessary and sufficient conditions for an almost-complex structure J on a manifold to arise from a complex structure on the manifold. An almost-complex structure J is said to be **integrable** if such a complex structure exists. Thus the Newlander-Nirenberg theorem gives necessary and sufficient conditions for an almost-complex structure on a manifold to be integrable. First we recall the notion of **Lie bracket** [u, v] of two vector fields on a manifold X. (Suppose that both u and v, and X itself, are C^∞.) Consider a point $w \in X$, and let ϕ and ψ be the local integral flows of ϕ and ψ near w. Thus the flow-lines of ϕ and ψ have the vector fields u and v as velocity vector fields, respectively. For each (small) $\epsilon > 0$, consider the curve ϕ_w starting at w, $\phi_w(t) = \phi(w,t)$, $0 \leqslant t \leqslant \epsilon$, and put $a = \phi_w(\epsilon)$. Next consider the flow-curve ψ_a starting at a, and put $b = \psi_a(\epsilon) = \psi(a,\epsilon)$. Define two further points c and d of X by $c = \phi(b,-\epsilon) = \phi_b(-\epsilon)$, and $d = \psi_c(-\epsilon) = \psi(c,-\epsilon)$. It follows that there is a curve $\gamma: [0,\eta] \to X$ given by $\gamma(\epsilon) = d$, $0 < \epsilon < \eta$, $\gamma(0) = w$, which is C^∞. The velocity vector $\gamma'(0)$ of γ at 0 is the value [u, v] (w) of the Lie-bracket vector field [u, v] at w.

In terms of a chart $\xi_\alpha: U_\alpha \to U'_\alpha$ on X, let $x = \xi_\alpha(w)$, $u(w) = [w,\lambda,\alpha]$, $v(w) = [w,\mu,\alpha]$. Then $[u,v](w) = [w,\nu,\alpha]$, where $\nu \in R^n$ ($n = \dim X$), and for all $s = 1, \ldots, n$,

$$\nu_s = \sum_{r=1}^{n} \left(\lambda_r \frac{\partial \mu_s}{\partial x_r} - \mu_r \frac{\partial \lambda_s}{\partial x_r}\right)$$

Using the Lie bracket (Fig. 2), we define the **torsion** of an almost-complex structure J. (The definition is an example of a more general concept of torsion which applies to any tensor field of contravariant and covariant order 1.) Let u and v be C^∞ vector fields on X, and J a C^∞ almost-complex structure on X. Then the torsion $\tau_J(u,v)$ of J on (u, v) is given by

$$\tau_J(u,v) = 2\{[J(u),J(v)] - [u,v] - J([u,J(v)]) - J([J(u),v])\}$$

If J is integrable, then J is just multiplication by i, so $J(u) = iu$, and $J(v) = iv$. Thus $\tau_J(u,v) = 0$ for all u, v, and so the tensor field τ_J is 0 in this case. The theorem of Newlander-Nirenberg asserts that $\tau_J = 0$ is also *sufficient* for the integrability of J. We can look at these facts from a slightly different point of view by considering the complexification $T_{\mathbb{C}}X$ of the tangent bundle of X. (Here, as elsewhere, we prefer to denote certain vector bundles by the symbol for their total space rather than that for their projection.)

Recall that the fibre of $T_{\mathbb{C}}X$ at $w \in X$ is $T_wX \times T_wX$, with multiplication by complex scalars given by writing $i(x,y) = (-y,x)$ and $(x,y) = x + iy$. The complex dimension of the fibre $T_wX \times T_wX$ is 2m, and so $T_{\mathbb{C}}X$, regarded as a C^∞ manifold, has dimension $6m = 3(2m) = 3 \dim X$. Thus multiplication of vectors in $T_{\mathbb{C}}X$ by i should not be confused with multiplication by i on a complex manifold, as described at the beginning of this section.

Now $T_{\mathbb{C}}X = T' \oplus T''$, where the fibres of T' and T'' are the eigenspaces of the eigenvalues $i, -i$ of $\tilde{J}$, where $\tilde{J}$ is the extension to $T_{\mathbb{C}}X$ of a given almost-complex (C^∞) structure J on X. As we saw before, the vectors in T' are those of the form $\alpha - iJ\alpha$, while those of T'' are those of the form $\beta + iJ\beta$. In fact, T' is just the subbundle of $T_{\mathbb{C}}X$ on which the almost-complex structures $\tilde{J}$ and i coincide.

We extend the definition of Lie bracket to sections of $T_{\mathbb{C}}X$ by linearity over $\mathbb{C}$. Thus if λ and μ are smooth ($= C^\infty$) sections of $T_{\mathbb{C}}X$, then $[\lambda,\mu]$ is a well defined C^∞ section of $T_{\mathbb{C}}X$.

Proposition. *A C^∞ almost-complex structure J on a manifold X is integrable iff, for any C^∞ sections λ,μ of T', the Lie bracket $[\lambda,\mu]$ is a section of T'.*

Proof. Let $\lambda = u - iJ(u)$, $\mu = v - iJ(v)$, where u,v are C^∞ vector fields on X. Then a straightforward calculation shows that if we write

$$\begin{aligned} \nu &= [\lambda,\mu] \\ &= [u - iJ(u), v - iJ(v)] \\ &= [u,v] - [J(u),J(v)] - i[J(u),v] - i[u,J(v)] \end{aligned}$$

then $\nu + iJ\nu = -(1 + iJ)\,\tau_J(u,v)$. But ν is a section of T', iff $\nu + iJ\nu = 0$, which proves the proposition, by the Newlander-Nirenberg theorem.

9. ALMOST-COMPLEX STRUCTURES ON SPHERES AND OTHER MANIFOLDS

We begin with some observations about surfaces – C^∞ 2-manifolds. Suppose that S is such a surface, and let S be oriented. Then S admits an almost-complex structure J. For suppose that ρ is a Riemannian structure on S (induced, perhaps, by an embedding of S in R^3). Then for each non-zero tangent vector u to S at $w \in S$, there is a unique tangent vector v to S at w such that v is perpendicular to u, and the ordered pair (u,v) is a right-handed system with respect to the chosen orientation, with $\|v\| = \|u\|$. That is to say, v is obtained by rotating u "anticlockwise" through $\pi/2$ in T_wS. Define $J_w : T_wS \circlearrowleft$ by $J_w(u) = v$. Then, trivially, we obtain an almost-complex structure on S.

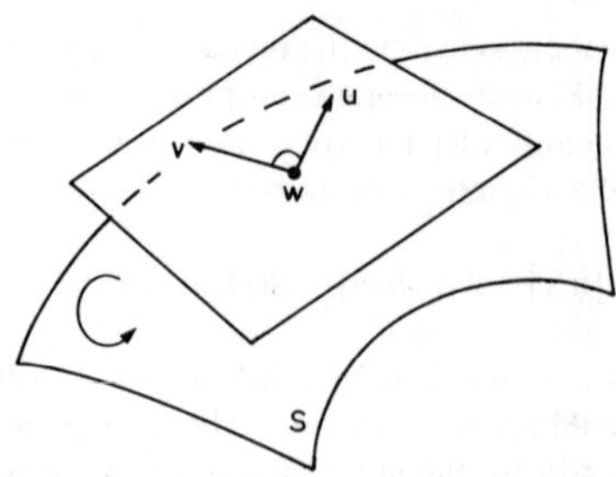

Now consider the torsion $\tau_J(\lambda,\mu)$ of any two C^∞ vector fields on S. In a neighbourhood of any point $w \in S$, we can choose a C^∞ vector field u such that the pair u, J(u) is a basis for the C^∞ vector fields on S in this neighbourhood U. That is, for any C^∞ vector field λ on U, there are C^∞ functions $\alpha: U \to R$, $\beta: U \to R$ with $\lambda = \alpha u + \beta J(u)$.

But

$$\tau_J(u,J(u)) = [u,J(u)] = 2\{[J(u),J^2(u)] - [uJ(u)] - J([u,J^2(u)]) - J([J(u),J(u)])\}$$

$$= 2\{[J(u),-u] - [u,J(u)] - J([u,-u]) - J([J(u),J(u)])\}$$

$$= 0$$

Hence $\tau_J(\lambda,\mu) = 0$ for any C^∞ vector fields λ,μ on S. That is, J is integrable.

This result can be established directly, without using the Newlander-Nirenberg theorem, by proving the existence of isothermal co-ordinates systems on S.

We turn now to the question of which spheres admit (a) almost-complex, and (b) complex structures. Of course we are talking about the even-dimensional spheres

$$S^{2m} = \{x \in R^{2m+1}: \|x\| = 1\}$$

In case m = 1, we already know that almost-complex structures exist, and that each such structure is integrable. An explicit example is given by the complex atlas constructed earlier on S^2, using stereographic projection. We recall that the complex 1-manifold so obtained is called the **Riemann sphere.**

Exercise. Let J be the almost-complex structure induced on S^2 by the inclusion $\iota: S^2 \hookrightarrow R^2$. That is, we give S^2 its standard orientation, and its standard Riemannian structure. Define $v = J(u)$ to be the tangent vector perpendicular to u, with $u \to v$ anticlockwise. As we have seen, J is integrable, so yields a complex structure $\mathscr{C}$ on S^2. We know further that $(S^2, \mathscr{S})$ is holomorphically equivalent to the Riemann sphere. Construct an explicit equivalence.

We next show that if S^{2m} admits an almost-complex structure, then m = 1 or m = 3. This is proved in two steps, embodied in the following two theorems. The first is an elementary geometrical result, while the second is a very deep topological theorem. We therefore prove only the former.

Theorem (Kirchhoff). *If S^{n-1} admits an almost-complex structure, then S^n is parallelizable.*

Proof. Let J be an almost-complex structure on S^{n-1}. Identify R^n with R^{n+1} under the inclusion $R^n \subset R^n \times R = R^{n+1}$, $x \mapsto (x,0)$, $x = (x_1, ..., x_n)$. Then each point $w \in S^n$ can be written uniquely in the form $\alpha x + \beta\nu$, where $\alpha > 0$ and $|\beta|$ are the lengths of the projections of w into R^n and R, $x \in S^{n-1}$, with the exceptions of the north pole $\nu = (0, ..., 0, 1)$ and the south pole $\sigma = (0, ..., 0, -1)$.

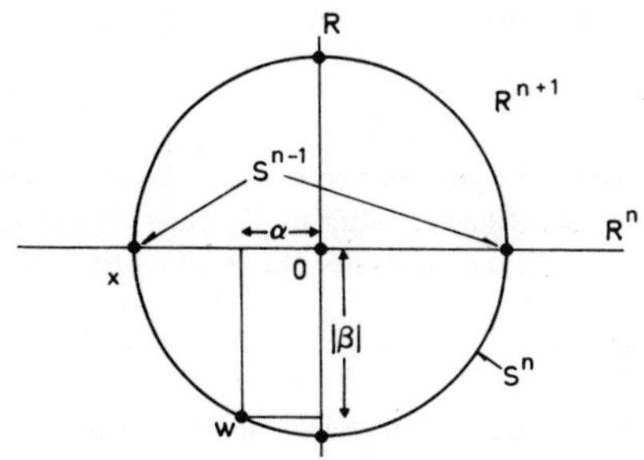

For such a point w, define a linear isomorphism $\theta_w : R^n \to T_w S^n$ as follows. Observe that for each $w \in S^n \setminus \{\sigma, \nu\}$ there is a direct sum decomposition of R^n into $T_x S^{n-1} \oplus L_x$, where L_x is the line generated by $x \in S^{n-1}$. Thus $p \in R^n$ may be written uniquely as $u \oplus tx$, where $u \in T_x S^{n-1}$ and $t \in R$. Put

$$\theta_w(x) = -\alpha\nu + \beta x$$

$$\theta_w(u) = \beta u + \alpha L_x(u)$$

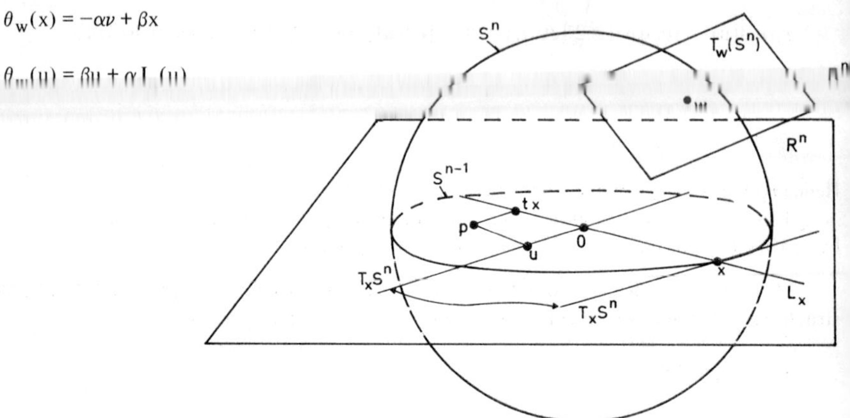

noticing that both $\theta_w(x)$ and $\theta_w(u)$ are perpendicular to w. (In these calculations, we have identified the linear spaces $T_x S^{n-1}$ and $T_w S^n$ with the linear subspaces of R^{n+1} parallel to their geometrical representations in R^{n+1}.)

As $\alpha \to 0$, $\beta \to \pm 1$, so we define $\theta_\nu : R^n \to T_\nu S^n = R^n$ and $\theta_\sigma : R^n \to T_\sigma S^n = R^n$ to be the identity 1_{R^n} and -1_{R^n} respectively.

Thus we have constructed an isomorphism from R^n onto $T_w S^n$ which varies continuously over the whole of S^n. Hence TS^n is vector-bundle isomorphic to $S^n \times R^n$, i.e. S^n is parallelizable.

Theorem. S^n *is parallelizable* $\Leftrightarrow$ n = 1, 3 *or* 7. (See, e.g., Husemoller [4].)

Corollary. *If* S^{2m} *admits an almost-complex structure, then* m = 1 *or* m = 3.

We now show how to construct an example of an almost-complex structure on S^6, using the algebra of Cayley numbers. First we review the salient features of this algebra.

We can regard the algebra of quaternions either as an algebra on R^4 in which addition is given by vector addition, and multiplication is determined by linear extension from the relations

$$i^2 = j^2 = k^2 = ijk = -1$$

where $1 = (1,0,0,0)$, $i = (0,1,0,0)$, $j = (0,0,1,0)$, $k = (0,0,0,1)$, or we can identify $R^4 = R^2 \times R^2 = \mathbb{C} \times \mathbb{C}$, and define addition as above, with multiplication given by

$$(z,w)\,(u,v) = (zu - \bar{v}w,\ vz + w\bar{u})$$

It is convenient to define the Cayley numbers or octonions as the algebra Ω given in terms of the quaternions H just as H is defined in terms of $\mathbb{C}$. Thus we identify R^8 with $R^4 \times R^4 = H \times$ and, taking addition as vector addition, we define the product of $(p,q) \in \Omega$ with $(r,s) \in \Omega$ by

$$(p,q)\,(r,s) = (pr - \bar{s}q,\ sp + q\bar{r})$$

where the **conjugate** $\overline{(z,w)}$ of the quaternion $(z,w) \in \mathbb{C} \times \mathbb{C}$ is defined by $\overline{(z,w)} = (\bar{z}, -w)$.

While the complex numbers form a field, the quaternion multiplication is not commutative and octonion multiplication is not even associative. However, for each $\zeta, \eta \in \Omega$, the relations

$$* \quad (\zeta\eta)\eta = \zeta(\eta\eta), \quad \zeta(\zeta\eta) = (\zeta\zeta)\eta$$

hold.

Exercise. Let the standard basis for R^8 be denoted by $1, i, j, k, e, i', j', k'$. Work out a set of relations between these vectors to determine octonion multiplication, analogous to those given above for quaternions.

Another useful way of talking about quaternions and Cayley numbers is to extend the terminology of "real and imaginary parts" familiar in the context of complex numbers. In the case of quaternions, we write $H = R^4 = R \times R^3$, and say that if $q = (x, y) \in R \times R^3$, then x is the **real** part of q, and y is its **imaginary** part. If $x = 0$, we say that q is **pure imaginary.**

The multiplication of pure imaginary quaternions is related to the standard inner product $\cdot$ and vector product $\times$ in R^3 by the equation

$$\dagger \quad (0, y)\,(0, y') = (-y \cdot y', y \times y')$$

Likewise, we can write $R^8 = R \times R^7$, and express $\zeta \in \Omega$ in the form $(\xi, \eta) \in R \times R^7$, saying that ξ and η are the **real** and **imaginary parts** of ζ, with ζ **pure imaginary** iff $\xi = 0$. Using the formula (†) as a guide, we then define a cross-product $\times$ in R^7 by putting

$$\# \quad (0, \eta)\,(0, \eta') = (-\eta \cdot \eta', \eta \times \eta')$$

where $\cdot$ denotes the standard inner product in R^7.

Exercise. Show that if η, η' and η'' are points of R^7, then $\eta \times \eta' = -\eta' \times \eta$, and $(\eta \times \eta') \cdot \eta'' = \eta \cdot (\eta' \times \eta'')$.
Deduce that

$$(0, \eta)\,(0, \eta) = (-\|\eta\|^2, 0) \in R \times R^7$$

We are at last in a position to apply all this to the 6-sphere

$$S^6 = \{\eta \in R^7 : \|\eta\| = 1\}$$

As we did above, we regard the linear space $T_\eta S^6$ as the linear subspace of R^7 parallel to the geometrical tangent space to S^6 at η. Hence

$$T_\eta S^6 = \{\zeta \in R^7 : \zeta \cdot \eta = 0\}$$

Now define $J_\eta : T_\eta S^6 \to T_\eta S^6$ by $J_\eta(\zeta) = \eta \times \zeta$. Then, using the results of the above exercise, we see that $J_\eta(\zeta) \cdot \eta = (\eta \times \zeta) \cdot \eta = -(\zeta \times \eta) \cdot \eta = -\zeta \cdot (\eta \times \eta) = 0$. So J_η is well-defined. Trivially, J_η is linear, so we have only to show that $(J_\eta)^2 = -1$.

Since $J_\eta(J_\eta(\zeta)) = \eta \times (\eta \times \zeta)$, we can write (bearing formula (#) in mind)

$$\begin{aligned}(0, J_\eta(J_\eta(\zeta))) &= (0, \eta)\,(0, \eta \times \zeta) + (\eta \cdot (\eta \times \zeta), 0)\\ &= (0, \eta)\,((0, \eta)\,(0, \zeta)) + (0, \eta)\,(\eta \cdot \zeta, 0)\\ &= ((0, \eta)\,(0, \eta))\,(0, \zeta) \quad \text{(by formula } (*))\\ &= (-\|\eta\|^2, 0)\,(0, \zeta) = (-1, 0)\,(0, \zeta) = (0, -\zeta)\end{aligned}$$

This proves that J is an almost-complex structure (in fact C^∞) on S^6.

Naturally, one should calculate the torsion tensor field τ_J. It turns out that $\tau_J \neq 0$, so J is not integrable. However, this does not prove that there may not be some other almost-complex structure on S^6 that is integrable.

Each almost-complex structure J on S^6 induces an almost-complex structure on an immersed oriented hypersurface of R^7, as follows. Let X be a C^∞-oriented 6-manifold, and $f: X \to R^7$ a C^∞ immersion. Then for each $w \in X$ there is a unique unit normal vector v to f(U) at f(w), where U is some open neighbourhood of x in X smoothly embedded by f in R^7, such that v forms a positively oriented frame in R^7 when added to the image under Tf of a positively oriented tangent frame to X at x. The translate $v' = v - f(x)$ of v to 0 is a point of S^6, and we may identify $Tf(T_x X)$ with $T_{v'}S^6$. Thus the endomorphism $J_{v'}$ of any almost-complex structure J on S^6 determines such an endomorphism on $Tf(T_x X)$ and hence another $\widetilde{J} = (Tf^{-1})J(Tf)$ on $T_x X$ itself. If J is C^∞, so is $\widetilde{J}$.

We remark also that the cotangent bundle T^*X of any C^∞ manifold X admits an almost complex structure (see the exercise at the end of § 7).

Despite the statement on p.10 of [9], it remains an unsolved problem (to the best of my knowledge) whether there is an integrable almost-complex structure on S^6. The reader of Chern's book [2] will know that van de Ven [12] has shown the existence of compact connected 4-manifolds that admit an almost-complex structure but no complex structure. In fact, there are infinitely many such manifolds, including

$$P_2(\mathbb{C}) \# (S^1 \times S^3) \# (S^1 \times S^3) \quad \text{and} \quad (S^2 \times S^2) \# (S^1 \times S^3) \# (S^1 \times S^3)$$

where $P_2(\mathbb{C})$ denotes the complex projective plane and # denotes the connected sum operation on manifolds. Van de Ven attributes to A. Howard the observation that if V is a non-singular algebraic surface of degree 4 in $P_3(\mathbb{C})$, then for any $k \geqslant 1$, $\#^{2k+1} V$ is a compact **simply-connected** 4-manifold which admits an almost-complex, but no complex, structure.

Thus for compact manifolds there can be a topological obstruction to deforming an almost-complex structure into a complex structure. For **open** manifolds (no compact component) the situation is rather different. Thus Landweber [7] (see also Brender [1]) has recently published proofs of the following results.

(1) Let X be an open 2m-manifold with $H^i(X, \mathbb{Z}) = 0$ for $i > m$. Then every almost-complex structure on X is homotopic to a complex structure.

(2) Let X be an open 2m-manifold with $H^i(X, \mathbb{Z}) = 0$ for $i \geqslant m$. Then there is a natural bijection from the set of homotopy classes of complex structures on X to the set of homotopy classes of almost-complex structures on X.

(3) Let X be a manifold whose stable tangent bundle admits a complex structure. Then $X \times R^n$ admits a complex structure, where $n = \dim_R X$.

For recent discussions of such integrability problems, see Pollack [10] or Kumpera and Spencer [6].

10. HERMITIAN AND KÄHLER MANIFOLDS

We have defined a Hermitian structure in a vector bundle $p: B \to X$ with fibre E as a section γ of the associated bundle $T_2^0 E$ which is symmetric, positive definite, and for which $\gamma_x(\lambda, \mu) = \gamma_x(J(\lambda), J(\mu))$ for all $x \in X$ and all $\lambda, \mu \in E_x$, where J is a given almost-complex structure in p.

In case X is a C^∞-manifold and $p: B \to X$ is the tangent bundle $\pi_X: TX \to X$, we say that, for a given almost-complex structure J on X, a Hermitian structure γ in TX is an **almost-Hermitian structure on** X, and we call (X,γ) an almost-Hermitian manifold with respect to J. If J is integrable, so that X is a complex manifold, then (X,γ) is said to be a **Hermitian manifold** (with respect to this complex structure).

If X is any almost-complex manifold, with almost-complex structure J, then X admits an almost-Hermitian structure with respect to J. We can prove this directly using a partition of unity on X, and bearing in mind that R^{2m} admits an almost-Hermitian structure with respect to J. Alternatively, we can use the fact that any smooth manifold admits a C^∞ Riemannian structure g (proved by using a partition of unity as above, or by using the fact that X can be smoothly embedded in some Euclidean space), to construct an almost-Hermitian structure γ by putting

$$\gamma_x(\lambda,\mu) = g_x(\lambda,\mu) + g_x(J(\lambda),J(\mu))$$

Thus
$$\begin{aligned}\gamma_x(J(\lambda),J(\mu)) &= g_x(J(\lambda),J(\mu)) + g_x(J^2(\lambda),J^2(\mu)) \\ &= g_x(J(\lambda),J(\mu)) + g_x(-\lambda,-\mu) \\ &= g_x(J(\lambda),J(\mu)) + g_x(\lambda,\mu) = \gamma_x(\lambda,\mu)\end{aligned}$$

Thus the mere existence of an almost-Hermitian structure on an almost-complex manifold imposes no further restrictions on the topology of the manifold. Nor does that of a Hermitian structure on a complex manifold. Thus the study of Hermitian geometry is a matter of exploring particular kinds of Hermitian and almost-Hermitian structures in the spirit of Riemannian geometry as a whole.

We now proceed to consider a special class of Hermitian structures, the existence of which imposes quite severe conditions on the topology of the manifold. To define this class of structure, we need the notion of **exterior derivative** for differential forms. This can be defined for differential forms on a smooth manifold, but not in vector bundles over any topological space. This is why we can talk of Hermitian structures in vector bundles, but now restrict ourselves to the tangent bundle of a smooth almost-complex manifold.

Suppose then that X is a C^∞ 2m-manifold. An exterior differential p-form ϕ can be expressed in the local co-ordinates x of some chart $\xi: U \to U'$ on X as

$$\phi(x) = \frac{1}{p!} \sum_{\alpha_1, \ldots, \alpha_p=1}^{2m} \phi_{\alpha_1 \ldots \alpha_p}(x)\, dx_{\alpha_1} \wedge \ldots \wedge dx_{\alpha_p}$$

We define the exterior derivative $d\phi$ of ϕ on U to be the $(p+1)$-form on U whose expression in ξ co-ordinates is

$$d\phi(x) = \frac{1}{p!} \sum_{\alpha_0, \alpha_1, \ldots, \alpha_p=1}^{2m} \left(\frac{\partial\phi(x)}{\partial x_{\alpha_0}} dx_{\alpha_0} \wedge dx_{\alpha_1} \wedge \ldots \wedge dx_{\alpha_p}\right)$$

which we can rewrite as

$$\frac{1}{(p+1)!} \sum_{\alpha_0, \alpha_1, \ldots, \alpha_p=1}^{2m} \psi_{\alpha_0 \alpha_1 \ldots \alpha_p}(x)\, dx_{\alpha_0} \wedge dx_{\alpha_1} \wedge \ldots \wedge dx_{\alpha_p}$$

Exercise. Express $\psi_{\alpha_0 \ldots \alpha_p}$ in terms of the derivatives of the functions $\phi_{\alpha_0 \ldots \alpha_p}$.

It is a routine exercise to verify that this definition is compatible with changes of local co-ordinates, and so $d\phi$ is a well defined $(p+1)$-form on X.

Exercise. Show that $d(d\phi) = 0$, for any p-form ϕ on X, $p \geqslant 0$. (A 0-form is just a C^∞ function f on X, and df is its differential. The relation $d(d\phi) = 0$ is fundamental and provides the link between the algebra of differential forms and the cohomology structure of X.)

Exercise. Let ϕ be a p-form on X, and ψ a q-form on X. Show that $d(\phi \wedge \psi) = d\phi \wedge \psi + (-1)^p \phi \wedge d\psi$.

Suppose now that J is an almost-complex structure on X, and γ an almost-Hermitian structure with respect to J. Then we can define an exterior 2-form ω on X by $\omega_x(\lambda, \mu) = \gamma_x(\lambda, J(\mu))$, and consider exterior derivative $d\omega$. If $d\omega = 0$, then γ is said to be an **almost-Kähler structure** on X with respect to J. If J is integrable, so that X is a complex manifold, then γ is said to be a **Kähler structure** on X, and (X, γ) is a **Kähler manifold.** We shall concentrate on the latter case, so from now on we suppose that X is a complex manifold with integrable almost-complex structure J, and that γ is a Kähler structure on X.

We consider the complexification $T_{\mathbb{C}}X$ of TX and its decomposition into the Whitney sum $T' \oplus T''$ by eigenspaces of J. If $z_1, \ldots, z_m$ are local complex co-ordinates for X on some neighbourhood $U \subset X$, then we can write any exterior form ϕ of type (p, q) as

$$p!\, q!\, \phi(z) = \sum_{\substack{\alpha_1 \ldots \alpha_p = 1 \\ \beta_1 \ldots \beta_q = 1}}^{m} \phi_{\alpha\bar{\beta}}(z) dz_{\alpha_1} \wedge \ldots \wedge dz_{\alpha_p} \wedge d\bar{z}_{\beta_1} \wedge \ldots \wedge d\bar{z}_{\beta_q}$$

The operator d can be extended to $T_{\mathbb{C}}X$ in an obvious way, and we introduce further operators $\partial, \bar{\partial}$ by

$$\partial\phi(z) = \frac{1}{p!\, q!} \sum_{\left.\begin{matrix}\alpha_0\, \alpha_1 \ldots \alpha_p \\ \beta_1\, \beta_2 \ldots \beta_q\end{matrix}\right\} = 1}^{m} \frac{\partial \phi_{\alpha\bar{\beta}}}{\partial z_{\alpha_0}} dz_{\alpha_0} \wedge dz_{\alpha_1} \wedge \ldots \wedge dz_{\alpha_p} \wedge d\bar{z}_{\beta_1} \wedge \ldots \wedge d\bar{z}_{\beta_q}$$

$$\bar{\partial}\phi(z) = \frac{1}{p!\, q!} \sum_{\left.\begin{matrix}\alpha_1 \alpha_2 \ldots \alpha_p \\ \beta_0 \beta_1 \ldots \beta_q\end{matrix}\right\} = 1}^{m} \frac{\partial \phi_{\alpha\bar{\beta}}}{\partial \bar{z}_{\beta_0}} d\bar{z}_{\beta_0} \wedge dz_{\alpha_1} \wedge \ldots \wedge dz_{\alpha_p} \wedge d\bar{z}_{\beta_1} \wedge \ldots \wedge d\bar{z}_{\beta_q}$$

Exercise. Show that $\partial\partial = 0$, $\bar{\partial}\bar{\partial} = 0$, $\partial + \bar{\partial} = d$, and $\partial\bar{\partial} = -\bar{\partial}\partial$.

$\left(\text{Remember that } \frac{\partial}{\partial z} = \frac{\partial}{\partial x} - i\frac{\partial}{\partial y}\ ,\ \frac{\partial}{\partial \bar{z}} = \frac{\partial}{\partial x} + i\frac{\partial}{\partial y}\right)$

In this context, we study the 2-form

$$\tilde{\omega}(z) = -2i \sum_{\alpha, \beta = 1}^{m} h_{\alpha\bar{\beta}}(z) dz_\alpha \wedge d\bar{z}_\beta$$

of the Hermitian structure γ, and note that $d\tilde{\omega} = 0$ since $d\omega = 0$. (If ϕ is of type (p, q), then $\partial\phi$ is of type (p + 1, q) and $\bar{\partial}\phi$ of type (p, q + 1).)

Exercise. Let

$$\tilde{\omega} = -2i \sum_{\alpha, \beta = 1}^{m} h_{\alpha\bar{\beta}}\, dz_\alpha \wedge d\bar{z}_\beta$$

be the 2-form of a Hermitian structure γ on a complex manifold X. Show that γ is a Kähler structure, i.e. $d\tilde{\omega} = 0$, iff for all $\alpha, \beta, \gamma = 1, \ldots, m$,

$$\boxdot \qquad \frac{\partial h_{\alpha\bar{\beta}}}{\partial z_\gamma} = \frac{\partial h_{\gamma\bar{\beta}}}{\partial z_\alpha} \qquad \text{or} \qquad \frac{\partial h_{\alpha\bar{\beta}}}{\partial \bar{z}_\gamma} = \frac{\partial h_{\alpha\bar{\gamma}}}{\partial \bar{z}_\beta}$$

We now examine the nature of the components $h_{\alpha\bar{\beta}}$ of the Hermitian matrix associated with a Kähler structure γ on X.

Suppose that V is an open subset of $\mathbb{C}^m$, and let $f: V \to R$ be a C^∞ real-valued function on V. Consider the matrix-valued function $H: V \to M_m$ given by

$$H(z) = \left[\frac{\partial^2 f(z)}{\partial z_\alpha \partial \bar{z}_\beta}\right] = [h_{\alpha\bar{\beta}}(z)]$$

Then $h_{\alpha\bar{\beta}} = \overline{h_{\beta\bar{\alpha}}}$, so H(z) is a Hermitian matrix.

Further, relations $\boxdot$ above are obviously satisfied, so if H(z) is positive-definite, then we can define a Kähler form on V by

$$\tilde{\omega} = -2i \sum_{\alpha, \beta = 1}^{m} \frac{\partial^2 f}{\partial z_\alpha \partial \bar{z}_\beta} dz_\alpha \wedge d\bar{z}_\beta$$

$$= -2i\, \partial\bar{\partial} f$$

Conversely, suppose that ϕ is an exterior 2-form on a simply-connected open set V, given by

$$\phi(z) = \sum_{\alpha,\beta = 1}^{m} h_{\alpha\bar{\beta}}(z) dz_\alpha \wedge d\bar{z}_\beta$$

with $d\phi = 0$. Then there is a real-valued C^∞ function $f: V \to R$ such that $\phi = \partial\bar{\partial} f$. We shall indicate the proof of this fact, as follows. We need two general theorems which we state without proof.

(i) Let θ be a C^∞ form of type (p,q) on $V \subset \mathbb{C}^m$, $q \geqslant 1$. Then $\bar{\partial}\theta = 0$ iff there is a C^∞ form ψ on V of type (p, q−1) with $\theta = \bar{\partial}\psi$, and $\partial\theta = 0$ iff there is a C^∞ form ζ on V of type (p−1, q) with $\theta = \partial\zeta$.

(ii) Let θ be a holomorphic p-form on $V \subset \mathbb{C}^m$, $p \geqslant 1$. Then there is a holomorphic (p−1)-fo ψ on V with $d\psi = \phi$ iff $d\phi = 0$.

Suppose then that $d\phi = 0$. Now $d = \partial + \bar{\partial}$, so $\partial\phi + \bar{\partial}\phi = 0$. But $\partial\phi$ is of type (2, 1), while $\bar{\partial}\phi$ is of type (1,2). So $\partial\phi = 0$ and $\bar{\partial}\phi = 0$. By (i), there is a form ψ on V of type (1, 0) such that $\phi = \bar{\partial}\psi$. Now we can write

$$\psi(z) = \sum_{\alpha=1}^{m} \psi_\alpha(z)\, dz_\alpha$$

for some C^∞ functions $\psi_\alpha \colon V \to R$.

Hence $0 = \partial\phi = \partial\bar{\partial}\psi = -\bar{\partial}\partial\psi$. But

$$\partial\psi(z) = \partial \sum_{\alpha=1}^{m} \psi_\alpha(z) dz_\alpha$$

$$= \sum_{\alpha,\beta=1}^{m} \frac{\partial\psi_\alpha(z)}{\partial z_\beta}\, dz_\beta \wedge dz_\alpha$$

$$= -\sum_{\alpha,\beta=1}^{m} \frac{\partial\psi_\alpha(z)}{\partial z_\beta}\, dz_\alpha \wedge dz_\beta$$

$$= \frac{1}{2} \sum_{\alpha,\beta=1}^{m} \left(\frac{\partial\psi_\alpha(z)}{\partial z_\beta} - \frac{\partial\psi_\beta(z)}{\partial z_\alpha} \right) dz_\beta \wedge dz_\alpha$$

Hence $0 = \bar{\partial}\partial\psi$ implies that for all $\alpha, \beta, \gamma = 1, \ldots, m$,

$$\frac{\partial}{\partial \bar{z}^\gamma} \left(\frac{\partial\psi_\alpha}{\partial z_\beta} - \frac{\partial\psi_\beta}{\partial z_\alpha} \right)(z) = 0$$

That is, $\partial\psi$ is a holomorphic 2-form on V. By (ii), there is a holomorphic 1-form η on V such that $\partial\psi = d\eta = \partial\eta$. It follows that $\partial(\psi - \eta) = 0$, and so, by (i), there is a C^∞ function $f\colon V \to R$ such that $-\psi + \eta = \partial f$.

We conclude that

$$\phi = \bar{\partial}\psi = \bar{\partial}(\eta - \partial f) = -\bar{\partial}\partial f = \partial\bar{\partial} f$$

Now suppose that $\tilde{\omega}$ is the 2-form of a Hermitian structure γ on a complex manifold X. Then it follows that γ is a Kähler structure iff for each complex chart $\xi: U \to U'$ on X there is a C^∞ real-valued function $f_\xi: U' \to R$ such that $\tilde{\omega} = -2i\,\partial\bar{\partial} f_\xi$ on U, with U′ a simply-connected open subset of $\mathbb{C}^m$.

We illustrate this result by constructing the standard Fubini-Study Kähler structure on complex projective m-space $P_m(\mathbb{C})$. Recall that for each $j = 0, 1, \ldots, m$, there is a chart $\xi_j: U_j \to \mathbb{C}^m$, where $U_j = \{[z]: z_j \neq 0\}$, and

$$\xi_j([z]) = \left(\frac{z_0}{z_j}, \ldots, \frac{\hat{z}_j}{z_j}, \ldots, \frac{z_m}{z_j}\right) = (w_1, \ldots, w_m) \in \mathbb{C}^m$$

and the set $\{\xi_0, \ldots, \xi_m\}$ is a complex atlas on $P_m(\mathbb{C})$.

Define $f_j: \mathbb{C}^m \to R$ by

$$f_j(w) = \log\left(1 + \sum_{r=1}^{m} |w_r|^2\right)$$

and hence define $K_j = f_j \circ \xi_j: U_j \to R$, $j = 0, \ldots, m$. Then we can write

$$K_j([z]) = \log\left(1 + \sum_{\alpha \neq j} \left|\frac{z_\alpha}{z_j}\right|^2\right)$$

$$= \log\left(\sum_{s=0}^{m} |z_s|^2\right) - \log|z_j|^2$$

Thus for each $p = [z] \in U_j \cap U_k$, we have

$$K_j([z]) - K_k([z]) = \log|z_k|^2 - \log|z_j|^2$$

$$= \log\left|\frac{z_k}{z_j}\right|^2 = \log|w_k|^2 \qquad (k < j, \text{ say})$$

$$= \log(w_k \bar{w}_k)$$

$$= \log w_k + \log \bar{w}_k$$

Hence $\partial\bar{\partial}(K_j - K_k)([z]) = \partial\bar{\partial}(\log w_k + \log \bar{w}_k) = 0$.

This shows that the form $\tilde{\omega} = -2i\,\partial\bar{\partial} K_j$, $j = 0, \ldots, m$, is well defined on the whole of $P_m(\mathbb{C})$, and, as we have already seen, $d\tilde{\omega} = 0$. We have therefore constructed a Kähler form, and so a Kähler metric, on $P_m(\mathbb{C})$.

11. COMPLEX SUBMANIFOLDS OF COMPLEX MANIFOLDS

Let S be a subset of a complex m-manifold X. We say that S is a **complex S-dimensional submanifold** of X iff for all $x \in S$ there is an open neighbourhood U of x in X and a chart ξ: $U \to U'$ in the complex structure of X such that $\xi(S \cap U) = \mathbb{C}^s \cap \xi(U) (= \mathbb{C}^s \cap U')$, where $\mathbb{C}^s$ is identified with the linear subspace of $\mathbb{C}^m$ generated by the first s standard basis vectors.

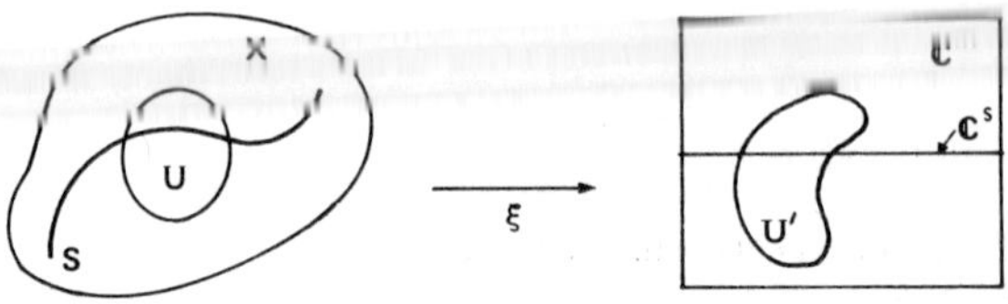

It follows that any complex submanifold S of X is itself a complex s-manifold.

Example. Show that $\mathbb{C}^s$ is a complex submanifold of $\mathbb{C}^m = \mathbb{C}^s \times \mathbb{C}^{m-s}$.

Exercise. Regarding $\mathbb{C}^s$ as a linear subspace of $\mathbb{C}^m$, show that the Grassmannian G(r,s−r) is a complex submanifold of G(r,m−r).

One way in which complex submanifolds arise is as follows. Let X and Y be complex manifolds, and let $f:X \to Y$ be a holomorphic map onto Y. Then for each $x \in X$, the derivative $Tf(x):T_xX \to T_yY$, $y = f(x)$, is a complex linear map (with matrix $Jf(x) = [\partial y_r/\partial x_s]$ with respect to local co-ordinate systems at x and y). Suppose that the rank of Tf(x) is equal to the dimension n of Y, for all $x \in X$. Then every contour $f^{-1}(y)$ of f, $y \in Y$, is a complex submanifold of X, of dimension m−n, where m = dim X.

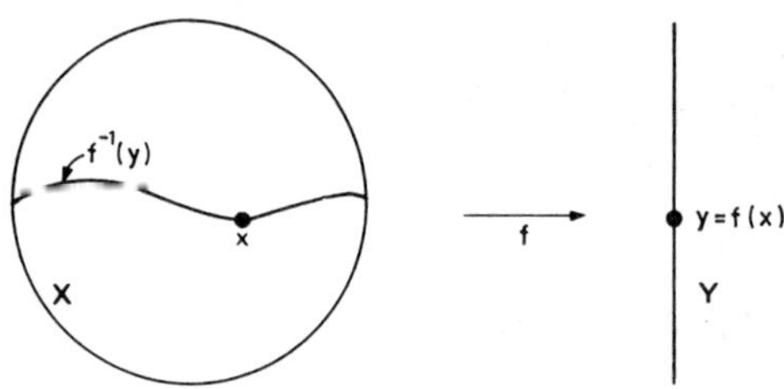

Exercise. Formulate a complex analogue of the implicit mapping theorem to obtain a localized version of the above situation.

An important example is the case where X is complex projective m-space $P_m(\mathbb{C})$. Suppose that $p_1, \ldots, p_k$ are homogeneous polynomial functions on $\mathbb{C}^{m+1}$, $p_r:\mathbb{C}^{m+1} \to \mathbb{C}$, r = 1, ..., k. Then there is a map $p:\mathbb{C}^{m+1} \to \mathbb{C}^k$ given by $p = (p_1, \ldots p_k)$. We study the zero contour $V = p^{-1}(0)$ of p. Since each p_r is homogeneous, for each $\lambda \in \mathbb{C}$, $\lambda V = \lambda$. That is, if $z \in V$, then the complex line generated by z ($\neq 0$) lies in V. Thus V is a complex cone with vertex $0 \in \mathbb{C}^{m+1}$. It follows that the set S of generators of this cone is a subset of $P_m(\mathbb{C})$. Such a set S is called an **algebraic variety** (or **complex** algebraic variety). Of course S need not be a complex submanifold of $P_m(\mathbb{C})$, but if the derivative of p has constant rank at all $z \in \mathbb{C}^{m+1}\backslash\{0\}$, then S *is* a complex submanifold of S, called a **non-singular** complex algebraic variety, or complex algebraic submanifold of $P_m(\mathbb{C})$.

Example. Let $\alpha_0, ..., \alpha_m \in \mathbb{C}$, and define $p:\mathbb{C}^{m+1} \to \mathbb{C}$ by

$$p(z) = \sum_{r=0}^{m} \alpha_r z_r$$

Then $Jp(z) = [\alpha_0\ \alpha_1\ ...\ \alpha_m]$ (a complex $1 \times (m+1)$ matrix). Thus if at least one $\alpha_r \neq 0$, the set S defined by $p^{-1}(0)$ as above is a complex algebraic submanifold of $P_m(\mathbb{C})$, of dimension $m-1$, a complex hypersurface of $P_m(\mathbb{C})$.

Exercise. Show that this hypersurface is complex isomorphic to $P_{m-1}(\mathbb{C})$.

Example. Let $p:\mathbb{C}^{m+1} \to \mathbb{C}$ be given by $p(z) = z_0^2 + ... + z_m^2$. Then $Jp(z) = [2z_0\ ...\ 2z_m]$, so the rank of $Jp(z)$ is 1 on $\mathbb{C}^{m+1}\backslash\{0\}$. Thus $p^{-1}(0)$ gives a complex hypersurface of $P_m(\mathbb{C})$ (a complex quadric).

It is an easy exercise to check from the definition of complex submanifold that if S is a complex submanifold of a Kähler manifold X, then the Riemannian structure induced on S by the Kähler structure of X is also a Kähler structure. Since we have given a Kähler structure to $P_m(\mathbb{C})$, it follows that every complex submanifold of $P_m(\mathbb{C})$ admits a Kähler structure. In particular, every non-singular complex algebraic variety admits a Kähler structure. In fact, every compact complex submanifold of $P_m(\mathbb{C})$ is an algebraic variety (Chow). The question therefore arises of whether there are compact Kähler manifolds that are not algebraic varieties. The answer is that there are many such manifolds. For instance, there are complex tori (admitting Kähler metrics) which are not algebraic, as in the paper of de la Harpe in these Proceedings.

We have indicated earlier that there are complex manifolds that do not admit any Kähler structure, and the final section of these notes will establish some of the topological conditions that any Kähler manifold must satisfy.

Much of what we have discussed may be expressed in the following sequence of implications, none of which is reversible (there are topological manifolds with no C^∞-structure; not every C^∞-manifold is orientable; some manifolds admit an almost-complex structure but no complex structure, and so on), except where indicated.

complex algebraic ⇒ Kähler ⇒ complex ⇒ almost-complex ⇒ even-dimensional / oriented / C^∞ ⇒ topological

complex ⇕ Hermitian; almost-complex ⇕ almost-Hermitian

Hermitian ⇒ almost-Hermitian

12. TOPOLOGY OF KÄHLER MANIFOLDS

Consider the cohomology groups of a complex manifold X over, say, the field $\mathbb{C}$ of complex numbers. Then for each $r = 0, 1, ..., 2m = \dim_R X$, the cohomology group $H^r(X, \mathbb{C})$ is a complex linear space. If we assume further that X is compact, then the dimension β_r of $H^r(X, \mathbb{C})$ is finite. This integer β_r is called the **r^{th} Betti number** of X.

Now the group $H^r(X, \mathbb{C})$ may be calculated from a knowledge of the exterior forms on X, using the theorem of de Rham. For a modern formulation of this theorem, it is necessary to make use of the theory of sheaves. In particular, we need the sheaves of germs of differential s-forms,

$s = 0, 1, \ldots, m$, on X. Roughly speaking, we consider the group $\Omega^r X$ of r-forms on X, and note that $d = d_r : \Omega^r X \to \Omega^{r+1} X$ is a homomorphism with $d^2 = 0$. Thus $\operatorname{im} d_{r-1} \subset \ker d_r$, and we can form the quotient $\ker d_r / \operatorname{im} d_{r-1}$. Then de Rham's theorem, in essence, says that this quotient group is isomorphic to $H^r(X, \mathbb{C})$.

A p-form ϕ for which $d\phi = 0$ is said to be **closed**, while a p-form which can be written as $d\psi$ for some (p-1)-form ψ is said to be **exact**. So we think of $H^r(X, \mathbb{C})$ as the group of closed r-forms divided by the group of exact r-forms.

Our aim in this section is to show that if X is a compact Kähler manifold of dimension m, then for all $k = 0, 1, \ldots, m$, the Betti number β_{2k} is non-zero (i.e. $\beta_{2k} \geqslant 1$). Thus we have to show that for each $k = 0, 1, \ldots, m$ there is a 2k-form on X which is closed but not exact.

We do this by considering the Kähler form ($\widetilde{\omega} =$) ω on X. This is a 2-form, so we can obtain a 2k-form $\omega^k = \omega \wedge \ldots \wedge \omega$ (k factors), $k = 1, \ldots, m$. (We need not worry about β_0, since $\beta_0 = \beta_{2m}$ anyway.) First we note that $d(\omega^k) = 0$ since $d\omega = 0$ (recall that $d(\omega \wedge \omega) = d\omega \wedge \omega + (-1)^2 \omega \wedge d\omega$ etc.). So ω^k is closed. We have only to show, therefore, that ω^k is not exact.

Next suppose that ω^m is exact, say $\omega = d\phi$ for some 1-form ϕ on X. Now by the generalized Stokes' theorem,

$$\int_X d\psi = \int_{\partial X} \psi$$

for any p-form ψ on X, $p \leqslant \dim X{-}1$. But in our case, $\partial X = \phi$, so $\int_X d\phi = \int_X \omega = 0$.

Further, if ω^k is exact, say $\omega^k = d\theta$, then $\omega^m = \omega^k \wedge \omega^{m-k} = d\theta \wedge \omega^{m-k} = d(\theta \wedge \omega^{m-k})$ since ω^k is closed. This shows that we have only to prove that $\int_X \omega^m \neq 0$ to establish that $\beta_{2k} \geqslant 1$, $k = 0, 1, \ldots, m$. We do this by proving that ω^m is a non-zero constant multiple of the volume 2m-form on X.

Consider then the expression for ω in local co-ordinates $z_1, \ldots, z_m$ on the domain U of some complex chart on X. Since we are not concerned with the precise expression for ω, we can ignore non-zero scalar multiples, and write

$$\omega = \sum_{\alpha, \beta = 1}^{m} h_{\alpha\bar{\beta}}(z)\, dz_\alpha \wedge d\bar{z}_\beta$$

Hence

$$\omega^m = \omega \wedge \ldots \wedge \omega \text{ (m factors)}$$

$$= \sum_{\substack{\alpha_1, \ldots, \alpha_m \\ \beta_1, \ldots, \beta_m}=1}^{m} h_{\alpha_1\bar{\beta}_1} h_{\alpha_2\bar{\beta}_2} \cdots h_{\alpha_m\bar{\beta}_m}\, dz_{\alpha_1} \wedge d\bar{z}_{\beta_1} \wedge \ldots \wedge dz_{\alpha_m} \wedge d\bar{z}_{\beta_m}$$

$$= \sum_{\alpha, \beta = 1}^{m} \epsilon(\alpha_1, \ldots, \alpha_m)\epsilon(\beta_1, \ldots, \beta_m) h_{\alpha_1\bar{\beta}_1} h_{\alpha_2\bar{\beta}_2} \cdots h_{\alpha_m\bar{\beta}_m}\, dz_1 \wedge d\bar{z}_1 \wedge \ldots \wedge dz_m \wedge d\bar{z}_m$$

where $\epsilon(\alpha_1, \ldots, \alpha_m) = 1, -1$ or 0 according as the map $(1, 2, \ldots, m) \mapsto (\alpha_1, \ldots, \alpha_m)$ is an even or an odd permutation, or is not bijective. So if we eliminate the last case and denote the symmetric group of degree m by S_m as usual, we can write

$$\omega^m = \sum_{\sigma, \tau \in S_m} \operatorname{sgn} \sigma \operatorname{sgn} \tau \, h_{\sigma_1 \bar{\tau}_1} \cdots h_{\sigma_m \bar{\tau}_m} \, dz_1 \wedge d\bar{z}_1 \wedge \ldots \wedge dz_m \wedge d\bar{z}_m$$

But

$$\sum_{\sigma \in S_m} \operatorname{sgn} \sigma \, h_{\sigma_1 \bar{\tau}_1} \cdots h_{\sigma_m \bar{\tau}_m} = \det \begin{bmatrix} h_{1\bar{\tau}_1} & \cdots & h_{1\bar{\tau}_m} \\ & & \\ h_{m\bar{\tau}_1} & \cdots & h_{m\bar{\tau}_m} \end{bmatrix}$$

$$= \operatorname{sgn} \tau \det \begin{bmatrix} h_{1\bar{1}} & \cdots & h_{1\bar{m}} \\ & & \\ h_{m\bar{1}} & \cdots & h_{m\bar{m}} \end{bmatrix} = (\operatorname{sgn} \tau)\, \Delta$$

where $\Delta > 0$.

Hence

$$\omega^m = \sum_{\tau \in S_m} (\operatorname{sgn} \tau)(\operatorname{sgn} \tau)\, \Delta = m!\, \Delta$$

and so

$$\int_X \omega^m = (-i)^m \, m! \; 2^m \operatorname{vol}(X) \neq 0$$

Since $dz \wedge d\bar{z} = (dx + i dy) \wedge (dx - i dy) = -2i \, dx \wedge dy$.

Example. Let X be the Hopf manifold discussed earlier. This is a complex manifold diffeomorphic to $S^1 \times S^{2m-1}$. If $m > 1$, then $\beta_{2k}(S^1 \times S^{2m-1}) = 0$ for $0 < k < m$. So these Hopf manifolds do not admit any Kähler structure.

13. CLOSING REMARKS

These notes are extremely elementary in the sense that we have not explored the interaction between connections on a manifold and Hermitian or Kähler structures on the manifold. In particular, we have not studied holomorphic sectional curvature on Kähler manifolds. Nor have we developed any of the sheaf theory that is needed to discuss most of the deeper results on complex manifolds – deformation of complex structures, for example. These and many other aspects of complex manifolds are discussed in the wide variety of books and research articles now available. Those listed here include some of the better known sources on which I have drawn freely in preparing these notes, and in which much more detailed bibliographies can be found.

REFERENCES

The reader's attention is drawn particularly to Refs [2], [5], [9] *and* [14], *which cover much of the material of this paper*

[1] BRENDER, A., Stable complex structures on real manifolds, J. Diff. Geom. **8** (1973) 579–588.

[2] CHERN, S.S., Complex Manifolds Without Potential Theory, Van Nostrand, Princeton (1967).

[3] HELGASON, S., Differential Geometry and Symmetric Spaces (Pure and Applied Mathematics Ser. Vol.12), Academic Press, New York (1962).

[4] HUSEMOLLER, D., Fibre Bundles, McGraw-Hill, New York (1966).

[5] KOBAYASHI, S., NOMIZU, K., Foundations of Differential Geometry, Vol.II, Wiley (Interscience) New York (1969).

[6] KUMPERA, A., SPENCER, D.C., Lie Equations: I – General Theory (Annals of Mathematics Studies No.73), Princeton Univ. Press (1972).

[7] LANDWEBER, P.S., Complex structures on open manifolds, Topology **13** (1974) 69–75.

[8] LICHNEROWICZ, A., Théorie Globale des Connexions et des Groupes d'Holonomie, Rome (1955).

[9] MORROW, J., KODAIRA, K., Complex Manifolds, Holt, Rinehart and Winston, New York (1971).

[10] POLLACK, A.S., The integrability problem for pseudogroup structures, J. Diff. Geom. **9** (1974) 355–390.

[11] UENO, K., Classification Theory of Algebraic Varieties and Compact Complex Spaces, Lecture Notes in Mathematics 439, Springer, Berlin (1975).

[12] VAN DE VEN, A., On the Chern numbers of certain complex and almost-complex manifolds, Proc. Natl Acad. Sci. USA, **55** (1966) 1624–1627.

[13] WEIL, A., Introduction à L'Etude des Variétés Kähleriennes, Hermann, Paris (1958).

[14] WELLS, R.O., Differential Analysis on Complex Manifolds, Prentice-Hall, Englewood Cliffs (1973).

TOEPLITZ MATRICES AND TOEPLITZ OPERATORS

H. WIDOM
Division of Natural Sciences,
University of California,
Santa Cruz, California,
United States of America

Abstract

TOEPLITZ MATRICES AND TOEPLITZ OPERATORS.

1. Introduction. 2. Toeplitz matrices and orthogonal polynomials. 3. Asymptotics of Toeplitz determinants and orthogonal polynomials. 4. Eigenvalues of Toeplitz matrices. 5. Hardy spaces. 6. Toeplitz operators. 7. Projection methods. 8. Classes of compact operators. 9. Applications to Toeplitz matrices.

1. INTRODUCTION

Toeplitz matrices and Toeplitz operators arise in and have application to a wide variety of fields (probability theory, numerical analysis, K-theory, complex analysis, statistical mechanics, etc.). These lectures will cover a selection of interrelated topics varying from classical to contemporary.

A Toeplitz matrix is a matrix (of finite or infinite order) whose j,k entry depends only on the difference $j-k$. A Toeplitz operator is an operator which, with respect to a certain basis for the space on which the operator acts, has as matrix a Toeplitz matrix. The distinction is clearly a vague one and we shall not adhere to it rigidly. For the most part we shall use the matrix terminology in the finite dimensional case, the operator terminology otherwise.

The entries of the matrices we shall consider arise as Fourier coefficients of an integrable function on the unit circle. If f is such a function its Fourier coefficients will be denoted by f_k:

$$f_k = \frac{1}{2\pi} \int_0^{2\pi} f(e^{i\theta})\, e^{-ik\theta}\, d\theta$$

The associated Toeplitz matrices are

$$T_n(f) = (f_{j-k}) \qquad 0 \leq j,\ k \leq n$$

$$T(f) = (f_{j-k}) \qquad 0 \leq j,\ k < \infty$$

Thus $T_n(f)$ is a square matrix of order $n + 1$ and $T(f)$ is a semi-infinite matrix. In the finite case $D_n(f)$ denotes the determinant of $T_n(f)$.

Toeplitz's work (about 1910) concerned what is now called the spectrum of the doubly-infinite matrix

$$(f_{j-k}) \qquad -\infty < j,\ k < \infty$$

in the case where f was the restriction to the unit circle of a function analytic in an annulus containing the circle. He showed the spectrum was exactly the range of $f(e^{i\theta})$.

2. TOEPLITZ MATRICES AND ORTHOGONAL POLYNOMIALS

The first really deep results concerning Toeplitz matrices and determinants were those of G. Szegö (beginning about 1920) and involved their connection with the theory of polynomials orthogonal on the unit circle. In this case f is a given nonnegative integrable function on the circle (not vanishing a.e.) and the sequence of orthogonal polynomials $P_0(z)$, $P_1(z), \ldots$ is obtained by applying the Gram-Schmidt orthogonalization process to the sequence $1, z, z^2, \ldots$ with respect to the inner product:

$$\langle\varphi,\psi\rangle = \frac{1}{2\pi}\int_0^{2\pi} \varphi(e^{i\theta})\,\overline{\psi(e^{i\theta})}\, f(e^{i\theta})d\theta$$

We assume the P_n normalized so that

$$\frac{1}{2\pi}\int_0^{2\pi} |P(e^{i\theta})|^2 f(e^{i\theta})d\theta = 1$$

and the coefficient of z^n in $P_n(z)$ is positive; these are the orthogonal polynomials associated with the weight function f.

One may think of $T_n(f)$ as acting on the space $\mathcal{P}_n$ of polynomials in z of degree at most n. The polynomial

$$P(z) = p_0 + p_1z+\cdots+ p_nz^n$$

is sent into the polynomial

$$Q(z) = q_0 + q_1z+\cdots+ q_nz^n$$

where

$$q_j = \sum_{k=0}^{n} f_{j-k}p_k$$

It is easy to check that, with (,) denoting the ordinary inner product,

$$(P,Q) = \frac{1}{2\pi}\int P(e^{i\theta})\,\overline{Q(e^{i\theta})}\, d\theta$$

one has

$$(T_n(f)\, P,Q) = \langle P,Q\rangle$$

This implies in particular that $T_n(f)$ is a positive definite Hermitian matrix.

<u>Theorem 2.1.</u> (a) $P_n = (D_n/D_{n-1})^{1/2}\, T_n(f)^{-1}z^n$.

(b) The highest coefficient of P_n equals $(D_{n-1}/D_n)^{1/2}$.

Theorem 2.1(a) and Cramer's rule show how to obtain P_n explicitly from the matrix $T_n(f)$; it is $(D_n D_{n-1})^{-1/2}$ times the determinant of the matrix obtained by replacing the last row of $T_n(f)$ by $1, z, \cdots, z^n$.

Extremal properties of D_n determine

$$\mu_n = \min \frac{1}{2\pi} \int_0^{2\pi} |Q(e^{i\theta})|^2 \, f(e^{i\theta}) \, d\theta$$

where Q runs over all monic polynomials of degree n.

<u>Theorem 2.2.</u> $\mu_n = D_n/D_{n-1}$ and this minimum is attained when $Q = (D_n/D_{n-1})^{1/2} \, P_n$.

Note that since $|e^{i\theta}| = 1$,

$$\mu_n = \min_{\alpha_1, \cdots, \alpha_n} \frac{1}{2\pi} \int |1 + \alpha_1 e^{-i\theta} + \cdots + \alpha_n e^{-in\theta}|^2 \, f(e^{i\theta}) \, d\theta$$

so that $\mu_{n+1} \leq \mu_n$ and

$$\mu = \lim_{n \to \infty} \mu_n$$

exists. An application of Jensen's inequality for analytic functions,

$$\frac{1}{2\pi} \int_0^{2\pi} \log |\varphi(e^{i\theta})| \, d\theta \geq \log |\varphi(0)|$$

(if φ is analytic for $|z| \leq 1$) gives the inequality

$$\mu_n \geq G(f)$$

where $G(f)$, the geometric mean of f, is defined by

$$G(f) = \exp \frac{1}{2\pi} \int_0^{2\pi} \log f(e^{i\theta}) \, d\theta \, .$$

Thus $\mu \geq G(f)$. The following will be established in the next section.

<u>Theorem 2.3.</u> $\mu = G(f)$.

3. ASYMPTOTICS OF TOEPLITZ DETERMINANTS AND ORTHOGONAL POLYNOMIALS

The main results are Theorem 2.3 and asymptotic formulas for $P_n(z)$ when $|z| \geq 1$. We consider first the case when f is strictly positive and very smooth.

Lemma 3.1. Suppose $f(e^{i\theta}) > 0$ and $f \in C^2$. Then there is a unique function F, continuous, nonzero and C^1 for $|z| \geq 1$, analytic for $|z| > 1$, satisfying

(a) $F(\infty) > 0$

(b) $|F(e^{i\theta})|^2 = f(e^{i\theta})$.

This function satisfies $F(\infty) = G(f)^{1/2}$.

By introducing polynomials obtained from the partial sums of the series for F^{-1} one proves

Theorem 3.2. Under the assumptions of Lemma 3.1 we have

$$\mu_n = G(f) + o(n^{-1})$$

and

$$\int |e^{-in\theta} P_n(e^{i\theta}) F(e^{i\theta})-1|^2 d\theta = o(n^{-1})$$

The last relation shows that on the unit circle $z^{-n} P_n(z)$ has mean square limit $F(z)^{-1}$. To show that this holds uniformly for $|z| \geq 1$ one uses

Lemma 3.3. Suppose each R_n is a polynomial of degree n in z and $|R_n(e^{i\theta})| \leqslant M$ for all θ. Then $|R_n'(e^{i\theta})| \leqslant cnM$ for all θ, where c is an absolute constant. (Actually one can take c = 1.)

The desired uniform asymptotic formula now follows:

Theorem 3.4. Under the assumptions of lemma 3.1

$$\lim_{n\to\infty} z^{-n} P_n(z) = F(z)^{-1}$$

uniformly for $|z| \geq 1$.

It is not hard to deduce from Theorem 3.2 that $\mu = G(f)$ in complete generality. Asymptotic results for P_n, for f general, are more subtle and depend on the construction of the analytic function F. This will be postponed to section 5.

4. EIGENVALUES OF TOEPLITZ MATRICES

Since $D_n(f)$ is just the product of the eigenvalues of $T_n(f)$ it would not be surprising if information about the distribution of these eigenvalues yields information about the asymptotic behavior of $D_n(f)$. What is surprising is that the reverse holds. From the limit relation (Theorems 2.2 and 2.3)

$$\lim_{n\to\infty} \frac{D_n(f)}{D_{n-1}(f)} = G(f)$$

one can deduce information about the eigenvalues. If we denote them by

$$\lambda_{n,o}, \lambda_{n,1}, \ldots, \lambda_{n,n}$$

then

$$\log D_n(f) = \sum_{k=0}^{n} \log \lambda_{n,k}$$

The limit relation above implies the slightly weaker relation

$$\lim_{n\to\infty} D_n^{1/n+1} = G(f)$$

and so

$$\lim_{n\to\infty} (n+1)^{-1} \sum_{k=0}^{n} \log \lambda_{n,k} = \frac{1}{2\pi} \int \log f(e^{i\theta})\,d\theta$$

In fact this holds for more general functions than the logarithm. We assume now that f is real and bounded and write

$$m = \text{ess inf } f(e^{i\theta}), \qquad M = \text{ess sup } f(e^{i\theta})$$

We have $m \leq \lambda_{n,k} \leq M$ for all n,k.

Theorem 4.1. If F is any continuous function on $[m,M]$

$$\lim_{n\to\infty} (n+1)^{-1} \sum_{k=0}^{n} F(\lambda_{n,k}) = \frac{1}{2\pi} \int_0^{2\pi} F(f(e^{i\theta}))d\theta$$

This theorem tells us that in some sense the eigenvalues are distributed like the values of f. The next theorem tells another way in which this is so. Denote by $N(a,b,n)$ the number of k such that

$$a < \lambda_{n,k} < b$$

and by $\Omega(a,b)$ the measure of

$$\{\theta: \ a < f(e^{i\theta}) < b\}$$

Theorem 4.2. Assume $\{\theta: f(e^{i\theta}) = a \text{ or } b\}$ has measure zero. Then

$$\lim_{n\to\infty} (n+1)^{-1} N(a,b,n) = (2\pi)^{-1} \Omega(a,b)$$

For example if f is the characteristic function of a semi-circle then for any $\varepsilon > 0$

$$\lim_{n\to\infty} (n+1)^{-1} N(-\varepsilon,\varepsilon,n) = \frac{1}{2}, \quad \lim_{n\to\infty} (n+1)^{-1} N(1-\varepsilon,1+\varepsilon,n) = \frac{1}{2}$$

One might wonder whether there are no eigenvalues, or very few, in the interval $[\frac{1}{3}, \frac{2}{3}]$, say. In fact there are unboundedly many.

Theorem 4.3. For any $\lambda \in [m,M]$ and any $\varepsilon > 0$

$$\lim_{n\to\infty} N(\lambda-\varepsilon, \lambda+\varepsilon, n) = \infty$$

This will be a consequence of results on Teoplitz operator we shall derive later on.

5. HARDY SPACES

These are spaces of functions on which infinite Toeplitz matrices are most usefully thought of as acting as operators. These may be thought of either as certain spaces of functions analytic in the unit disc or as certain (isomorphic) spaces of functions defined on the circle.

First we recall some facts from Fourier analysis. If f is a function in L_1 (of the circle) with Fourier coefficients f_k, one defines the Abel means

$$f_r(e^{i\theta}) = \sum_{k=-\infty}^{\infty} f_k\, r^{|k|}\, e^{ik\theta} \qquad (0 \leq r < 1)$$

There is the Poisson integral representation

$$f_r(e^{i\theta}) = \int_0^{2\pi} P(r,\theta-\varphi)\, f(e^{i\varphi})\, d\varphi$$

where

$$P(r,\theta) = (2\pi)^{-1} \frac{1-r^2}{1-2r\cos\theta + r^2}$$

These Abel means have the following properties:

(a) $\|f_r\|_p \leq \|f\|_p \qquad (1 \leq p \leq \infty)$

(b) If $f \in L_p$ with $p < \infty$ then $\lim_{r \to 1} \|f - f_r\|_p = 0$

(c) If $f \in L_1$ then $\lim_{r \to 1} f_r(e^{i\theta}) = f(e^{i\theta})$ almost everywhere

<u>Definition</u>. If $f \in L_p$ then we say $f \in H_p$ if $f_k = 0$ for $k < 0$.

The spaces H_p clearly decrease with increasing p.

<u>Theorem 5.1</u>. If $f \in H_p$, $g \in H_q$ $(p^{-1} + q^{-1} = 1)$ then $fg \in H_1$.

For each $f \in H_p$ there is an associated function

$$F(z) = \sum_{k=0}^{\infty} f_k z^k$$

analytic for $|z| < 1$. Clearly $F(re^{i\theta}) = f_r(e^{i\theta})$. We say that F belongs to H_p (there should be no confusion in the use of the same notation since one applies to functions on the circle, the other to functions on the disc) and has boundary function f. It follows from (a) above that $F \in H_p$ satisfies, for some $M > 0$,

$$\|F(re^{i\theta})\|_p \leq M \quad \text{for all} \quad r < 1$$

<u>Theorem 5.2.</u> A function F, analytic in $|z| < 1$, belongs to H_p if and only if it satisfies an inequality of the above form.

An important fact concerning functions f in H_1 is that they cannot be very small without vanishing identically. "Very small" refers to the integrability of $\log|f|$. Write

$$\log^+ x = \max(o, \log x), \quad \log^- x = \min(o, \log x)$$

Since $\log^+ x \leq x$ and $f \in L_1$ we clearly have

$$\int \log^+ |f(e^{i\theta})| \, d\theta < \infty$$

What is interesting is the integrability of $\log^-|f|$.

<u>Theorem 5.3</u> If $f \in H_1$ and f is not a.e. equal to zero then $\log^-|f|$ is integrable.

<u>Corollary</u>. If $f \in H_1$ then $f = 0$ either almost everywhere or almost nowhere.

We now indicate the construction of the analogue of the function F of Lemma 3.1 in case f is merely nonnegative, integrable, and satisfies

$$\int \log f(e^{i\theta}) \, d\theta > -\infty$$

Let u denote the Poisson integral of $\frac{1}{2} \log f(e^{-i\theta})$

$$u(re^{i\theta}) = \frac{1}{2} \int P(r,\theta-\varphi) \log f(e^{-i\varphi}) \, d\varphi$$

Then $u(z)$ is harmonic for $|z| < 1$ so $v(z) = u(z^{-1})$ is harmonic for $|z| > 1$. Let w denote its harmonic conjugate, which is uniquely determined when the condition $w(\infty) = 0$ is imposed. Finally, set

$$F(z) = \exp\{v(z) + i\, w(z)\}$$

Then $F(z)$ belongs to H_2 of $|z| > 1$ (i.e. $F(z^{-1})$ belongs to H_2 of $|z| < 1$), its boundary function $F(e^{i\theta})$ satisfies

$$|F(e^{i\theta})|^2 = f(e^{i\theta}) \quad \text{a.e.}$$

and $F(\infty) = G(f)^{1/2}$. In case f satisfies the conditions of Lemma 3.1, F is the function of the conclusion of that lemma.

Theorem 5.4. As $n \to \infty$ we have

(a) $\lim \|e^{-in\theta} P_n(e^{i\theta}) F(e^{i\theta})-1\|_2 = 0$

(b) $\lim z^{-n} P_n(z) = F(z)^{-1}$ uniformly for $|z| \geq R > 1$

6. TOEPLITZ OPERATORS

The space H_2 (of the circle) is a Hilbert space (being a subspace of L_2) with orthonormal basis $e^{ik\theta}$ $(k = 0, 1, 2, \ldots)$. Given a function f belonging to L_∞ of the circle, the Toeplitz operator $T(f)$ will denote the operator which, with respect to that basis, has matrix (f_{j-k}) $(0 \leq j, k < \infty)$.

If $\varphi \in H_2$ then the jth Fourier coefficient of $T(f)(\varphi)$ is

$$\sum_{k=0}^{\infty} f_{j-k}\, \varphi_k = \sum_{k=-\infty}^{\infty} f_{j-k}\, \varphi_k$$

since $\varphi_k = 0$ when $k < 0$. The right side is just the jth Fourier coefficient of the product $f\varphi$. Hence

$$(*) \qquad T(f)(\varphi) = P(f\varphi)$$

where P denotes the projection of L_2 onto H_2. This implies in particular the inequality

$$\|T(f)\| \leq \|f\|_\infty$$

where the left side is the ordinary operator norm.

We shall be concerned first with questions related to the inversion of $T(f)$: how to invert it in practice (i.e. how explicitly to solve $T(f)(\varphi) = \psi$) and how to determine from the properties of f whether or not $T(f)$ is invertible in principle. Since for any constant λ one has

$$T(f) - \lambda I = T(f-\lambda)$$

(I is the identity operator) the latter is equivalent to determining $\sigma(T(f))$, the spectrum of $T(f)$, the set of λ for which $T(f) - \lambda I$ is not invertible.

If it were not for the projection P appearing on the right side of (*) one would have

$$T(f)\, T(g) = T(fg)$$

Although this is generally false it does hold in certain very useful cases. We use the notation $\overline{H}_p$ to denote the set of complex conjugates of functions in H_p of the circle.

<u>Theorem 6.1</u>. If $f \in \overline{H}_\infty$ or $g \in H_\infty$ then
$$T(fg) = T(f)\, T(g)$$

This simple theorem allows us explicitly to invert a large class of Toeplitz operators.

<u>Definition</u>. A factorization $f = gh$ is called a Wiener-Hopf factorization if g and g^{-1} belong to $\overline{H}_\infty$, h and h^{-1} belong to H_∞.

Wiener and Hopf used a similar factorization to solve a certain class of integral equations — continuous analogues of infinite Toeplitz matrix equations.

<u>Theorem 6.2</u>. If $f = gh$ is a Wiener-Hopf factorization then $T(f)$ is invertible and

$$T(f)^{-1} = T(h^{-1})\, T(g^{-1})$$

A sufficient condition for f to possess a Wiener-Hopf factorization is that it have a logarithm with an absolutely convergent Fourier series, for then one may take

$$g(e^{i\theta}) = \exp \sum_{k=-\infty}^{-1} (\log f)_k e^{ik\theta}$$

$$h(e^{i\theta}) = \exp \sum_{k=0}^{\infty} (\log f)_k e^{ik\theta}$$

For f even to have a continuous logarithm it is necessary that f be continuous and satisfy

$$(**) \qquad f(e^{i\theta}) \neq 0, \qquad \underset{0 \leq \theta \leq 2\pi}{\Delta} \arg f(e^{i\theta}) = 0$$

The Wiener-Lévy theorem states that if f satisfies (**) and in addition has an absolutely convergent Fourier series then so does any continuously defined log f.

Thus if f is continuous, satisfies (**), and has an absolutely convergent Fourier series then T(f) is invertible. Actually, as we shall see later, the last condition is unnecessary.

Theorem 6.3. If T(f) is invertible then $f^{-1} \in L_\infty$.

Corollary. $\sigma(T(f))$ contains R(f), the essential range of f.

To find a set containing $\sigma(T(f))$ one uses the fact that if an operator A on Hilbert space has norm less than one then I + A is necessarily invertible. This implies

Theorem 6.4. $\sigma(T(f))$ is contained in the smallest convex set containing R(f).

These last two results are about all that can be said concerning the relationship between $\sigma(T(f))$ and $R(f)$ in general. We turn now to certain cases where the spectrum may be determined precisely. The result for the first of these was already mentioned; it follows easily from Theorems 6.1 and 6.4.

<u>Theorem 6.5</u>. If f is continuous then $T(f)$ is invertible if and only if (**) holds.

If f is continuous and nonzero the quantity

$$(2\pi)^{-1} \underset{0 \le \theta \le 2\pi}{\Delta} \arg f(e^{i\theta})$$

is called the "winding number" of f and equals the net number of times the curve described by $f(e^{i\theta})$ winds around the origin as $e^{i\theta}$ traverses the unit circle counterclockwise.

<u>Corollary</u>. If f is continuous then $\sigma(T(f))$ consists of $R(f)$ together with those points $\lambda \notin R(f)$ such that $f-\lambda$ has nonzero winding number.

There is an extension of this which holds if f is continuous except for jump discontinuities. In this case the curve described by $f(e^{i\theta})$, which has gaps, must have added to it the line segments running from $f(e^{i\theta_k-})$ to $f(e^{i\theta_k+})$ for all points of discontinuity $e^{i\theta_k}$.

The next theorem describes the spectrum of a self-adjoint Toeplitz operator.

<u>Theorem 6.6</u>. If f is real, $\sigma(T(f)) = [\text{ess inf } f, \text{ ess sup}$

Finally, H_∞ .

Theorem 6.7. If $f \in H_\infty$ then $T(f)$ is invertible if and only if $f^{-1} \in H_\infty$.

Corollary. Suppose $f \in H_\infty$ with corresponding analytic function $F(z)$ $(|z| < 1)$. Then $\sigma(T(f))$ is the closure of the range of F.

Observe that in all cases for which the spectrum of $T(f)$ has been determined, f continuous except perhaps for jump discontinuities, f real, f belonging to H_∞, the spectrum can easily be verified to be a connected set. In fact the spectrum is always connected.

7. PROJECTION METHODS

Given an invertible operator A on a Hilbert space H and a sequence of projections P_n converging strongly to I(i.e. $P_n x \to x$ for each $x \in H$), one says that the projection method $\{P_n\}$ is applicable to A if the operators $P_n A P_n$, thought of as acting on $P_n H$, are invertible for sufficiently large n and

$$(P_n A P_n)^{-1} P_n \to A^{-1}$$

strongly as $n \to \infty$.

This concept is of particular interest in the study of Toeplitz operators since

$$T_n(f) = P_n T(f) P_n$$

where $T_n(f)$ is thought of as acting on $\mathscr{P}_n$ (the space of polynomials in $e^{i\theta}$ of degree at most n) and P_n is the projection from H_2 to $\mathscr{P}_n$.

There is one general result, quite a simple one, concerning projection methods which will be of interest to us.

<u>Definition</u>. A sequence of operators $\{A_n\}$ (acting on perhaps different Hilbert spaces) is called uniformly invertible if A_n is invertible for sufficiently large n and the norms $\|A_n^{-1}\|$ are bounded as $n \to \infty$.

<u>Theorem 7.1</u>. For any operator A on a Hilbert space H and any sequence of projections P_n converging strongly to I, if the sequence $\{P_n A P_n\}$ is uniformly invertible then A is itself invertible and the projection method $\{P_n\}$ is applicable to A.

Theorem 4.3 is a simple consequence of this result and Theorem 6.6. Another consequence is a limit theorem for Toeplitz determinants.

<u>Theorem 7.2</u>. If the sequence $\{T_n(f)\}$ is uniformly invertible then

$$\lim_{n\to\infty} \frac{D_{n-1}(f)}{D_n(f)}$$

exists and is equal to the constant term (zero'th Fourier coefficient) of $T(f)^{-1} 1$.

The identification of this number with the reciprocal of a suitably interpreted geometric mean of f will be discussed in section 9.

We mention here one case where uniform invertibility occurs; the proof is like that of Theorem 6.4.

Theorem 7.3. If $\mathcal{R}e\, f \geq c > 0$ then $\{T_n(f)\}$ is uniformly invertible.

8. CLASSES OF COMPACT OPERATORS

Compact operators on Hilbert space play a fundamental role in several aspects of Toeplitz theory. We set down here their basic properties.

Given an operator A on Hilbert space, one defines the s-numbers $s_n(A)$ $(n = 1, 2, \ldots)$ by

$$s_n(A) = \inf_{F_n} \|A - F_n\|$$

where F_n runs through all operators of rank (dimension of range) at most $n-1$. Thus

$$s_1(A) \geq s_2(A) \geq \cdots$$

and $s_1(A) = \|A\|$. These s-numbers have the following easily established properties:

(a) $s_n(ABC) \leq \|A\|\; s_n(B)\; \|C\|$

(b) $s_{n+m-1}(A+B) \leq s_n(A) + s_m(B)$

(c) $s_{n+m-1}(AB) \leq s_n(A) \quad s_m(B)$

Definition. A is called compact if $\lim_{n\to\infty} s_n(A) = 0$. If A is compact we define for $1 \leq p \leq \infty$

$$\|A\|_p = \{\sum_{n=1}^{\infty} s_n(A)^p\}^{1/p}$$

$$S_p = \{A: \|A\|_p < \infty\}$$

In particular S_∞ is the set of compact operators and $\|A\|_\infty$ is just the operator norm $\|A\|$. It follows easily from (b) above that S_p is a linear space and from (a) that it is closed under multiplication by any bounded operator. Less obvious but true is that $\| \|_p$ is a norm on S_p making it into a Banach space in which the operators of finite rank form a dense subspace.

In case H is finite dimensional and A is positive semi-definite, $s_n(A)$ is exactly the nth largest eigenvalue of A. If A is arbitrary it follows from the definition of s_n that if

$$A = U(A^*A)^{1/2}$$

is the polar decomposition of A, with U unitary, then

$$s_n(A) = s_n((A^*A)^{1/2})$$

and so equals the nth largest eigenvalue of $(A^*A)^{1/2}$

The spaces S_p are in some ways analogous to the Lebesgue spaces L_p. In particular Schwarz's inequality holds.

<u>Theorem 8.1</u>. If A, $B \in S_2$ then $AB \in S_1$ and $\|AB\|_1 \leq \|A\|_2 \|B\|_2$.

The spaces S_1 and S_2 are of special importance. An operator in S_1 is said to be of "trace class" (since for such operators there can be defined, as we shall see, a well-defined trace) and an operator in S_2 is said to be "Hilbert-Schmidt". In the finite-dimensional case, with $A = (a_{j,k})$,

$$\|A\|_2^2 = \|A^*A\|_1 = \operatorname{tr} A^*A = \sum_{j,k} |a_{j,k}|^2$$

There is no such simple representation of $\|A\|_1$. However, there is the important inequality

$$\|A\|_1 \geq \Sigma \, |a_{i,i}|$$

In particular, in the finite dimensional case,

$$|\mathrm{tr}\ A| \leq \|A\|_1$$

This inequality allows one to define the trace of an arbitrary operator in S_1 of a Hilbert space H.

Theorem 8.2. Let P_n be a sequence of finite rank projections on H which tends strongly to I as $n \to \infty$. Then if $A \in S_1$

$$\lim_{n\to\infty} \ \mathrm{tr}\ P_n A\ P_n$$

exists and is independent of the choice of the sequence $\{P_n\}$.

The limit is of course denoted by tr A. It is a continuous linear functional on S_1 and moreover

$$\mathrm{tr}\ AB = \mathrm{tr}\ BA$$

if $A \in S_1$, B arbitrary.

Similarly one can define the determinant of an operator differing from the identity by an operator in S_1.

Theorem 8.3. Let P_n, A be as in Theorem 8.2. Then

$$\lim_{n\to\infty} \det P_n(I+A)\ P_n$$

exists and is independent of the choice of $\{P_n\}$.

This limit is denoted by det(I+A) and is also a continuous function on S_1. It is this determinant which arises in the finer study of the limiting behavior of the Toeplitz determinants $D_n(f)$.

An important property of S_p is that multiplication by an operator from this space transforms strong convergence (which is actually quite weak) into convergence in the norm of S_p (which is very strong).

Theorem 8.4. Suppose $A \in S_p$, $B_n \to B$ strongly and $C_n^* \to C^*$ strongly. Then

$$\lim_{n\to\infty} \|B_n A C_n - BAC\|_p = 0$$

9. APPLICATIONS TO TOEPLITZ MATRICES

One would infer from the results of the last few sections that if T(f) is compact then interesting results concerning $T_n(f)$ could be obtained. Actually T(f) is compact only if f is identically zero and the applications arise in a slightly more subtle way.

We define the Hankel operator H(f) on H_2 as the operator which, with respect to the standard basis $\{e^{ik\theta}\}$ (k=0,1,...) has matrix

$$(f_{j+k+1}) \qquad 0 \le j, k < \infty$$

Theorem 9.1. If f is continuous then $H(f) \in S_\infty$. If $\sum_{k=0}^{\infty} k|f_k|^2 < \infty$ then $H(f) \in S_2$.

Thus Hankel operators are very nice, and they are closely connected with Toeplitz operators.

We use the following notation: For a function f defined on the unit circle, $\tilde{f}(e^{i\theta}) = f(e^{-i\theta})$.

A trivial computation establishes the following.

<u>Theorem 9.2.</u> For any $f, g \in L_\infty$ we have

$$T(fg) - T(f)\, T(g) = H(f)\, H(\tilde{g})$$

<u>Corollary.</u> If f or g is continuous then $T(fg) - T(f)\, T(g)$ is compact. If $\sum_{-\infty}^{\infty} |k|\, |f_k|^2 < \infty$ and $\sum_{-\infty}^{\infty} |k|\, |g_k|^2 < \infty$ then this difference belongs to S_1.

For applications to Toeplitz matrices we shall use a finite-dimensional analogue of Theorem 9.2. As before, think of $T_n(f)$ as acting on $\mathcal{P}_n$ and denote by P_n the projection from H_2 to $\mathcal{P}_n$.

<u>Theorem 9.3.</u> If $f \in H_\infty$ or $g \in \overline{H}_\infty$ we have

$$T_n(fg) - T_n(f)\, T_n(g) = P_n H(f)\, H(\tilde{g})\, P_n$$

Two applications of this identity together with Theorems 8.4 and 9.1 yield the following.

<u>Theorem 9.4.</u> Let $f \in L_\infty$, $g \in H_\infty$, $h \in \overline{H}_\infty$. Then if g and h are continuous

$$\lim_{n\to\infty} \{T_n(g)\, T_n(f)\, T_n(h) - P_n\, T(g)\, T(f)\, T(h)\, P_n\} = 0$$

in operator norm. If $\sum_{-\infty}^{\infty} |f_k| + \{\sum_{-\infty}^{\infty} |k|\, |f_k|^2\}^{\frac{1}{2}} < \infty$, and similarly for g and h, then the conclusion holds in the norm of S_1.

From this fact and Theorem 8.3 one deduces the applicability of the projection method if f is continuous:

Theorem 9.5. If f is continuous and satisfies

$$f(e^{i\theta}) \neq 0, \qquad \underset{0 \leq \theta \leq 2\pi}{\Delta} \arg f(e^{i\theta}) = 0$$

then the sequence $\{T_n(f)\}$ is uniformly invertible.

Corollary. Under the assumptions of the theorem we have

$$\lim_{n\to\infty} \frac{D_n(f)}{D_{n-1}(f)} = G(f) = \exp\left\{\frac{1}{2\pi}\int_0^{2\pi} \log f(e^{i\theta})\,d\theta\right\}$$

where in the integral on the right any continuous $\log f$ is used.

Thus Szegö's limit theorem (Theorem 2.3) has been extended to the case of nonreal f, albeit with a continuity assumption. One can obtain a stronger limiting result if more is known about f. Let us assume that f satisfies, in addition, the conditions

$$(*) \qquad \sum_{k=-\infty}^{\infty} |f_k| < \infty, \quad \sum_{k=-\infty}^{\infty} |k|\, |f_k|^2 < \infty$$

Then f^{-1} satisfies these conditions also and it follows from the second part of the corollary to Theorem 9.2 that $T(f)\,T(f^{-1})$ differs from I by an operator in S_1. Hence its determinant is defined.

Theorem 9.6. If f satisfies the assumptions of Theorem 9.5 and in addition (*) holds, then

$$\lim_{n\to\infty} \frac{D_n(f)}{G(f)^{n+1}} = \det T(f)\,T(f^{-1})$$

This result, under the assumption that f is real and satisfies a certain smoothness condition, but with a different expression for the limit, was also first obtained by Szegö. To obtain Szegö's expression for the limit, one uses the following formula relating the determinant of a multiplicative commutator with the trace of an additive commutator.

Theorem 9.7. If A and B are operators on Hilbert space such that $AB - BA \in S_1$ then $e^A e^B e^{-A} e^{-B}$ differs from I by an operator of S_1 and

$$\det e^A e^B e^{-A} e^{-B} = \exp \operatorname{tr}(AB-BA)$$

This is applied to operators A and B such that

$$e^A = T(g), \quad e^B = T(h)$$

where $f = gh$ is a Wiener-Hopf factorization and one obtains the formula

$$\det T(f)\, T(f^{-1}) = \exp \sum_{k=1}^{\infty} k(\log f)_k (\log f)_{-k}$$

which is Szegö's expression for the limit.

SECRETARIAT OF THE COURSE

DIRECTORS

A. Andreotti	Department of Mathematics, School of Science, Oregon State University, Corwallis, Oregon 97331, United States of America and Istituto Matematico, Università di Pisa, Via Derna 1, Pisa, Italy
J. Eells	Mathematics Institute, University of Warwick, Coventry CV4 7AL, Warwickshire, United Kingdom
F. Gherardelli	Istituto Matematico "U. Dini", Università di Firenze, Viale Morgagni 67/A, Firenze, Italy

EDITOR

Miriam Lewis	Division of Publications, IAEA, Vienna, Austria

The following conversion table is provided for the convenience of readers and to encourage the use of SI units.

FACTORS FOR CONVERTING UNITS TO SI SYSTEM EQUIVALENTS *

SI base units are the metre (m), kilogram (kg), second (s), ampere (A), kelvin (K), candela (cd) and mole (mol).
[For further information, see International Standards ISO 1000 (1973), and ISO 31/0 (1974) and its several parts]

Multiply		*by*	*to obtain*
Mass			
pound mass (avoirdupois)	1 lbm	= 4.536×10^{-1}	kg
ounce mass (avoirdupois)	1 ozm	= 2.835×10^{1}	g
ton (long) (= 2240 lbm)	1 ton	= 1.016×10^{3}	kg
ton (short) (= 2000 lbm)	1 short ton	= 9.072×10^{2}	kg
tonne (= metric ton)	1 t	= 1.00×10^{3}	kg
Length			
statute mile	1 mile	= 1.609×10^{0}	km
yard	1 yd	= 9.144×10^{-1}	m
foot	1 ft	= 3.048×10^{-1}	m
inch	1 in	= 2.54×10^{-2}	m
mil (= 10^{-3} in)	1 mil	= 2.54×10^{-2}	mm
Area			
hectare	1 ha	= 1.00×10^{4}	m^2
(statute mile)^2	1 mile^2	= 2.590×10^{0}	km^2
acre	1 acre	= 4.047×10^{3}	m^2
yard^2	1 yd^2	= 8.361×10^{-1}	m^2
foot^2	1 ft^2	= 9.290×10^{-2}	m^2
inch^2	1 in^2	= 6.452×10^{2}	mm^2
Volume			
yard^3	1 yd^3	= 7.646×10^{-1}	m^3
foot^3	1 ft^3	= 2.832×10^{-2}	m^3
inch^3	1 in^3	= 1.639×10^{4}	mm^3
gallon (Brit. or Imp.)	1 gal (Brit)	= 4.546×10^{-3}	m^3
gallon (US liquid)	1 gal (US)	= 3.785×10^{-3}	m^3
litre	1 l	= 1.00×10^{-3}	m^3
Force			
dyne	1 dyn	= 1.00×10^{-5}	N
kilogram force	1 kgf	= 9.807×10^{0}	N
poundal	1 pdl	= 1.383×10^{-1}	N
pound force (avoirdupois)	1 lbf	= 4.448×10^{0}	N
ounce force (avoirdupois)	1 ozf	= 2.780×10^{-1}	N
Power			
British thermal unit/second	1 Btu/s	= 1.054×10^{3}	W
calorie/second	1 cal/s	= 4.184×10^{0}	W
foot-pound force/second	1 ft·lbf/s	= 1.356×10^{0}	W
horsepower (electric)	1 hp	= 7.46×10^{2}	W
horsepower (metric) (= ps)	1 ps	= 7.355×10^{2}	W
horsepower (550 ft·lbf/s)	1 hp	= 7.457×10^{2}	W

* Factors are given exactly or to a maximum of 4 significant figures

Multiply			*by*	*to obtain*
Density				
pound mass/inch^3	1 lbm/in^3	=	2.768×10^4	kg/m^3
pound mass/foot^3	1 lbm/ft^3	=	1.602×10^1	kg/m^3
Energy				
British thermal unit	1 Btu	=	1.054×10^3	J
calorie	1 cal	=	4.184×10^0	J
electron-volt	1 eV	$\simeq$	1.602×10^{-19}	J
erg	1 erg	=	1.00×10^{-7}	J
foot-pound force	1 ft·lbf	=	1.356×10^0	J
kilowatt-hour	1 kW·h	=	3.60×10^6	J
Pressure				
newtons/metre^2	1 N/m^2	=	1.00	Pa
atmosphere[a]	1 atm	=	1.013×10^5	Pa
bar	1 bar	=	1.00×10^5	Pa
centimetres of mercury (0°C)	1 cmHg	=	1.333×10^3	Pa
dyne/centimetre^2	1 dyn/cm^2	=	1.00×10^{-1}	Pa
feet of water (4°C)	1 ftH_2O	=	2.989×10^3	Pa
inches of mercury (0°C)	1 inHg	=	3.386×10^3	Pa
inches of water (4°C)	1 inH_2O	=	2.491×10^2	Pa
kilogram force/centimetre^2	1 kgf/cm^2	=	9.807×10^4	Pa
pound force/foot^2	1 lbf/ft^2	=	4.788×10^1	Pa
pound force/inch^2 (= psi)[b]	1 lbf/in^2	=	6.895×10^3	Pa
torr (0°C) (= mmHg)	1 torr	=	1.333×10^2	Pa
Velocity, acceleration				
inch/second	1 in/s	=	2.54×10^1	mm/s
foot/second (= fps)	1 ft/s	=	3.048×10^{-1}	m/s
foot/minute	1 ft/min	=	5.08×10^{-3}	m/s
mile/hour (= mph)	1 mile/h	=	4.470×10^{-1}	m/s
			1.609×10^0	km/h
knot	1 knot	=	1.852×10^0	km/h
free fall, standard (= g)		=	9.807×10^0	m/s^2
foot/second^2	1 ft/s^2	=	3.048×10^{-1}	m/s^2
Temperature, thermal conductivity, energy/area·time				
Fahrenheit, degrees −32	°F − 32		$\frac{5}{9}$	°C
Rankine	°R		$\frac{5}{9}$	K
1 Btu·in/ft^2·s·°F		=	5.189×10^2	W/m·K
1 Btu/ft·s·°F		=	6.226×10^1	W/m·K
1 cal/cm·s·°C		=	4.184×10^2	W/m·K
1 Btu/ft^2·s		=	1.135×10^4	W/m^2
1 cal/cm^2·min		=	6.973×10^2	W/m^2
Miscellaneous				
foot^3/second	1 ft^3/s	=	2.832×10^{-2}	m^3/s
foot^3/minute	1 ft^3/min	=	4.719×10^{-4}	m^3/s
rad	rad	=	1.00×10^{-2}	J/kg
roentgen	R	=	2.580×10^{-4}	C/kg
curie	Ci	=	3.70×10^{10}	disintegration/s

[a] atm abs: atmospheres absolute;
atm (g): atmospheres gauge.

[b] lbf/in^2 (g) (= psig): gauge pressure;
lbf/in^2 abs (= psia): absolute pressure.

HOW TO ORDER IAEA PUBLICATIONS

An exclusive sales agent for IAEA publications, to whom all orders and inquiries should be addressed, has been appointed in the following country:

UNITED STATES OF AMERICA	UNIPUB, P.O. Box 433, Murray Hill Station, New York, N.Y. 10016

In the following countries IAEA publications may be purchased from the sales agents or booksellers listed or through your major local booksellers. Payment can be made in local currency or with UNESCO coupons.

ARGENTINA	Comisión Nacional de Energía Atómica, Avenida del Libertador 8250, Buenos Aires
AUSTRALIA	Hunter Publications, 58 A Gipps Street, Collingwood, Victoria 3066
BELGIUM	Service du Courrier de l'UNESCO, 112, Rue du Trône, B-1050 Brussels
CANADA	Information Canada, 171 Slater Street, Ottawa, Ont. K1A 0S9
C.S.S.R.	S.N.T.L., Spálená 51, CS-110 00 Prague Alfa, Publishers, Hurbanovo námestie 6, CS-800 00 Bratislava
FRANCE	Office International de Documentation et Librairie, 48, rue Gay-Lussac, F-75005 Paris
HUNGARY	Kultura, Hungarian Trading Company for Books and Newspapers, P.O. Box 149, H-1011 Budapest 62
INDIA	Oxford Book and Stationery Comp., 17, Park Street, Calcutta 16; Oxford Book and Stationery Comp., Scindia House, New Delhi-110001
ISRAEL	Heiliger and Co., 3, Nathan Strauss Str., Jerusalem
ITALY	Libreria Scientifica, Dott. de Biasio Lucio "aeiou", Via Meravigli 16, I-20123 Milan
JAPAN	Maruzen Company, Ltd., P.O.Box 5050, 100-31 Tokyo International
NETHERLANDS	Marinus Nijhoff N.V., Lange Voorhout 9-11, P.O. Box 269, The Hague
PAKISTAN	Mirza Book Agency, 65, The Mall, P.O.Box 729, Lahore-3
POLAND	Ars Polona, Centrala Handlu Zagranicznego, Krakowskie Przedmiescie 7, Warsaw
ROMANIA	Cartimex, 3-5 13 Decembrie Street, P.O.Box 134-135, Bucarest
SOUTH AFRICA	Van Schaik's Bookstore, P.O.Box 724, Pretoria Universitas Books (Pty) Ltd., P.O.Box 1557, Pretoria
SPAIN	Diaz de Santos, Lagasca 95, Madrid-6 Calle Francisco Navacerrada, 8, Madrid-28
SWEDEN	C.E. Fritzes Kungl. Hovbokhandel, Fredsgatan 2, S-103 07 Stockholm
UNITED KINGDOM	Her Majesty's Stationery Office, P.O. Box 569, London SE1 9NH
U.S.S.R.	Mezhdunarodnaya Kniga, Smolenskaya-Sennaya 32-34, Moscow G-200
YUGOSLAVIA	Jugoslovenska Knjiga, Terazije 27, YU-11000 Belgrade

---

Orders from countries where sales agents have not yet been appointed and requests for information should be addressed directly to:

Division of Publications
International Atomic Energy Agency
Kärntner Ring 11, P.O.Box 590, A-1011 Vienna, Austria

76- 06934